Diffusion and Use of Geographic Information Technologies

NATO ASI Series

Advanced Science Institutes Series

A Series presenting the results of activities sponsored by the NATO Science Committee, which aims at the dissemination of advanced scientific and technological knowledge, with a view to strengthening links between scientific communities.

The Series is published by an international board of publishers in conjunction with the NATO Scientific Affairs Division

A Life Sciences **B Physics**	Plenum Publishing Corporation London and New York
C Mathematical and Physical Sciences **D Behavioural and Social Sciences** **E Applied Sciences**	Kluwer Academic Publishers Dordrecht, Boston and London
F Computer and Systems Sciences **G Ecological Sciences** **H Cell Biology** **I Global Environmental Change**	Springer-Verlag Berlin, Heidelberg, New York, London, Paris and Tokyo

NATO-PCO-DATA BASE

The electronic index to the NATO ASI Series provides full bibliographical references (with keywords and/or abstracts) to more than 30000 contributions from international scientists published in all sections of the NATO ASI Series.
Access to the NATO-PCO-DATA BASE is possible in two ways:

– via online FILE 128 (NATO-PCO-DATA BASE) hosted by ESRIN, Via Galileo Galilei, I-00044 Frascati, Italy.

– via CD-ROM "NATO-PCO-DATA BASE" with user-friendly retrieval software in English, French and German (© WTV GmbH and DATAWARE Technologies Inc. 1989).

The CD-ROM can be ordered through any member of the Board of Publishers or through NATO-PCO, Overijse, Belgium.

Series D: Behavioural and Social Sciences - Vol. 70

Diffusion and Use of Geographic Information Technologies

edited by

Ian Masser

Department of Town and Regional Planning,
University of Sheffield,
Sheffield, U.K.

and

Harlan J. Onsrud

National Center for Geographic Information and Analysis,
Department of Surveying Engineering,
University of Maine,
Orono, Maine, U.S.A.

Springer-Science+Business Media, B.V.

Proceedings of the NATO Advanced Research Workshop on
Modeling the Diffusion and Use of Geographic Information Technologies
Sounion, Greece
April 8–11, 1992

ISBN 978-94-010-4778-4 ISBN 978-94-011-1771-5 (eBook)
DOI 10.1007/978-94-011-1771-5

Printed on acid-free paper

TABLE OF CONTENTS

Preface

This book is based on the premise that research necessary to advance the utility of geographic information systems must extend far beyond concerns with technical issues. The search for formalisms and generalizable principles relative to the behavior and needs of individuals, organizations and institutions is just as important in enabling optimal use of geographic information innovations.

This book consists of papers prepared by participants in the NATO Advanced Research Workshop (ARW) on "Modeling the Diffusion and Use of Geographic Information Technologies." The workshop was held April 8-11, 1992, in Sounion, Greece. The idea for the workshop and this volume arose from discussions among U.S. and European researchers who had been involved in carrying out studies on the use and diffusion of geographic information innovations and actively involved in critically analyzing each other's work. We felt the time was ripe for reporting studies recently accomplished by the National Center for Geographic Information and Analysis (U.S.A.), the Economic and Social Research Council (U.K.), and others relative to these research topics. A workshop would allow contact among and international comparisons with those who were working independently on similar problems with similar or alternative approaches. It would also allow the bringing together of scholars in technology diffusion, management information systems, and sociology with scholars from the GIS community. Through this interdisciplinary approach the intent was to provide constructive criticism of the methods used and results achieved in the geographic information use and diffusion studies to date and to map directions for extending research into priority areas.

We were very fortunate to have Dr. Lyna Wiggins and Dr. Nicos Polydorides as part of the initial organizing team to share their ideas, provide scholarly insights, and to help identify and contact key researchers throughout the U.S. and Europe for this effort.

Our hope is that for those organizations who may gain from the adoption of geographic information systems (GIS) or those who wish to optimize their use of geographic information innovations, the conceptual frameworks, studies, and experiences related in this volume will help to advance the knowledge base which might allow them to adopt sooner, implement the technology more effectively, and use such innovations more efficiently.

Ian Masser and Harlan Onsrud
University of Sheffield, U.K. and University of Maine, U.S.A.

Acknowledgments

This book is the result of contributions by many persons. It was a pleasure and honor to exchange thoughts on the use and diffusion of geographic information technologies with each of the scientists at the workshop. We thank them all for their openness, their willingness to engage in constructive debate, and their continued good humor during the entire workshop. We gained immensely from the plenary session and small group debates. It is our hope and belief that all the participants similarly came away with greater insights into the use and diffusion of geographic information technologies and the formalisms that may be applied to advance understanding in the field.

The support of our two colleagues on the organizing committee cannot be stressed enough. Dr. Lyna Wiggins contributed to the inception of the idea for the workshop and provided thoughtful suggestions regarding both scholarly and management issues throughout the project. Dr. Nicos Polydorides handled the local arrangements and picked a delightful oceanside site for the workshop with a majestic view of the Temple of Poseidon. His telling of Greek myths, their relation to modern day problems of geography, and his efforts in introducing us to "authentic" Greek culture and cuisine created a wonderful atmosphere that allowed the intense intellectual work to proceed. Without their contributions, the workshop would not have been possible nor would it have been as productive and enjoyable.

Another important contributor to the workshop and this book is Kathleen Hornsby, of the NCGIA at the University of Maine. She not only handled much of the administrative work, but also dealt with budgeting issues, compiled the various contributions, and formatted them for this book. Similarly, Chris Holmes helped in communicating with the European participants and dealing with the budget in Europe.

We are grateful to the Scientific Affairs Division of the North Atlantic Treaty Organization (NATO) for the grant which made the ARW and this book possible. The National Center for Geographic Information and Analysis (supported by U.S. National Science Foundation grant SES-88-10917), the Regional Research Laboratories (U.K. Economic and Social Research Council), and the Urban and Regional Spatial Analysis Network for Education and Training - URSA-NET (Greece - COMETT program of the European Economic Community) each supplied additional funding for the workshop and contributed in numerous other ways to its organization.

Finally, we would like to thank the authors and publishers of the following works for permitting us to include excerpts from these works: From Richard E. Walton, *Up and Running: Integrating Information Technology and the Organization*. Boston, MA: Harvard Business School Press, 1989, p.5. Reprinted by permission. From *The Diffusion of Innovations*, Third Edition, by Everett M. Rogers. Copyright 1962, 1971, 1983 by The Free Press, A Division of Macmillan, Inc. Reprinted with the permission of the publisher.

DIFFUSION AND USE OF GEOGRAPHIC INFORMATION TECHNOLOGIES: AN INTRODUCTION

IAN MASSER
Department of Town and Regional Planning
University of Sheffield
Sheffield, S10 2TN
United Kingdom
and
HARLAN J. ONSRUD
Department of Surveying Engineering, 5711
University of Maine
Orono, ME 04469-5711, USA

The roots of the present geographic information systems (GIS) phenomenon go back at least a quarter of a century. A special issue of *The American Cartographer* in 1988 devoted to "The transition from analog to digital representation of space" drew attention to the historical development of GIS in relation to the cartographic and surveying professions. In his introduction to that issue, Roger Tomlinson traces the shifting emphasis from paper maps and statistical data in hard copy form to geographically referenced data in machine readable form (1988, p249). As a result he argues that "the classic steps of observation, measurement, description, explanation, forecast and decision can be grouped into the more familiar categories of data gathering, data handling, data analysis models (mental and formal) and decision making" that are the essential ingredients of modern geographic information systems.

While many of the concepts supporting GIS emerged from the 60's and '70s, the technology to support GIS has only become available during the last five to ten years. This has generated a multi million dollar hardware and software industry whose sales have been expanding world-wide by 15-20% each year. The eventual public and private investment in improved geographic information handling capabilities is being estimated in terms of many billions of dollars.[1] The GIS phenomenon is now generating its own specialist literature. The first academic journal to be entirely devoted to GIS, the *International Journal of Geographic Information Systems* began publication as recently as 1987 and the first trade journal, *GIS World* started in 1988. In the same year the National Science Foundation established its National Center For Geographic Information and Analysis (NCGIA).

Generally North America has led the way in geographic information technological developments but there are considerable variations in the time lag between it and other countries. British experience, for example, closely parallels that of North America, with the publication in 1987 of the landmark Chorley Report on *Handling Geographic Information* and the establishment of the Economic and Social Research Council's Regional Research

I. Masser and H. J. Onsrud (eds.), Diffusion and Use of Geographic Information Technologies, 1–8.

Laboratories earlier in the same year. On the other hand, the first European GIS conference did not take place until 1990 and the level of awareness in some European countries is still relatively low.

The recent growth and diffusion of GIS raises a number of important questions regarding its potential impact. For example:

- What are the main features of the GIS diffusion process?
- What criteria should be used to measure the impacts of GIS?
- To what extent is the adoption and utilization of GIS facilitated or impeded by the institutional and organizational context within which it takes place?
- How is GIS affecting organizations, institutions, and decisions?
- How do national cultural factors affect the diffusion process? and,
- For those wishing to embrace geographic information innovations, what mechanisms might be used at the regional, national and international levels to facilitate their diffusion?

Questions such as these prompt a further question. To what extent should GIS be regarded as a unique phenomenon? Or put another way, what insights might be gained about GIS diffusion from the findings of research on the diffusion of other technologies? It may be argued that the diffusion of geographic information innovations is very different from many other innovations. The differences arise from the complexity in using the innovation and the substantial capital and human investments necessary to optimize use of these innovations. It has also been argued that the adoption and use of GIS has been different from that of other information systems due to inherent difficulties in handling and analyzing geographic data relative to other forms of data.

In his "Reflections on 25 Years of GIS," David Sinton presents his views on what is unique about GIS. He argues that "the essential nature of geographic data creates both a fundamental theoretical problem and a nasty technical problem that uniquely define the industry" (Sinton,1992,p.2). The essence of the theoretical problem is the existence of inter-dependence among geographic entities which violates conventional statistical assumptions about independence. The nasty technical problem is that geographic data is inherently irregular and variable in nature. Its variability largely precludes the use of standard data base management packages as a GIS solution and its irregularity is such that it cannot be represented by abstract engineering data models.

The Advanced Research Workshop

Regardless of the validity of these claims, it may be argued that a great deal can be learned from the experience of other technological innovations in so far as the analysis of GIS diffusion processes is concerned. Past studies of other innovations are likely to be of considerable importance in tracing the past, present and future development of the field. With these considerations in mind Ian Masser, Harlan Onsrud, Nicos Polydorides, and Lyna Wiggins, submitted a proposal to the North Atlantic Treaty Organization (NATO) for an Advanced Research Workshop on "modeling diffusion and use of geographic information technologies". In this proposal they argued that the time was ripe for such a workshop because quantitative evidence was beginning to accumulate about the characteristics of GIS diffusion in various application fields in different countries. Much of the GIS user analysis, diffusion, and impact assessment work was taking place in isolation from classic diffusion research. There had been only very limited contacts with colleagues involved in parallel investigations in the management information systems and organizational behavior fields. The basic objectives of the Advanced Research Workshop were therefore to bring together those most directly involved to review the current state of the art in GIS diffusion research and at the same time to reconsider current research methodology with a view to developing a stronger inter-disciplinary perspective in future research.

The Advanced Research Workshop was held at Sounion in Greece from 8th to 11th April 1992. It involved thirty participants from thirteen different countries. This volume contains the papers presented at the workshop together with lead-in commentaries by rapporteurs at each of the workshop sessions and some of the more extended commentaries on papers by participants. The papers have been grouped into four sections which are preceded by the key note address by Everett Rogers. These sections reflect the questions raised above:

1. Research methodology and interdisciplinary perspectives.
   What can GIS researchers learn from research elsewhere on organizational behavior and technological innovation?
2. Assessing and modeling the diffusion of geographic information innovations.
   What are the main features of the GIS diffusion process?
   What criteria should be used to measure the organizational and institutional impacts of GIS?
3. Assessing cultural and institutional issues.
   To what extent is the adoption and utilization of GIS facilitated or impeded by the institutional and organizational context within which it takes place?
   How do national cultural factors affect the diffusion process?
4. Mechanisms for facilitating the diffusion of GIS technology.
   What mechanisms might be used at the regional, national and international levels to facilitate the diffusion of GIS?

Key Elements of Diffusion Research

In his key note address Everett Rogers reviews the evolution of the diffusion of innovations models over the last twenty five years and considers its appropriateness for the analysis of GIS diffusion. He argues that most of the key elements of this model, the S shaped curve form, the notion of the critical mass which is required before innovation can take off in terms of its widespread adoption, and the importance that must be attached to the activities of a limited number of champions in the early phases, can be used without any difficulty in the analysis of GIS diffusion.

In the light of the discussion contained in the opening paragraphs of this introduction it is worth noting that the historical development of GIS over the last quarter of a century has many of the hallmarks of the classic diffusion of innovations model as described by Rogers. A long period of relatively slow gestation is followed by a period of rapid take off as the diffusion process moves up the S Shaped curve and a critical mass of operational knowledge is established to sustain the momentum of these developments. There are also numerous examples of classic diffusion "champions" throughout the GIS literature.

However, Rogers points out at least two important differences between GIS and most innovations for which the classic diffusion model has been found useful in the past. The first difference reflects the extent to which GIS is an evolving technology which is constantly changing over time. Under these circumstances reinvention is an important characteristic of the diffusion process. Adopting agencies may go through several cycles of technological change during the course of implementing their systems. Secondly, Rogers points out that GIS are acquired by organizations rather than individuals as is the case in much of the classic diffusion research. Because of their greater complexity there are very large differences between the way the same technology is utilized in different organizations. Consequently there is a need for a more sophisticated treatment of organizational and institutional behavior than has been the case in many past diffusion studies. The themes outlined by Rogers provide a valuable starting point for the discussion of the other papers presented at the workshop.

Research Methodology And Interdisciplinary Perspectives

The first group of papers deal with the development of research methodology and the introduction of interdisciplinary perspectives in GIS research. The essential question underlying this group of papers is: what can GIS researchers learn from research elsewhere on organizational behavior and technological innovation?

In the first paper Kenneth Eason takes up the organizational behavior theme identified by Rogers and presents some findings from action research on the human and organizational problems associated with the introduction of advanced information technology. In the second paper, Paul Goodman presents a framework for understanding the successful implementation of new information technology. William Huxhold also considers the role of human and organizational factors in the implementation of advanced information systems and reflects on a number of theories about the nature of GIS in local government that have developed from research and development activities in the broader information systems field. The last paper by Gary Moore directly draws upon the findings from diffusion research in the management information systems field.

Taken as a whole, these papers and the commentary on Goodman's paper by Peter Burrough provide a number of interdisciplinary perspectives which essentially confirm the usefulness of the diffusion of innovations approach. They also identify some of the main issues that must be taken into account when incorporating organizational factors into the research framework.

Modeling The Diffusion Of Geographic Information Innovations

The five papers in the next section present some of the main findings of recent research on GIS diffusion. As with Huxhold's analysis, much of the discussion in these papers focuses around GIS applications in local government. There are two main reasons for this. First, local government is one of the largest users of GIS in most of the countries represented at the workshop. In addition, the fact that there are a large number of local government agencies carrying out similar functions makes them particularly appropriate for diffusion research. Second, local government agencies represent a number of interesting challenges for GIS because of their need for political accountability, their organizational complexity and the wide range of potential applications within the one organization. For these reasons local government provides an ideal starting point for modeling the diffusion of geographical innovations.

The first two papers by Ian Masser and Heather Campbell, discuss some of the findings from a major long term study of the impacts of GIS on British local government that is underway at the University of Sheffield. In the third paper, Lyna Wiggins draws on findings from a nation-wide survey of URISA members and a state-wide survey of cities and towns to examine in some depth the adoption and use of GIS in American local government jurisdictions. The fourth paper by Jeffrey Pinto and Harlan Onsrud reports the findings of a nation-wide survey of local government users of GIS in the U.S. in which statistical correlations between various factors and the ultimate level of use of GIS are discerned. The last paper by Peter Zwart looks at diffusion in general rather than specific terms and argues that the diffusion of GIS technology will only be complete when those in the community gaining advantage from its use, know it so well that they stop being aware of it.

Together these papers and the commentary by Michael Wegener provide a good cross section of the current state of the art in GIS diffusion research. The findings from the empirical studies generally support many of the assumptions underlying the diffusion model while at the same time highlight the diversity of implementation within similar organizations. Several of the papers highlight strengths and weaknesses of various research methods, often through self criticism, and some conclude with recommendations for increased case study research to complement and put into perspective the findings of survey research.

Assessing Cultural And Institutional Issues

The third section of the book contains four papers assessing the influence of cultural and institutional issues on the diffusion of GIS in different parts of the world. They address questions such as, how do national cultural factors affect the diffusion process, and to what extent is the adaptation and utilization of GIS facilitated or impeded by the institutional and organizational context within which it takes place?

In the first paper of this section Michael Wegener and Hartwig Junius draws upon German experience to argue that national land information traditions determine the institutional environment and procedures into which GIS technology must be integrated and these traditions can severely restrict the transfer of "universal GIS software" developed in different national settings. Massimo Rumor lists a number of negative factors which he sees as inhibiting the diffusion of GIS in Italy but also illustrates how a great deal can be achieved at the local level if there is political commitment and a strong local champion. Antonio Arnaud examines similar impediments to the diffusion of GIS in Portugal, stresses the key role that municipalities must play in establishing a network of data suppliers in that country, and stresses the need for both project and institutional GIS champions. The final paper in this section by Timothy Cartwright considers the potential of geographic information systems technology in less developed countries and itemizes the considerable practical difficulties in using GIS technology in low-technology information- poor environments.

Together, the four papers in this section present a variety of perspectives on the influence of institutional and cultural factors on the diffusion of GIS technology. In particular they reemphasize the significance of the conclusions reached in the Chorley Report that the availability of GIS technology is "a necessary but not a sufficient condition for the take up of GIS to increase rapidly" (DOE, 1987, P.10). Like the Chorley Report these papers show that "some of the barriers are human and institutional, some are administrative and legal, some are financial, and some others are of a technical nature" (p.11).

Mechanisms For Facilitating The Diffusion of GIS Technology

The fourth section of the book examines some of the institutional mechanisms that have been devised with a view to facilitating the diffusion of GIS technology. It consists of four case studies of what might be regarded as current best practice in this respect at the regional, national and international levels respectively. The last paper returns to the theme that continued advancement of technological capabilities is needed in addition to improved institutional mechanisms in order to expand the use and diffusion of geographic information handling capabilities through broader cross sections of society.

Gerry Post and John McLaughlin outline the main stages in the development of a spatial data marketplace in Maritime Canada, and they describe current efforts toward development and maintenance of a spatial data infrastructure for the public and private sectors in the region. Henk Otten's paper describes the experiences of the Dutch National Centre for Expertise in Geographic Information (NEXPRI) and illustrates how the wide range of meta information services provided by this national resource have greatly facilitated the adoption and implementation of GIS in that country. In the third paper, Nicos Polydorides reviews the successes and challenges of the Urban and Regional Systems Analysis - Network for Education and Training (URSA-NET) project which is funded by the European Community COMETT program and which is coordinated from Athens, Greece. Finally, Wim Douven, Michel Grothe, Peter Nijkamp, and Henk Scholten urge the integration of GIS capabilities with planning models as an area of technological advancement which should greatly expand the social uses to which geographic information technologies may be effectively applied.

As a group the four papers in this section illustrate a number of alternative strategies that can be used to facilitate the diffusion of GIS technology at the regional, national and international levels. They also highlight four important issues that must be tackled in the process: the establishment of regional, national or international spatial data infrastructures, the creation of meta databases, and the value of education and training programs, and the value of continued technological advancements.

The Future Research Agenda

As a result of the discussions that took place during the Advanced Research Workshop at Sounion, a number of fundamental questions and research topics emerged which are discussed in the final chapter of this book. These extend the current research agenda in a number of important directions. In general, the research agenda emphasizes the need for a more systematic approach to the study of GIS diffusion. First, research efforts must be expanded to include longitudinal studies of implementation as well as studies of adoption and non-adoption. Second, there is a need for more research on criteria that can be used to measure the use and value of geographic information as the relative paucity of existing measures presents a major barrier to further research. In addition to evaluating impacts on organizations and institutions, impacts on society need to be critically evaluated. Third, greater emphasis must be placed on both institutional and cultural issues in future research given the extent to which they influence GIS diffusion. In the process, special attention needs to be given to the comparative evaluation of national and cross national experiences. Finally, the mechanisms that have been or might be used to overcome these institutional and cultural barriers at the regional, national and international levels need to be more critically examined. The case studies described in the last section of the book give some indication of the diversity of approaches that are currently being employed. A more comprehensive analysis which results in a taxonomy of these approaches would be of considerable value for those involved in such ventures.

In conclusion, then, it is felt that the Advanced Research Workshop achieved its objectives and that the papers presented there represent a valuable picture of the current state of the art of research on diffusion and use of geographic information systems technologies. Valuable insights were obtained by the introduction of an interdisciplinary perspective into the discussion. Similarly, the inclusion of several case studies in Section 4 of this book introduce a policy dimension to the discussion that is otherwise largely missing. As a result, this volume sets out a research agenda for the next five to ten years of GIS diffusion research.

Endnote

1. In 1988, Daratech's "GIS Markets and Opportunities" reported revenues of $527.5 million from sales of GIS hardware, software, and services for North American companies (as quoted in *Photogrammetric Engineering and Remote Sensing*, December 1990, p.1603) and Matrix Consultants of Boston estimated total GIS revenues at $750 million for that same year (as quoted in *GIS World*, Nov/Dec 1989). The *International GIS Sourcebook* reported a worldwide industry revenue of $2.4 billion dollars in 1990, based primarily on U.S. sources, and predicts that the total U.S. federal GIS commitments by the year 2000 will be at least $6 billion (1991-1992 International GIS Sourcebook, p.4). *The Economist* reported the GIS industry's revenues rose in America to around $3.5 billion in 1991 (*The Economist*, March 21, 1992, p. 69). GIS Research Corporation projects $75 to $95 billion being spent in the U.S. alone on GIS through the year 2000 (as reported in Forest, Montgomery & Juhl (1990) *Intelligent. Infrastructure Workbook: A Management Level Primer on GIS*.)

References

Department of the Environment (1987) Handling Geographic Information, *Report to the Secretary of State for the Environment of the Committee of Enquiry into the Handling of Geographic Information*, Lord Chorley (Chairman), Her Majesty's Stationery Office, London, 208pp.

Sinton, D. (1992) Reflections on 25 Years of GIS, insert in *GIS World*, February 1992.

Tomlinson, R.F. (1988) The Impact of the Transition from Analogue to Digital Cartographic Representation, *The American Cartographer*, 15(3) 249-262.

THE DIFFUSION OF INNOVATIONS MODEL

EVERETT M. ROGERS
Center for Advanced Study in the Behavioral Sciences
202 Junipero Serra Blvd.
Stanford, CA 94305, U.S.A.
and
Annenberg School for Communication
University of Southern California
Los Angeles, CA 90089-0281, U.S.A

ABSTRACT. The main elements in the diffusion of innovations model are explained, stressing the role of space in diffusion. Special aspects of the diffusion of geographic information technologies are discussed. Particular attention is given (1) to re-invention of the innovation by users, and (2) the "critical mass" in the rate of adoption. The classical diffusion model grew out of applied research carried out some 50 years ago, that was conducted with normative value-orientations among U.S. farmers (for instance, the pro-innovation bias). The diffusion model assumes that diffusion is an essentially social process, featuring the individual-to-individual transmission of subjective evaluations of an innovation, based primarily on personal experiences with the new idea. Today, more than 4,000 diffusion research publications have appeared, authored by scholars in each of the social science disciplines. These scholars display general agreement about the main elements in the diffusion process, and about the appropriate research methodologies for carrying out such studies. Research on the spread and adoption of geographic information technologies have, and are, reinvigorating the diffusion research tradition.

1. Introduction

This paper describes the diffusion of innovations model, assesses the current state-of-the-art of diffusion theory, and synthesizes empirical investigations of diffusion.[1] The main bases for writing the present paper are (1) my 38 years of experience in conducting diffusion research, (2) my ongoing review and synthesis of the research literature on diffusion, which I actively gather, and (3) my reading of several reviews and critiques of diffusion research, written from a variety of disciplinary viewpoints.[2] My viewpoint is that of a sympathetic insider to the diffusion research field, with an attempt to present a balanced perspective, including a critical view of diffusion work.

One way to understand the evolution of diffusion research is to divide this history into three eras, beginning with the pre-paradigm period of the pre-1960s.

I. Masser and H. J. Onsrud (eds.), Diffusion and Use of Geographic Information Technologies, 9–24.

2. Era #1: Steps Toward a Paradigm

The diffusion model started from humble origins. The first important diffusion investigations concerned the spread of agricultural innovations among U.S. farmers. This research was sponsored by the agricultural R&D units that created the innovations (state agricultural experiment stations attached to land-grant universities), and by the state agricultural extension services responsible for diffusing the new ideas to American farmers. A normative orientation characterized these early diffusion studies. It was assumed that the innovations should be diffused and adopted by all farmers, that the rate of adoption should be more rapid, and that the innovations should be neither re-invented nor rejected, nor discontinued. These early agricultural diffusion studies were mainly conducted by rural sociologists in colleges of agriculture. An intellectual emphasis was placed upon determining the social characteristics of individuals who were relatively earlier and later to adopt the agricultural innovations, and upon the role of the social relationships among farmers in a community which served as interpersonal channels for diffusion.

2.1 THE HYBRID CORN STUDY

The prototype early diffusion study is the Ryan and Gross (1943) investigation of the spread of hybrid seed corn among Iowa farmers. The hybrid corn study, more than any other diffusion study, influenced the methodology, theoretical framework, and interpretations of later diffusion scholars. Ryan and Gross did not explicitly state the diffusion paradigm that was to emerge around 1960, but their hybrid corn study contained the intellectual seeds for the diffusion model. In her investigation of the invisible college of rural sociologists studying diffusion as of the mid-1960s, Crane (1972), p. 74) identified the scholars who first utilized a new concept and/or methodological tool in studying diffusion. The Ryan and Gross (1943) study is credited with launching 15 of the 18 most widely used intellectual innovations in the rural sociology diffusion research tradition. By any measure, the Ryan and Gross study was a landmark. We are still standing in its intellectual shadow, mainly for better, but in certain respects, for worse.

Information is patterned matter-energy that affects the probabilities available to an individual making a decision (Rogers and Kincaid, 1981). A lack of information about an innovation is seldom an important factor retarding an individual or an organization from adopting a new idea. Often a more important problem is coping with information overload, so as to find the most relevant information in deciding to adopt or reject the innovation. A key factor in deciding to adopt an innovation is to find trustworthy information about it, so as to deal with the uncertainty surrounding the innovation. The credibility of an information source, as perceived by an individual, is thus crucial. Most individuals evaluate technological innovations through the personal experiences of adopters who are much like themselves. If the near-peer with whom an individual discusses an innovation is a satisfied adopter, this interpersonal network influence is a force for adoption; if the near-peer is a dissatisfied adopter, rejector, or discontinuer, the individual is influenced negatively toward the innovation. Social modeling of the innovation behavior of others is a fundamental process underlying the diffusion of innovations (Bandura, 1977).

2.2 THE SOCIAL NATURE OF DIFFUSION

The Coleman and others' (1966) study of the diffusion of a new antibiotic drug (tetracycline) among physicians in four Illinois communities, conducted in the mid-1950s, marked a turning point in recognizing the social nature of the diffusion process. James Coleman and his colleagues in the Bureau of Applied Social Research at Columbia University focused on the doctor-to-doctor contagion process through which personal evaluations of the new drug were communicated in a medical community. They found an S-shaped diffusion curve characterized the behavior of earlier and later adopters and found a rather limited role of commercial detailmen (that is, salesmen) and pharmaceutical advertising in medical journals. Mainly, doctors depended on other doctors for personal opinions about whether to adopt tetracycline. They did not view commercial sources and channels of communication as credible (these channels/sources, however, did create awareness-knowledge of the new drug).

A seeming paradox concerns the introduction and early diffusion of an innovation in a system, and the norms[3] of the system. Initially, the norms usually oppose the innovation, which explains why innovators (the very first individuals to adopt) are viewed by members of a system as social deviants at the time that they adopt a new idea. The innovation gradually fits with the norms of the system, especially when the opinion leaders in the system adopt (these leaders generally are early adopters, the second category to adopt a new idea, rather than innovators). When these influentials adopt, the norms of the system become favorable to the innovation. Once an innovation becomes compatible with a system's norms (usually at a rate of adoption somewhere around 10 to 20 percent), it is then difficult or impossible to halt the further diffusion of the innovation. Thus for those in charge of promoting the diffusion of an innovation in a system, the strategic goal is to reach the point at which the "critical mass" occurs (Rogers, 1991).

2.3 LACK OF INTELLECTUAL INTEGRATION

Although the various elements that were later to constitute the diffusion model were emerging in the years prior to 1960, at this stage diffusion studies mainly consisted of unintegrated empirical findings. A common theoretical framework was not yet widely accepted by the various sets of diffusion scholars. Each disciplinary research tradition had little understanding of diffusion study in other traditions. For example, Coleman and others (1966) expressed their surprise in finding out about the rural sociology diffusion research tradition some time after they had gathered the medical drug diffusion data. Neither the rural sociologists nor the medical diffusion scholars knew about the spatial diffusion studies that had been conducted by Professor Torsten Hägerstrand (1952) at Lund University in Sweden. His pioneering work, based on a sophisticated model for using Monte Carlo simulation techniques, was to remain largely unknown in the United States for 13 years (Hägerstrand, 1969).

Today spatial diffusion research, mainly conducted by geographers, has advanced considerably, and has infected the entire geography discipline: "Within the whole range of geographical research, the study of spatial diffusion occupies a central place. The dynamics by which a phenomenon originally located at one point becomes transferred to another is a question which is as difficult to answer as it is easy to pose" (Cliff and others, 1981, p. 191). For example, spatial diffusion scholars found that measles epidemics in Iceland were best explained by a mixed model containing both (1) a hierarchical element, in which an innovation spreads down the urban-size hierarchy, and (2) a contagious element, in which an innovation spreads outwardly from earlier adopting centers to other areas (Cliff and others, 1981, p. 184). Clearly, space is an important factor in the diffusion of every innovation, even though diffusion scholars other than geographers have not paid sufficient attention to spatial variables. Individuals talk to other individuals about an innovation in a social process that is space-dependent: We talk to others who are close to us.

The most up-to-date count of the number of diffusion publications by discipline/research tradition as of 1982 (Rogers, 1983, pp. 52-53), shows the following:

1.	Rural sociology	26%
2.	Communication	12%
3.	Education	11%
4.	Marketing	10%
5.	Sociology (other than rural sociology and medical sociology	10%
6.	Public health/medical sociology	7%
7.	Economics (including agricultural economics)	6%
8 .	Anthropology	4%
9.	Geography	4%
10.	Psychology	2%
11 .	Political science	1%
12.	Other disciplines	7%
	Total	100%

Despite their lack of intellectual integration by 1960, about 500 diffusion researches had been completed (Figure 1). And the diffusion approach had spread to most of the social sciences, although at this time it had only achieved a very small beachhead in some disciplines.

3. Era #2: The Diffusion Paradigm Emerges

The decade of the 1960s was a period of intellectual coming-together for innovation diffusion research as the "dominant paradigm" was formed. The 1960s were also a time of proliferation in the number of diffusion publications. Figure 1 shows this sharp, four-fold increase during the 1960s, from about 500 diffusion publications in 1960 to around 1,900 in 1970. But the most important trend during Era #2 was the infection of every social science discipline by the diffusion paradigm, and the intellectual integration that resulted.

One reason for the unusual number of scientific disciplines engaged in diffusion research traces to the fact that no single discipline can offer a complete explanation of

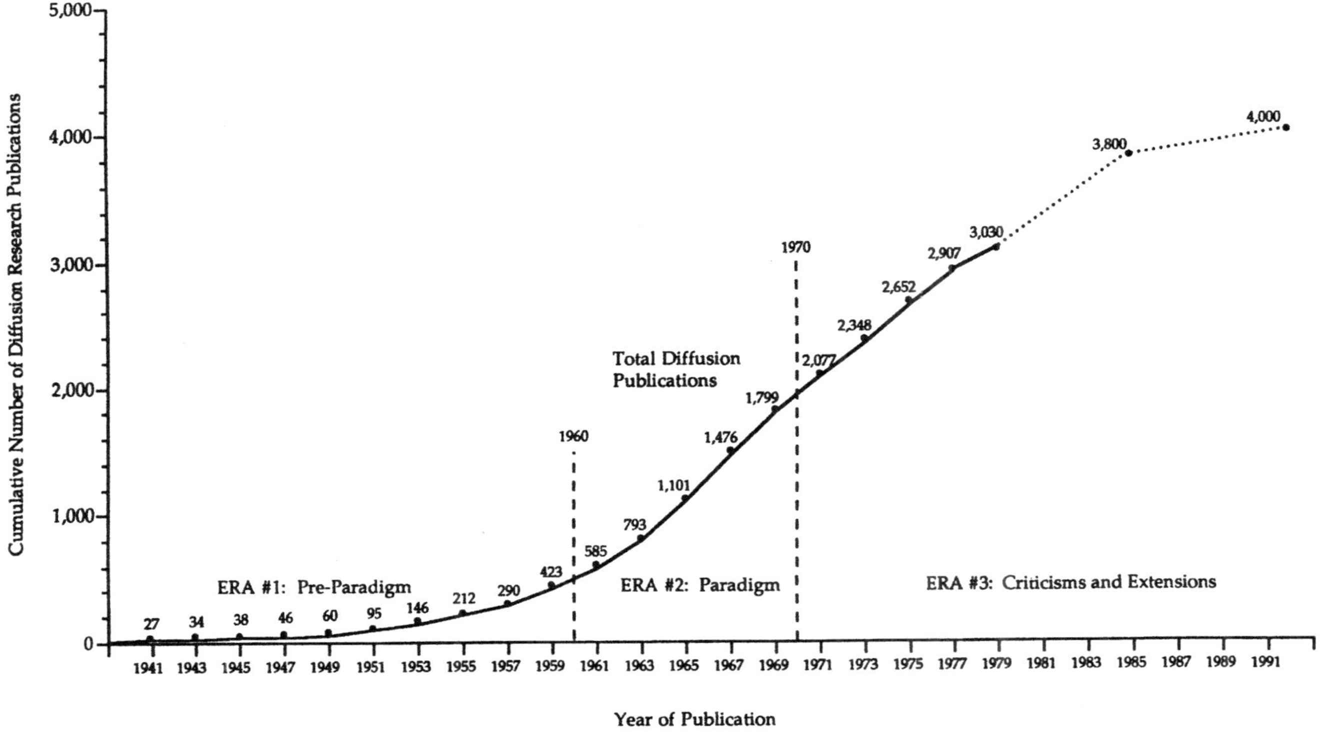

Figure 1. Cumulative Number of Diffusion Research Publications by Year.
***Source*: Based on Rogers (1983, p. 47).**

diffusion behavior. Certainly the diffusion process involves communication of information about the innovation, so it is understandable why communication researchers study diffusion. Interpersonal social relationships and such social structural factors as system norms are centrally involved in diffusion, so sociologists also focus on diffusion for example (Coleman and others, 1966). Social modeling of one individual's innovation behavior by another individual lies at the heart of the diffusion process, so certain psychologists (especially social psychologists) study diffusion (for example, Bandura, 1977 and 1986, pp. 142-181). Adoption decisions inevitably involve economic choices by individuals and organizations, so economists investigate diffusion (for example, Griliches, 1957). Spatial distance impedes the person-to-person diffusion of innovations in a system, and so geographers have conducted diffusion studies (for instance, Hägerstrand, 1969; Brown, 1981). Anthropologists investigate cultural variables in the adoption of new ideas, and the consequences of innovation (for example, Pelto, 1973). Political scientists analyze the spread of new laws and of new government programs among state legislators in America (for instance, Walker, 1966; Gray, 1973). Innovation occurs in organizations (as well as on the part of individuals as decision-making units), and so organization scholars also study innovation diffusion (for example, Zaltman and others, 1973). Because the diffusion approach offers one means of understanding social change, a topic central to every social science discipline, every social science has a piece of the action.

Innovation diffusion has emerged as one of the most multidisciplinary research topics in the social sciences today. Paisley's (1984) citation analysis shows that a book about innovation diffusion is the second most widely-cited publication cited in all social science journals published in the past decade (first place went to a research methodology textbook about experimentation).

The diffusion of innovations is a research front that is unique (1) in the proliferation of studies that have been completed by social scientists, and (2) in the wide interest that it has captured among social science disciplines. "The diffusion model is a conceptual paradigm with relevance for many disciplines" (Rogers, 1983, p. 89). The 1960s decade was particularly important in the history of diffusion because of the emergence of a common paradigm that was generally accepted across the dozen or so social science disciplines concerned with diffusion.

3.1 PARADIGMS AND INVISIBLE COLLEGES

A given field of scientific research usually begins with a major breakthrough or reconceptualization, called a "revolutionary paradigm" by Kuhn (1970), that provides a new way of looking at some phenomenon. A paradigm is a scientific approach to some phenomena that provides model problems and solutions to a community of scholars. The new paradigm typically sets off considerable intellectual effort as scientists are attracted to advance the new conceptualization with their research or to disprove certain aspects of the new paradigm. Gradually, a scientific consensus about the paradigm is developed, and the invisible college (the informal network of researchers who form around an intellectual paradigm to study a common topic) declines in scientific interest as fewer findings of an exciting nature are turned up. These are the usual stages in the normal growth of science,

Kuhn (1970) claims.

Research on the diffusion of innovations followed the rise-and-fall stages described by Kuhn (1970) rather closely (Crane, 1972), although the final stage of demise has not begun (except in certain of the disciplines that study diffusion). The hybrid corn diffusion study by Bryce Ryan and Neal Gross (1943), described previously, began a new approach to the study of diffusion, that was soon followed up by an increasing number of scholars. However, in the late 1980s the rate of increase in the number of diffusion publications per year began to level off somewhat (see Figure 1). Rural sociology, the specialty in which diffusion study began 50 years ago, lost interest in investigating the diffusion process in recent decades, as most of the interesting problems had been explored. But in very recent years, some rural sociologists have again begun to conduct diffusion inquiries, this time of conservation and other environment-related innovations.

An invisible college centered on an intellectual paradigm provides the typical scientist with the information that he or she needs to reduce the uncertainty of the research process. Of the many alternative directions that a research project might pursue, a paradigm guides a researcher toward one general approach. Thus, the paradigm and the invisible college of scientists that follow the paradigm provide a researcher with a source of intellectual security in the uncertain world of a research front. A paradigm can also prevent seeing important dimensions of the process of study, as a result of trained incapacity. So a paradigm provides a means of scientific simplification, for better or worse.

3.2. WHAT IS THE DIFFUSION PARADIGM?

The main elements in the diffusion of new ideas are: (1) an innovation, (2) which is communicated through certain channels, (3) over time, (4) among the members of a social system (Figure 2). An innovation is an idea, practice, or object perceived as new by an individual or other unit of adoption (Rogers, 1983). The characteristics of an innovation, as perceived by the members of a social system, determine its rate of adoption. Five attributes of innovations are: (1) relative advantage, (2) compatibility, (3) complexity, (4) trialability, and (5) observability.

A communication channel is the means by which messages get from one individual to another. Mass media channels are usually more effective in creating awareness-knowledge of innovations, whereas interpersonal channels are more effective in forming, and in changing, attitudes toward a new idea, and thus indirectly influencing the decision to adopt or reject a new idea. Most individuals evaluate an innovation that they are considering adopting, not on the basis of scientific research by experts, but through the subjective evaluations of near-peers who have previously adopted the innovation (as discussed previously).

Time is involved in diffusion in (1) the innovation-decision process, (2) innovativeness, the degree to which an individual or other unit of adoption is relatively earlier in adopting new ideas than are other members of a social system, and (3) an innovation's rate of adoption. The innovation-decision process is the mental process through which an individual or other decision-making unit passes from first knowledge of an innovation, to forming an attitude toward the innovation, to a decision to adopt or reject, to

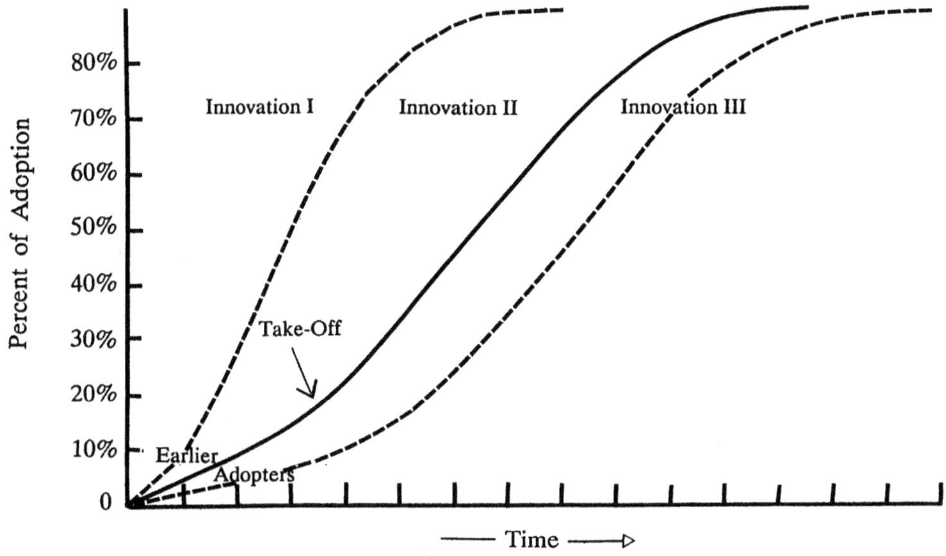

Figure 2. Diffusion is the process by which an innovation is communicated through certain channels over time among the members of a social system.

implementation of the new idea, and to confirmation of this decision (Figure 3). Innovativeness is the most frequently-studied issue in diffusion research. The purpose of such a study is to determine factors related to innovativeness.

Rate of adoption is the relative speed with which an innovation is adopted by members of a social system; Figure 2 shows three innovations that vary in their relatively more rapid, and relatively slower, rate of adoption. Innovations that diffuse more rapidly are generally characterized by greater relative advantage, more compatability, less complexity, more trialability, and greater observability .

A social system is a set of interrelated units that are engaged in joint problem-solving to accomplish a common goal. The units in a system may be individuals, firms, or families. A system has structure, the patterned arrangements of the units in a system, and this structure provides regularity and predictability to behavior in a system, including the adoption of innovations.

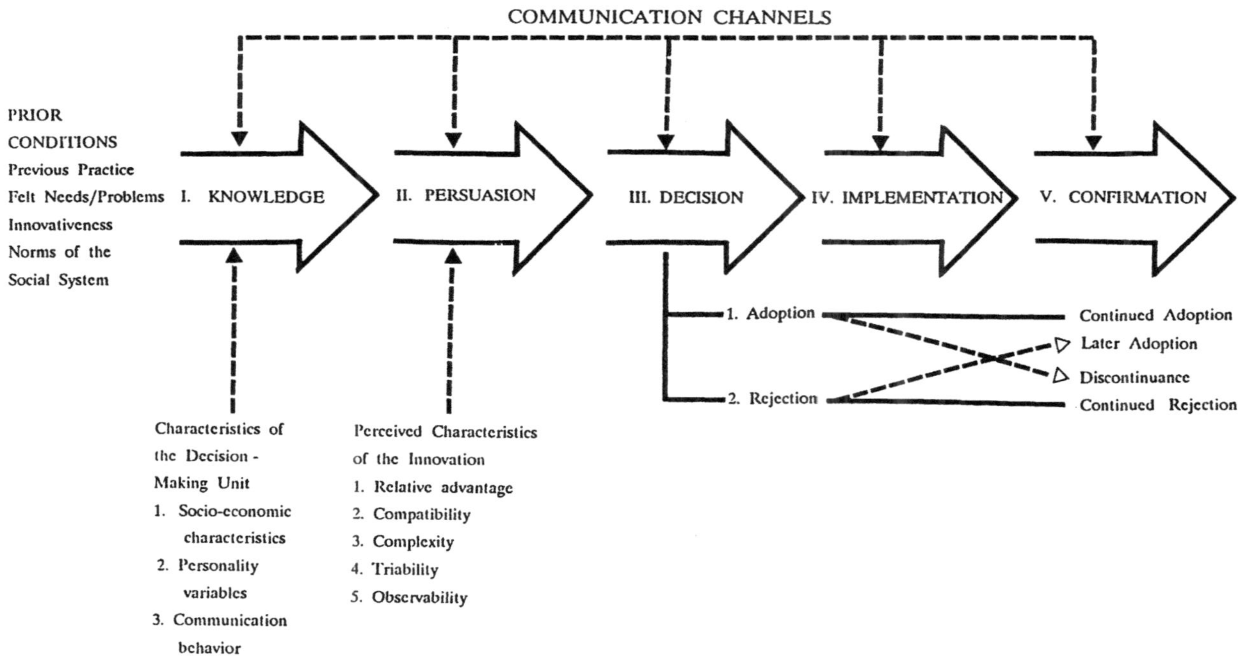

Figure 3. Stages in the innovation-decision process (Rogers 1983, p. 165).

These four main elements in diffusion, the key definitions of concepts, and the theoretical relationships among these conceptual variables constitute the diffusion paradigm. It grew gradually out of numerous empirical investigations, which pointed to the regularities of innovation diffusion behavior across a wide variety of innovations (e.g., hybrid corn, snowmobiles, microcomputers), types of respondents (e.g., farmers, doctors, managers of firms), and different national and cultural contexts.

Until 1960, most diffusion studies were conducted in the United States. Thereafter, diffusion research gained attention in Europe and during the 1960s, in many Third World nations. The diffusion paradigm fit well with the rural, health, nutrition, and family planning development programs being carried out in Latin American, Africa, and Asia. While the diffusion paradigm seemed applicable, especially at first, in Third World nations, the development programs occasionally outran the diffusion model, forcing certain modifications in the paradigm.

4. Era #3: Introspective Criticism and Extensions of the Paradigm

From 1970 to the present, the number of diffusion publications has more than doubled from about 1,900 to over 4,000 (see Figure 1). Until 1970, almost no intellectual criticism of the diffusion paradigm had appeared. That changed as both diffusion scholars themselves and outsiders offered critical perspectives of the diffusion paradigm. Era #3 is also a time in which applications of the basic paradigm were extended in various ways.

4.1 CRITICISMS OF DIFFUSION RESEARCH

Although diffusion research has made numerous important contributions to our understanding of human behavior change, its potential might have been even greater had it not been characterized by certain shortcomings. Every field of scientific research makes certain simplifying assumptions about the complex reality that it studies. Such assumptions are built into the intellectual paradigm that guides the scientific field. Often these assumptions are not recognized, even as they affect such important matters as what is studied and what is ignored, and which research methods are favored and which are not followed. So when a scientist follows a theoretical paradigm, he or she puts on a set of intellectual blinders that help the researcher to avoid seeing much of reality, while focusing narrowly on phenomena of central interest.

What are the main criticisms of diffusion research?

1. The pro-innovation bias is the implication in much diffusion research that an innovation should be diffused and adopted by all members of a social system, that it should be diffused more rapidly, and that the innovation should be neither re-invented nor rejected nor discontinued. Seldom is the pro-innovation bias straightforwardly stated in diffusion publications . Rather, the bias is assumed and implied. This lack of recognition of the pro-innovation bias makes it especially troublesome, and potentially dangerous in an intellectual

sense. The bias leads diffusion researchers to ignore the study of ignorance about innovations, to underemphasize the rejection or discontinuance of innovations, to overlook re-invention, and to fail to study anti-diffusion programs designed to prevent the diffusion of "bad" innovations (like drugs, or cigarette consumption, or aircraft skyjacking, for example).

2. Individual-blame is the tendency to hold an individual responsible for his or her problems, rather than the system of which the individual is a part. In other words, an individual-blame orientation implies that "if the shoe doesn't fit, there's something wrong with your foot." An opposite point of view would blame the system, not the individual; it might imply that the shoe manufacturer or the marketing system could be at fault for a shoe that does not fit.

When coupled with the pro-innovation bias, the individual-blame viewpoint assumes that an individual is at fault for not adopting an innovation. Perhaps on closer analysis it would become apparent that the system is at fault for not providing an innovation more appropriate to the individual's needs, and so the individual may be quite justified in rejecting the new idea. Or perhaps the innovation is appropriate, but it is simply not available to the potential adopter; an example would be the Indian farmer who cannot adopt a new rice variety because it is not available in his village.

As in the case of the pro-innovation bias in diffusion research, perhaps one of the most important ways to guard against the individual-blame bias is to be aware that it may exist. To what extent does diffusion research have an individual-blame bias? It is difficult to assess in a precise way, but on careful reading there is a certain flavor of individual-blame in many diffusion publications. An individual-blame orientation is not, in and of itself, always inappropriate. Perhaps individual-level variables are the most appropriate to investigate in a particular diffusion study. But in many cases such a "psychological approach," centering on individual-level variables, is not a complete explanation of the diffusion behavior being investigated. Diffusion scholars should keep a more open mind toward system-blame explanations of diffusion behavior than they have in the past.

4.2 THE INNOVATION PROCESS IN ORGANIZATIONS

During the first several decades of studying the diffusion of innovations, this research focused mainly on individuals as the units of adoption. Then it was realized that many innovations are adopted not by individuals, but by organizations. This shift in the unit of adoption is particularly important in the case of the new communication technologies, which frequently are adopted by organizations: For instance, microcomputers in schools, and electronic mail and teleconferencing in business firms.

Studying the innovation process in organizational settings represents a shift from the classical diffusion model in important ways. For example, the innovation process in organizations is usually considered successful if it leads to implementation (including routinization of the new idea), not just to the adoption decision per se. The individual plays an active, creative role in the innovation process by matching the innovation with a perceived organizational problem, and perhaps in re-inventing the innovation. An innovation should not be conceived as a fixed, invariant, and static element in the innovation process, but as a flexible and adaptable idea that is consecutively defined and redefined as the innovation process gradually unfolds, often leading to re-invention.

4.3 NETWORKS OF DIFFUSION

Era #3 is also the period when diffusion investigations began to focus on the interpersonal networks through which an innovation spreads (for example, Rogers and Kincaid, 1981). The crucial role of such networks has been realized since the Coleman and others' (1966) drug diffusion research, but precise means of analyzing networks were not available until the 1970s and 1980s. Network analysis methods are today used to turn up occasionally surprising findings, such as Granovetter's (1973) theory of "the strength of weak ties" (that new ideas often flow among individuals who are strangers or casual acquaintances, rather than close friends). At least in some situations, one is more likely to learn about a new idea from brief contacts with numerous acquaintances than from frequent contact with a small circle of close friends (Bandura, 1986, p. 152).

Network analyses of innovation diffusion help provide an improved understanding of the micro-level of behavior. Diffusion network studies focus on the interpersonal information-exchanges through which a new idea spreads. Such research has helped us to see how and why the S-shaped diffusion curve takes off at about 10 or 15 or 20 percent adoption when the "critical mass" occurs.

4.4 NEW COMMUNICATION TECHNOLOGIES

What is different about the diffusion of the new media from that generally found for other new ideas? A "critical mass" of adopters of an interactive communication technology is necessary for the utility of the new idea to be sufficient for an individual to adopt. The usefulness of a new communication system increases for all adopters with each additional adopter. An extreme illustration is provided by the first individual to adopt a telephone in the 1870s; this interactive technology had zero utility until a second individual adopted.

The critical mass dimension of interactive communication technologies is a crucial factor in their adoption and use. The critical mass aspect of the new media traces from their interactive quality. If these innovations were not themselves a very special kind of communication channel, the critical mass dimension would not be so important. In the first stages of an innovation's diffusion, the lack of critical mass is a negative influence slowing the rate of adoption. Then, once a critical mass of adopters is using a new interactive technology, its further diffusion proceeds rapidly (Figure 4).

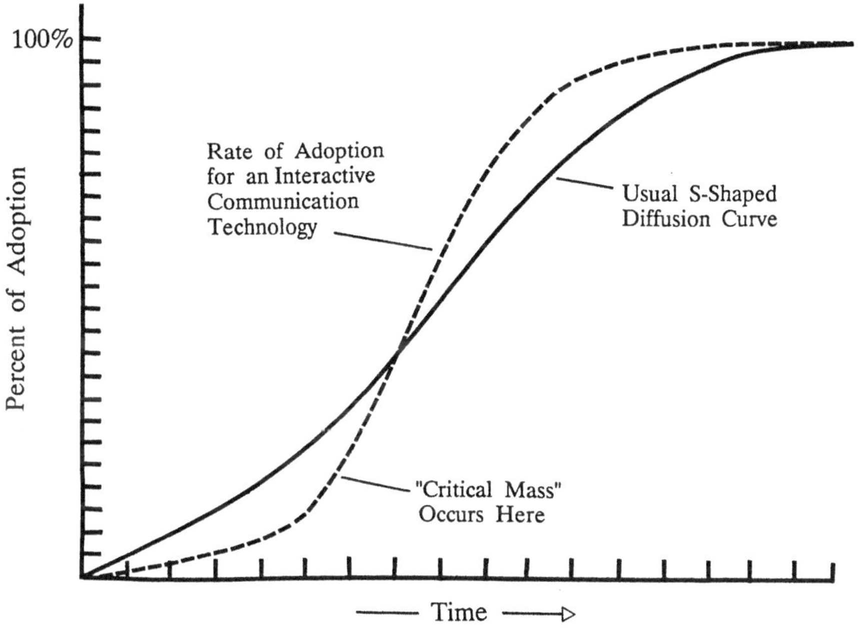

Figure 4. The rate of adoption (1) for the usual innovation, and (2) for an interactive communication technology, showing the "critical mass"

4.5 DISTINCTIVE ASPECTS OF GIS DIFFUSION

Certain dimensions of GIS affect their rate of diffusion:

1) Geographical information systems are computer-based tools for analysis, and can be applied by users to an extremely wide range of problems. Thus, GIS innovations are frequently re-invented as they diffuse. Such a high degree of re-invention creates problems for scholars who seek to study the diffusion of GIS, as each "adoption" may represent a quite different behavior. The re-inventability of GIS also speeds the rate of adoption of these innovations.

2) The commercial products available for GIS are constantly changing as more advanced technologies rapidly replace previous generations of the innovation. Such a rapidly changing technology for GIS may slow the rate of adoption of these innovations (a) as adopters know that their investment today to acquire a GIS product might be made obsolete tomorrow, and (b) as potential adopters must continually learn how to use more advanced systems.

3) GIS are currently characterized by a lack of user-friendliness, so that adopters must devote considerable efforts in order to learn how to use them effectively. Such complexity of GIS slows their rate of adoption.

5. Conclusions

The first important studies of the diffusion of innovations were conducted fifty years ago. Now, some 4,000 diffusion researches later, we understand a great deal about how new ideas spread and are adopted, why some innovations diffuse more rapidly than others, which individuals are relatively early versus late to adopt a new idea, and the role of different sources and channels of communication at stages in the innovation process. A paradigm for diffusion study formed about 30 years ago, leading to a proliferation in the number of diffusion studies conducted per year, a rate of research activity that eventually exhausted itself and slowed down. In very recent years the promising areas for diffusion inquiry, in my opinion, are (1) the diffusion of new information technologies of an interactive nature, (2) in organizations, (3) with attention to the process of re-invention, and (4) the role of the critical mass . I have just described the exact intellectual intersection at which study of the diffusion of geographic information technologies is located.

6. Endnotes

1. The present paper draws upon various of the author's previous writings about diffusion, especially from Chapter 1 of The Diffusion of Innovations, Third Edition by Everett M. Rogers. copyright c 1962, 1971, 1983 by The Free Press, a Division of Macmillan, Inc. Reprinted with the permission of the publisher.

2. Including communication (McAnany, 1984; Rogers, 1962 and 1983; Rogers with Shoemaker, 1971), geography (Brown, 1981), marketing (Robertson, 1971; Gatignon and Robertson, 1985), psychology (Bandura, 1986, pp. 142-181), and political science (Downs and Mohr, 1976).

3. Norms are the established behavior patterns for the members of a social system (Rogers, 1983, p. 27). They define a range of tolerable behavior and serve as a guide for the members of a system.

7. References

Bandura, Albert (1977) Social Learning Theory, Prentice-Hall, Englewood Cliffs, New Jersey.

Bandura, A. (1986) Social Foundations of Thought and Action: A Social Cognitive Theory, Prentice-Hall, Englewood Cliffs, New Jersey.

Brown, Lawrence A. (1981) Innovation Diffusion: A New Perspective, Methuen, New York.

Coleman, James, Menzel, Herbert, and Katz, Elihu (1966) Medical Innovation: A Diffusion Study, Bobbs-Merrill, New York.

Cliff, A. D., Haggett, P., Ord, J. K., and Versey, G. R. (1981) Spatial Diffusion: An Historical Geography of Epidemics in an Island Community, Cambridge University Press, New York.

Crane, Diana (1972) Invisible Colleges, University of Chicago Press.

Downs, George W., and Mohr, Lawrence B. (1976) Conceptual issues in the study of innovations, *Administrative Science Quarterly* 21: 700-714.

Gatignon, Hubert, and Robertson, Thomas S. (1985) A propositional inventory for new diffusion research, *Journal of Consumer Research* 11: 849-867.

Granovetter, Mark S. (1973) The strength of weak ties, *American Journal of Sociology* 78: 1360-1380.

Gray, Virginia (1973) Innovation in the states: A diffusion study, *American Political Science Review* 67: 1174-1185.

Griliches, Zvi (1957) Hybrid corn: An exploration in the economics of technological change, *Econometrica* 25: 501-522.

Hägerstrand, Torsten (1952) The Propagation of Innovation Waves, Lund, Sweden, University of Lund, Lund Studies in Geography 4.

Hägerstrand, Torsten (1969) Innovation Diffusion as a Spatial Process, University of Chicago Press.

Kuhn, Thomas S. (1970) The Structure of Scientific Revolutions, University of Chicago Press.

McAnany, Emile G. (1984) The diffusion of innovation: Why does it endure? *Critical Studies in Mass Communication* 1: 439-442.

Paisley, William (1984) Communication in the communication sciences, In B. Dervin and M.J. Voigt (eds.), Progress in the Communication Sciences, Ablex, Norwood, New Jersey.

Pelto, Pertti J. (1973) The Snowmobile Revolution: Technology and Social Change in the Arctic, Cummings, Menlo Park, California.

Robertson, Thomas S. (1971) Innovative Behavior and Communication, Holt, Rinehart and Winston, New York.

Rogers, Everett M. (1962) Diffusion of Innovations, Free Press, New York.

Rogers, Everett M. (1986) Communication Technology, Free Press, New York.

Rogers, Everett M. (1991) The "critical mass" in the diffusion of interactive technologies in organizations, In Kraemer, K. L. (ed.), *The Information Systems Research Challenge: Survey Research Methods*, Harvard Business School Press, Boston, 3: 245-263.

Rogers, Everett M., and Kincaid, D. Lawrence (1981) Communication Networks: Toward a New Paradigm for Research, Free Press, New York.

Rogers, Everett M., and Shoemaker, F. Floyd (1971) Communication of Innovations: A Cross-Cultural Approach, Free Press, New York.

Ryan, Bryce, and Gross, Neal C. (1943) The diffusion of hybrid seed corn in two Iowa communities, *Rural Sociology* 8: 15-24.

Walker, Jack L. (1966) The diffusion of innovations among the American states, *American Political Science Review* 63: 880-899.

Zaltman, Gerald, Duncan, Robert, and Holbek, Jonny (1973) Innovations and Organizations, Wiley, New York.

SECTION 1:
RESEARCH METHODOLOGY AND INTERDISCIPLINARY PERSPECTIVES

GERARD RUSHTON
Department of Geography
The University of Iowa, 316 JH
Iowa City, IA 52242-1316, USA

Research in the adoption of information technologies covers many perspectives from the study of the behavior of key actors in their organizational setting to action-oriented research where interventions designed to encourage acceptance and accelerate diffusion are introduced and their effects studied. The four papers that follow illustrate these perspectives. They are written by authors whose background is in information technology. Two of them, Eason and Goodman, have not previously been associated with GIS systems so their questions apply to the kinds of questions they would ask about the adoption of any advanced information technology. Eason asks what would be necessary to obtain more effective diffusion of GIS technology? He attempts to identify barriers to the adoption of GIS technology and asks how these might be removed. His premise is that a major obstacle to diffusion is the lack of effective means of gaining organizational and user acceptance of technical systems. Adopting the approach of Walton (1989) he asks how users and their organizations can coalesce behind an alignment of strategies for implementing advanced information technologies, and acquire widespread commitment, support, ownership, competence and mastery of the new systems. The integration of the technical and social conditions that preserve this alignment is, Eason maintains, fragile and techniques for sustaining integration are therefore of prime importance. Since these "user-centered design" approaches have only been implemented in ad hoc situations, Eason advocates an action-research program in which the approaches are diffused and, concurrently, research is conducted on the efficacy of the different steps in the process. The main body of Eason's paper describes ten principles for user-centered systems implementation. He concludes by identifying issues where research is needed and by calling for an action-oriented research agenda in which research organizations participate in the technology diffusion process and systematically examine the results of different techniques for promoting user-centered implementations of complex technical systems.

Like Eason, Goodman also argues for a dynamic process of redesign of technology implementation mechanisms although he also argues that much of this redesign can be accomplished prior to adoption through techniques such as simulation. He argues for the broadest possible context for considering interventions that may help or hinder implementation. When the unique characteristics of the new technology are combined with the unique characteristics of the adopting organization, it is unlikely, Goodman argues, that any single combination of values of variables will predict successful adoption. Rather, one should not

I. Masser and H. J. Onsrud (eds.), Diffusion and Use of Geographic Information Technologies, 25–26.

expect to know, a priori, exactly which variables should be manipulated in which way to effect adoption. Instead, he suggests that consideration of seven critical issues are essential in implementing new information technologies; defining the appropriate social system for implementing new technology, defining the meaning of success, understanding the object and socially constructed characteristics of new information technology, and the roles of the socialization, commitment, rewards, and learning processes on the implementation of new technology. Goodman believes that each of the principles he espouses is relevant to GIS implementation. Following the Goodman paper is a response by Peter Burrough. In his commentary, Burrough argues that knowledge of both what a given technology does and how it does it should be presumed to be of critical importance in the rapid appraisal and strategic use of new technologies and for reducing the risks associated with implementation.

Huxhold's paper focuses on the use of GIS in local governments in the United States. He advocates use of information engineering tools in designing GIS use in local governments. Such an approach, which he asserts has never been attempted in a GIS in U.S. local government, assumes that the organization adopts enterprise-wide goals that focus on the information needs of all users, rather than meeting sub-goals associated with individual offices or departments, as currently prevails. He ascribes the low rate of adoption of GIS in U.S. local governments--which he estimates at just over 2% in 1991--to typical organizational structures in which department and office heads are appointed by and report to different governmental entities. The greater part of his paper describes a typical evolution of computing management in local governments. He argues that knowing the computing management phase of any local government entity (Kraemer et. al. 1989) is a good predictor of its likely success in adopting GIS technology.

Moore, representing the field of management information systems, is a member of the recently formed Diffusion Interest Group in Information Technology (DIGIT). This group investigates processes by which individuals decide to adopt or reject personal use of information technology with the goal of influencing the shape of the diffusion curve. In this spirit, Moore empirically examines the adoption of information technology by end-users. He tests the validity of Ajzen and Fishbein's "theory of reasoned action" in which an individual's behavior is a function of his or her intentions to perform that behavior and that one's intentions are a function of one's own attitude and one's subjective norm toward the behavior. By interviewing managers and end users of GIS in oil exploration companies and by applying the theory of reasoned action to GIS diffusion, Moore concludes that three perceptions of adopting GIS appear to be predominant in the decision to acquire GIS; namely compatibility, relative advantage, and trialability.

Taken together, these four papers introduce many of the principles that are investigated further in the chapters that follow.

References

Kraemer, K., King, J., Dunkle, D., and Lane, J. (1989) *Managing Information Systems: Change and Control in Organizational Computing*, Jossey-Bass Inc., San Francisco, California.

Walton, R.E. (1989) *Up and Running*. Harvard Business School Press, Boston.

GAINING USER AND ORGANISATIONAL ACCEPTANCE FOR ADVANCED INFORMATION SYSTEMS

KENNETH D EASON
HUSAT Research Institute
Department of Human Sciences
Loughborough University of Technology
Loughborough, Leics, LE11 3TU,
United Kingdom

ABSTRACT. An argument is advanced that the implementation of advanced information systems in organisations is often limited by the non-acceptance of the technology by individual users and the organisation at large. The dominant design method for advanced information systems is technocentric and does not facilitate systematic attention to user and organisational issues. The paper advances a model of user-centred design as an alternative and, by reference to research and practical experience of systems development, discusses the methods and research necessary to institutionalise and test such an approach. The paper examines ten user-centred design principles in four groups. First, the requirements to achieve system integration between business objectives, organisational development and technical innovation. Second, the structures necessary for users and technical staff to work together and third the processes by which they can address and evaluate user needs and future system options. Finally, the processes needed for successful implementation and user mastery of new opportunities. The paper concludes that to advance and test this kind of methodology an action research approach is necessary in which current practice is encoded for general use in order that systematic evaluations can be undertaken.

1. The Up-take of Advanced Information Systems

Studies of the spread of Geographical Information Systems have largely been informed by models of the diffusion of innovations (Rogers 1983). Such studies treat an organisation as a single unit and adoption is the point at which the unit takes up the innovation. As Rogers (1992) points out, in a large organisation successful innovation also involves an internal implementation process in which the many agents in the organisation incorporate the innovation in their work. This process can be highly problematic! This paper considers the internal processes that are necessary to gain user and organisational acceptance for an innovation and begins by considering evidence from studies of the implementation of other forms of advanced information system.

The pace of development of information technology has been bewilderingly rapid. It has created wave upon wave of technological advance which, in principle, could be of enormous significance to the operation of business and industry. Most organisations have, however, not been able to take full advantage of these new possibilities. They suffer from what has become known in Europe as the up-take problem; they may try to apply the

I. Masser and H. J. Onsrud (eds.), Diffusion and Use of Geographic Information Technologies, 27–44.

technology but rarely manage to implement it fully. A common conclusion is that about 20% of applications are successful and meet their intended objectives (see, for example, Mowshowitz 1976). In many circumstances the application may never be implemented or may be rejected. Probably the most common outcome is partial use of the technology; use by a limited part of the potential user population and of only a small part of the available functionality. This lack of internal diffusion has been noted in many application areas; office automation (Pye et al 1986), decision support systems (McCosh 1984), expert systems (Berry and Hart 1990) and applications for small and medium sized enterprises (Wroe 1986).

The reasons for slow and difficult internal diffusion have been widely researched and many contributing factors have been identified. In general the problem is that the impetus for the development is usually technical and insufficient attention is paid to the organisational needs the technology might serve or to the specific needs of individual users.

The purpose of this paper is to explore what would be necessary to obtain more effective diffusion. The route to this end requires answers to four questions:

1. What are the current methods of design and implementation which lead to inadequate diffusion?
2. What alternative methods have been or could be proposed which might lead to more effective diffusion?
3. What are the barriers to the widespread adoption of these alternative methods?
4. What research is appropriate to test whether the assumptions that underpin the provisional answers to these questions are justified?

The answers provided below to the first three questions are based on existing research and experience with some alternative methods. The answer to the final question is an attempt to set a research agenda which might carry our knowledge further and assist the adoption of new methods in the application of Geographical Information Systems. The work to be reported is largely research on the application of other forms of advanced information systems with only marginal reference to research on Geographical Information Systems. This paper is therefore an attempt to abstract lessons from other research which can be applied to both practice and research in Geographical Information Systems.

It should be noted that the following analysis focuses upon methods used to implement major systems within organisations. In these cases information systems are designed or purchased which are intended to meet corporate needs and to serve many departments. Implementation has therefore to meet the requirements of many different users and the system has to be tightly coupled to the organisational structures and processes. There are many information systems that do not have such widespread organisational ramifications. A secretary planning to use a word processing package on a personal computer or a manager wanting to use an external information service to plan foreign travel do not cause major organisational disruption. There may be issues of perceived benefit, ease of use and training etc. but these will be localised effects. Major corporate use of a GIS is, however, likely to require more substantial organisational planning.

2. Existing Methods of Design and Implementation

The dominant method by which major information systems are designed and implemented in organisations has been characterised as rational and technocentric. It has the following major characteristics:

1. **Technology 'Push'.** The aim is to find an application for a particular form of technology. At its worst it is a solution looking for a problem. The volume edited by Berry and Hart (1990) gives examples of a number of expert system applications which were initially very poor 'solutions'. Only when the 'problem' had been fully understood could an appropriate solution be identified and, in a number of instances, most of the 'expert' elements of the system disappeared. Internal diffusion demands 'organisational pull' to understand the need as well as 'technology push' to provide the solution.

2. **The Waterfall or Sequential Model of Development.** Most systems development methods exhibit a linear or 'waterfall' structure in which each of a series of steps have to be completed before the next can begin. In general, analysis precedes design which precedes implementation. This seemingly obvious and immutable structure has a number of serious disadvantages. For example, it presumes it is possible to analyse requirements before a solution is designed. Yet many users can only be specific about requirements when they physically see what can be implemented especially when it is a technology of which they have no experience.

3. **The Dominance of Technical Specialists.** The people engaged in developments are dominantly technical specialists, hardly surprising if the development is conceived as the design and implementation of a technical system. However, this dominance has profound ramifications for potential users and stakeholders who are not technical specialists. What roles can and should they play in the development and how can they develop any sense of ownership and commitment? It is easy to slip into the classical resistance to change paradigm in which 'they' are imposing a solution upon 'us'.

4. **The Rational Technical Language.** Development programmes utilize a specific form of representation which emphasises logic and rationality and best suits the coding of technical solutions. This makes it difficult, however, for non-technical specialists to understand the nature and implications of technical proposals. It also makes it difficult to express all the needs and issues that may exist in the organisation which do not lend themselves to this kind of representation. As a result these issues, however significant they are, may never be systematically treated in the design processes.

These characteristics do not make it easy for user and organisational issues to play a major part in design and implementation. As a consequence user and organisational issues are often barriers to diffusion when attempts are made to implement a solution. The system may not be acceptable to the organisation or its potential users. These problems are well documented but the technical methods are deeply embedded and are sustained despite the evidence of their inadequacy.

3. Towards Alternative Methods of Systems Development

The implication of this analysis is that a major obstacle to diffusion is the lack of effective means of gaining organisational and user acceptance of technical systems. This paper examines alternative systems development methods which explicitly seek organisational and user acceptance for new proposals.

Many authors have identified the problems described above and have specified the qualities that a systems development would need if it were to create systems that would be accepted in organisations, for example, Mumford and Weir (1979), Tapscott (1982), Pava (1983), Hirschheim (1985), Long (1987), Eason (1988) and Walton (1989). There is quite widespread agreement amongst these authors upon the general model needed for system development and we will therefore concentrate upon summarising two overall approaches. The first is the Walton (1989) model for the integration of information technology in organisations. This model derives from an analysis of the successful features of 19 cases from the USA and the UK. It emphasises the three key ingredients shown in Figure 1.

Key Ingredients	**Phase One** Generating the Context for IT		**Phase Two** Designing an IT System		**Phase Three** Putting the IT System into Practice
Alignment	Vision alignment with business, organisation and technology strategies	→	Systems design aligned with vision	→	Operational use of system aligned with vision
Commitment/ Support/ Ownership	High organisational commitment; stakeholder support for IT	→	System designed to tap and promote user ownership	→	Users feel strong ownership for system
Competence/ Mastery	General task competence and IT literacy	→	System designed to use and promote mastery	→	Users mastering the system

Figure 1. A Model for System Integration (Walton 1989)[1]

The aim of this approach is to produce a technical system which is integrated with organisational development strategies both of which serve the objectives of the business. The key ingredients are the alignment of these strategies, the commitment, support and ownership of the future users of the system and the development of competence and mastery of the new work system by those who will work with it. The framework demonstrates how these ingredients have to be present for each stage of development if the operational system is to be acceptable to and usable by the members of the organisation.

The model developed within the HUSAT Research Institute (Eason 1982, 1988) has similar objectives and ingredients. The basic tenet is that successful systems development depends upon a user-centred design process for the creation of a socio-technical system to serve organisational and human objectives. The fundamental structure is given in Figure 2. This approach derives from a number of surveys of design practice, eg. Bjørn-Andersen, Eason and Robey (1986), Harker, Olphert and Eason (1990), and the analysis of 12 cases in which the HUSAT Research Institute have implemented and tested many components of the model (Harker and Eason, in press).

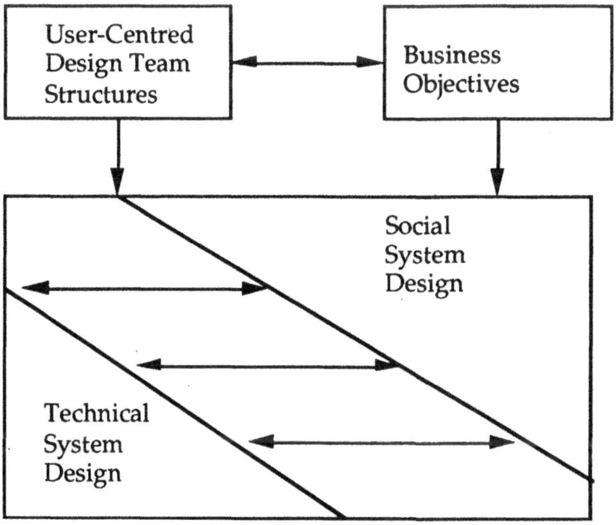

Figure 2. User-Centred Design of a Socio-Technical System

In this model a user-centred approach is adopted in which teams of users and technical specialists work together on the construction of a future system to meet business objectives. The future system has to be a compatible social system and technical system. The major problem to be overcome is the continuing tendency of the social and technical system to 'split' (Klein and Eason 1991). The laws and principles of the two kinds of system are quite different and the methods of constructing them are owned by different specialists. At all stages of development they are likely to separate and different, perhaps conflicting objectives may be pursued. The techniques for sustaining integration (represented by the horizontal lines in Figure 2) are therefore of prime importance.

The two models, in common with others, emphasise the need to move away from technical design towards socio-technical design and to give users a much more central role in the development process. The underlying hypothesis is that this approach will create systems which are more successful and acceptable than those developed by the technocentric process. An obvious research objective is to test this hypothesis; to examine a comparable range of systems developed by these two different types of approach and to test the degree to which integrated and acceptable systems result. Unfortunately such a research programme is not easy to carry out. The methods embodied in the technical systems design process are much more fully developed and dominant in the design community. The methods of the user-centred approach to socio-technical systems design are more embryonic and whilst there are examples of their use they are often partial approaches and are either the results of the work of their originators or are the consequence of strong users taking control of development and creating informal, local practices to serve their purposes. A comparison would be between the mature and the immature. 'Pure' examples of the two approaches would also be hard to find. An additional complication is the diversity of application settings; ideally the two approaches would be used in similar organisations to create new systems of similar kinds on similar scales. Since any development must use the best known methods to meet the needs of the client, it is most unlikely that such a comparison will be possible.

4. Operationalising, Institutionalising and Testing the User-Centred Approach

4.1 A PROGRAMME FOR THE DIFFUSION OF THE USER-CENTRED APPROACH

For the reasons given above the strategy being employed to develop and test the user-centred approach within the HUSAT Research Institute has, for a research programme, a number of unusual attributes. Thus far the main vehicles for its development have been survey and case study analysis of technocentric approaches and action research studies in which members of the Institute have participated in systems development processes and facilitated user-centred methods wherever possible. It is, of course, difficult in these circumstances to separate the effects of the method from the effects of the individual facilitator who is often also the evaluator of the effect. To implement the user-centred approach more fully and to test it more systematically requires three kinds of action:-

1. The Operational Development of User-Centred Design Methods. The many concepts and methods which are part of the user-centred approach need formalising and documenting in order that they can be made available for other system developers to make use of them.

2. The Institutionalisation of User-Centred Design. The methods that are being developed for a user-centred approach do not constitute an independent and comprehensive way of developing systems. They have to be related to the many existing methods for technical systems construction, for establishing business

objectives etc. and tailored to the particular circumstances of the development organisation. The methods have therefore to be developed in such a way that they can be institutionalised into all stages of the other development methods which are in widespread use.

3. The Evaluation of Effectiveness of Methods. Once a fully-fledged, operational set of user-centred design methods is embodied within a widely used systems development methodology it will become possible to more systematically research whether their claims with respect to integration and acceptability are justified. Before this stage, however, each method has a rationale and, once a method is in use by people other than its originators, it can be tested for effectiveness for its specific purpose. The evaluation process has therefore to operate on both the parts and the whole of the user-centred approach.

Together these activities constitute a broad based action-research programme in which researchers have to disseminate the results of earlier work in order to test whether their conclusions are justified. In effect, there has to be a diffusion of the user-centred design approach before it can be tested as a vehicle for the diffusion of advanced information systems.

Recent work has begun the process of operationalising user-centred methods and undertaking evaluations at each stage of development. In the next sections some of the concepts and methods under development are presented and the research issues associated with them are outlined. The range of work to be done is considerable and the methods and issues discussed are illustrative rather than comprehensive.

Part of the process of operationalising user-centred design as an approach has been to articulate a set of principles. These principles are then served by the specific methods of the approach. The ten principles are grouped into the following four categories.

1. Systems Integration. Overarching principles which relate to the prime objective of aligning business, organisational and technical strategies.

2. User-Centred Design Structures. Principles for the construction of system development teams which would establish and sustain a partnership between users and technical specialists.

3. User-Centred Design Processes. Principles by which users and designers can formulate requirements and test alternative solutions.

4. System Implementation. Principles to govern the manner in which change is undertaken to ensure users maintain mastery over the work organisation and systems they utilise.

Below the principles, methods and research issues for each of these categories are presented.

4.2 SYSTEMS INTEGRATION

Figure 3 presents two principles for systems integration and identifies some relevant methods and research issues.

System Integration Principles

1. Socio-Technical Design. The goal is the progressive development of socio-technical systems to serve business objectives.
2. Systems for Diverse Goals. User-centred design works for the development of integrated solutions which have the flexibility to serve diverse and changing goals.

Concepts and Methods	Research Issues
1. Iterative Design.	1. The compatibility of waterfall and iterative design methods.
2. Evolutionary Design.	2. Controlled development and evolving requirements.
3. Minimal Critical Design.	

Figure 3. Systems Integration

The first principle is a repeat of the assertion that systems integration depends upon the bringing together of social and technical systems to serve business objectives. It incorporates the concept that it is a 'progressive' process, not achieved in a single implementation, for the reason stated in the second principle; the goals the system serves will be diverse and changing. Organisations face the challenge of continuing change in their environments and internal systems must be capable of flexibility and evolutionary change. Within any organisation at any point in time there will be an array of goals that are being pursued, many of them in conflict, and positioning any new systems with respect to them will be a major task. Integrating a system to serve a stable and unitary goal can be a difficult task; to achieve this end for a diverse, perhaps poorly articulated, perhaps conflicting and certainly changing, set of goals stretches all existing methodologies.

It might be presumed that organisational implementation of a major Geographical Information System would be a corporate objective with integration across the organisation with respect to data sets a necessary process to achieve benefits. As Campbell (1992) has shown from her survey of UK Local Authorities there are many organisations in which separate corporate and departmental GISs are being developed. The goals of corporate integration and local autonomy are difficult to reconcile. A user-centred approach has to accept the reality of organisational conflict and there are a number of strategies by which an appropriate degree of integration can be defined and sought.

The concepts and methods in Figure 3 which relate to system integration principles concern the overall strategy and in particular the sequence of activities in the development process. An iterative process is necessary to move between the business, organisational and technical aspects of design and ensure each part remains compatible with the others. Evolutionary design is necessary to keep track of the evolving requirements that may result from the changing circumstances in which the organisation has to function. The concept of minimal critical design is in particular a useful strategy by which differences between corporate and departmental needs can be addressed. It is a way of limiting the overall system to a minimum critical specification so as to preserve flexibility in the system in order that it can be adaptive and responsive to locally changing requirements. This is a concept from socio-technical systems theory (Cherns 1976) which has been regularly applied to social system design but has not been fully explored in technical systems design.

There are a number of large scale research issues here which touch on the fundamental nature of the design process. One is whether the requirements for evolutionary and iterative components in the design process can ever be compatible with the waterfall character of technocentric design. In the spirit of institutionalising user-centred methods within the current design processes, Damodaran et al (1988), found it possible and effective to embed a wide variety of methods in the structured design method, SSADM. A fundamental problem remains, however, when an agreed set of enduring requirements cannot be established in the first phase of design. There are alternative approaches based upon prototyping and the progressive development of systems but when these routes are taken the major question is whether uncontrolled and undisciplined developments result. The challenge is to undertake controlled developments that can cope with evolving and therefore unknowable requirements.

4.3 PRINCIPLES FOR USER-CENTRED DESIGN STRUCTURES

Achieving a socio-technical system solution requires inputs from more than technical specialists and the design team that creates the new system has to be broad based. The three principles in Figure 4 establish the requirements for these team structures. First, if there is to be a commitment from the future user community, they must play a significant role in the design process. Second the involvement must not be limited to users and technical designers but must include all other significant stakeholders, for example, buyers, customers, suppliers etc., if acceptance is to be assured.

Finally, the knowledge and expertise necessary to construct a socio-technical system goes beyond information technology and requires the disciplines of business studies, organisational theory, personnel management etc. It will be necessary to integrate the contributions of many disciplines if an integrated system is to result.

There are many concepts and methods that concern the composition and operation of design teams. A preliminary requirement in obtaining a broad representation is an analysis of the stakeholders in any change process (Mitroff 1980) and a method for undertaking such an analysis is included within the user-centred tool set developed in the ESPRIT HUFIT project (Galer et al 1992).

The application of such techniques is an important way of overcoming a very common problem; design teams tend to concentrate attention upon a limited set of focal 'users' and may omit consideration of others who are critical to the success of a system. In a study by

Billingham (1990) of a GIS. applied in a Water Authority designers focused upon the small team who had to create and run the database. These 'users' were important to the operation of the system but the users who determined the success of the system were the inspectors and planners in the field who were to use the new maps. They were not consulted and rejected the new service when it was offered to them.

Principles for User-Centred Design Structures

3. Developing User Ownership. The design process has to develop a sense of ownership and commitment towards the integrated system amongst the user community.

4. Involving All Stakeholders. Design must be a partnership between users and designers and must seek the involvement of all types of stakeholders.

5. Integrating Contributions Across Disciplines. The development of an integrated solution requires the integration of contributions from all relevant disciplines.

Concepts and Methods	**Research Issues**
Stakeholder Analysis. User Representative Structures. Shared Representations of Future Systems.	Power and Influence in Systems Design. The Management of Large Scale Participation. Effective Inter-disciplinary Design.

Figure 4. User-Centred Design Structures

When the range of users and stakeholders is known, the problem is to create temporary organisational structures in which they can influence the system developed. Eason (1988) describes many such structures and the processes by which appropriate structures may be constructed which will yield a sense of ownership and enable users to contribute their knowledge.

The significant research issues in this area relate to the fundamental problem that the design process is both a formal process of analysis and design and a political process in which a solution is sought which will meet many, often conflicting, needs and goals. Acceptance that there will be conflicts between the requirements of the stakeholders is a pre-requisite to creating systems that will gain widespread organisational acceptance. Without it the system may serve one set of powerful 'winners' at the expense of rejection by others. The question of who controls the design process is central for many investigators. Ehn (1988), for example, describes participative processes in which the workforce take control in place of management. In other circumstances the power issues may lie between users and technical specialists or between head office and user departments. The design of team structures to cope with formal design and the management of power relations remains an important research topic. Case studies in which user communities are successfully engaged in systems design, such as those by Mumford (1983) and Hedberg (1980) are usually small scale (less than 100 users) and another significant research problem is how to design participative structures when there are thousands of users.

The fifth principle highlights a different set of concepts, methods and research issues. The contributors from different disciplines speak, in effect, different languages and the reports of misunderstandings when, for example, technical staff present proposals to users, are legion. Popular methods for achieving a common language are to create simulations or prototypes early in the design process in order that the concrete and real characteristics of the future system can be made evident to users. Whilst this appears a fruitful approach it needs systematic research because, as Harker (1990) points out static, technical prototypes may represent only a few of the relevant features of the future socio-technical system.

Principles for User-Centred Design Processes

6. User Requirements and Values. User-centred design should make explicit user requirements which include the goals and values of the organisation and the users within it, the nature of the tasks to be undertaken and the characteristics of users.

7. Representing Solutions. Informed decisions require the representation throughout design of concrete visions of socio-technical opportunities.

8. Reviewing Options. Users need time, resources and structures in which to gather data about the implications of options and to review options against their requirements.

Concepts and Methods	Research Issues
Functional and Non-Functional Requirements Capture. From Scenarios to Trials. Usability and Acceptability Evaluation.	Systems analysis or requirements generation. Realistic representations of future systems. The validity of early evaluations.

Figure 5. User-Centred Design Processes

4.4 PRINCIPLES FOR USER-CENTRED DESIGN PROCESSES

The next set of principles concern the processes users and designers will need to work through if there is to be a partnership which generates and evaluates future options in the design process. Existing methods emphasise formal analysis by systems staff, the progressive detailing of solutions and the evaluation by users at the time of implementation if at all. These principles support a process in which analysis, design and evaluation occur continuously and are a product of close designer-user cooperation. Principle 6 stresses the need for a much broader range of requirements to be considered than the normal functional requirements; they need to include the values, needs, goals and characteristics of different kinds of users and the organisation at large. Principle 7 repeats the point made in relation to Principle 5, that throughout the process future options have to be represented in concrete forms that all participants can understand. This provides an opportunity for a continuous process of evaluation and Principle 8 indicates that, to be effective, this process has to give users the time and resources to explore the broader implications of options in order to examine their ability to meet requirements.

The concepts and methods needed to support this approach to the design process all have as their objective the engagement of users in the design process. There is a need therefore for analysis techniques which enable users to specify their requirements. This raises an important research issue. Current systems analysis techniques are based upon the concept that analysis of work systems can capture the facts that will establish system requirements. The user-centred approach takes the view that requirements emerge when users consider alternative futures and develop a consensus of which goals and values should be pursued. Research is needed to establish an effective balance between these positions and how the two kinds of requirement relate to one another. In the CEC ESPRIT project ORDIT (Harker et al 1990) we have been developing techniques for the generation and representation of non-functional (organisational) requirements which raise important questions about the relation between functional and non-functional requirements.

Reference has already been made to the concepts and methods needed to represent possible future systems in ways that all stakeholders can appreciate. The aim is an early and realistic representation of the future which can facilitate knowledgeable selection before major investments are made. The techniques suggested include scenario building, envisionment techniques, static and dynamic prototyping, technology demonstrations, simulation building and experiments, pilot and trial implementations etc. There are a number of research issues that surround the use of these techniques; which technique for what purpose, how realistic do they have to be and, above all, how necessary is it for them to be socio-technical representations rather than the more usual technical representations?

The purpose of developing representations of systems is to facilitate evaluations of the match with user requirements and to do this as early in the design process as possible. Again there are many techniques to assist this process; expert evaluations, assessments against guidelines and standards, user cost-benefit evaluations, usability audits, acceptability audits etc. The main research questions concern the validity of predictions made at an early stage; do these evaluations identify the critical issues which might affect later implementation?

4.5 PRINCIPLES FOR USER-CENTRED SYSTEM IMPLEMENTATION

The final pair of principles summarise the many issues surrounding the successful implementation of the planned system. Principle 9 highlights the fact that technical system implementation also involves organisational change and that this change is best accomplished by an evolutionary strategy which provides users with opportunities to adapt, evaluate and shape the future course of implementation. Principle 10 focuses on the new learning needed to understand the new socio-technical system and to exploit the new opportunities it offers. It proposes that user support processes are needed which will give continuous point-of-need help because users will need to progressively develop their new competencies.

There are a number of studies which suggest these issues are particularly pertinent to the implementation of GISs because they involve the adoption of new working practice if their benefits are to be realised and, in the current manifestations, they are rarely easy to use. In the factor analysis study conducted by Onsrud and Pinto (1992) in US local Government the three factors most closely correlated with user satisfaction were access to learning, ease of use and the effects of use (adverse and beneficial) on individual users. Zwart and Coleman (1992) report a study of GIS usage which shows that most end users need and use a restricted set of facilities which is only a small percentage of functionality available on a large scale GIS. They point to the emergence of limited functionality desktop mapping systems as evidence of users seeking easy-to-use systems for dedicated rather than comprehensive purposes. Whatever the technical systems provision it is likely that end users will need help and support to develop the competence to make good use of it.

Principles for User-Centred Systems Implementation

Principle 9: Organisational Adaptation and Evaluation. Implementation requires local adaptation through incremental steps and user reviews to provide for evolutionary growth.

Principle 10: User Support. Implemented systems must contain point-of-need support mechanisms to ensure users can progressively develop the competence to exploit the technological facilities provided.

Concepts and Methods	Research Issues
System Implementation Strategies; from 'big bang' to phased implementation.	Evolutionary design and the parallel undertaking of design and implementation.
The planning of Organisational Change.	The emergence of new organisational forms through technological innovation.
User Support Tools and Strategies.	In-system help vs. human support structures.

Figure 6. User-Centred System Implementation

There are many concepts and methods already available in the organisational change literature to support the planning of changes in the social system, for example, techniques for new team building (Bass and Barrett, 1981). The many techniques by which technical systems can be introduced have been less well presented in the literature although Eason (1988) outlines six strategies ranging from the 'big bang' approach to evolutionary implementation. The latter provides more time and opportunity for users to learn and to influence later stages of development.

One of the research issues emerging in system implementation is the movement away from 'one shot implementation', ie. design before implementation, in favour of evolutionary implementation in which first stage implementation is run in parallel with the design of the second stage implementation etc. In theory such an evolutionary process permits users to influence later developments on the basis of experience of early forms of the system. But how can early forms be designed that will be compatible with later but unknown forms? And how can evolutionary technical system delivery be matched with a process of planned organisational change?

Another research issue is the exploration of the new organisational forms that are made possible by the use of technological innovations, for example, working from home, (Kinsman 1987), or the establishment of networked organisations etc. (Savage, 1990). A major problem is that most staff find it difficult to conceive and evaluate the different organisational forms which may be possible and desirable and as a result the amount of organisational change is often quite conservative and may inhibit the achievement of major benefits. The processes by which the staff of organisations can both conceive of major organisational and technological opportunities and plan an implementation programme to bring about the necessary changes are little understood.

Another set of concepts and methods that already exists in some profusion are those to support the new learning users would require in order to exploit the opportunities provided by the new system. The majority of concepts and methods are concerned with the formal training of users in the use of the new technical system or with the design of manuals to support the system. However research shows that users mostly learn after implementation, that they need to learn new approaches to their work as well as how to operate a new technical system and that they prefer to learn from other human beings (Damodaran 1986). In practice systems that are implemented usually develop an array of human-based support mechanisms that operate after implementation, for example, help desks, information centres, local experts, liaison officers etc. By contrast, greater effort is being put into the provision of support facilities within systems, for example, help facilities, demonstration programs etc. It is quite clear that users need continuing and wide ranging learning facilities if they are to make good use of new opportunities. It is not at all clear which mechanisms are appropriate for which kind of learning.

5. Conclusions; Action Research for the Development of User-Centred Design

Our conclusions after being involved in research and in active roles in the development and implementation of advanced information systems can be summarised as follows:-

1. Successful diffusion within an organisation is often limited because there is insufficient organisational and user acceptance of the planned innovation.
2. Gaining organisational and user acceptance depends upon
   (a) aligning the strategies for technical and organisational development to business objectives,
   (b) gaining user commitment through engagement in the design process and
   (c) ensuring users develop mastery of new facilities and work processes.
3. There are few techniques in regular use to support these three objectives and therefore the user-centred design philosophy they represent is not in widespread use in a systematically applied manner. It is therefore difficult to test the

proposition that user-centred design is an effective method for the diffusion of information technology.

The result of this analysis is that, as a Research Institute interested in the development of user-centred approaches, we find it necessary for us to play a pro-active role. We are therefore engaged in a process of formalising current best practice in user-centred design, institutionalising this practice in the normal design processes of organisations and testing whether the effects are as predicted. In a form of action research we aim to test specific hypotheses, for example, about the effects of different ways of representing future systems, which can contribute to the goals of user-centred design. This is a process which goes well beyond the normal research agenda. It actually embodies a complete cycle of research, encoding research results in practice, disseminating and institutionalising the practice, testing the results of use and utilising the results to revise and develop the practice. It is our belief that, difficult as it may be to complete this cycle, the advancement of approaches which are contrary to the dominant approach of the age, the technocentric approach, demands this range of contributions.

6. Endnote

1. Figure 1 is from Richard E. Walton, *Up and Running: Integrating Information Technology and the Organization.* Boston, MA: Harvard Business School Press, 1989, p.5. Reprinted by permission.

7. References

Bass, B.M. and Barrett, G.V. (1981) *People, Work and Organizations.* Allyn and Bacon, Boston.

Berry, D. and Hart, A. (1990) *Expert systems: Human Issues.* Chapman and Hall, London.

Billingham, R. (1990) The Implementation of Geographical Information Systems in Organisations. M.Sc. Dissertation, Loughborough University of Technology, U.K.

Bjørn-Andersen, N., Eason, K.D. and Robey, D. (1986) *Managing Computer Impact.* Ablex, Norwood, NJ.

Campbell, H.J. (1993) GIS Implementation in British Local Government. In Masser, I. and Onsrud, H.J. (eds.) *Use and Diffusion of Geographic Information Technologies.* Kluwer, Dordrecht, The Netherlands.

Cherns, A.B. (1976) The principles of socio-technical design. *Human Relations* 29: 783-792.

Damodaran, L., Ip, K. and Beck, M. (1988) Integrating human factors principles into a structured design methodology: a case study in the UK Civil Service. In Bullinger, H.-J., Protonotarios, E.N., Bouwhuis, D. and Reim, F. (eds.) *EURINFO '88* North-Holland, Amsterdam.

Damodaran, L. (1986) User Support. In Bjørn-Andersen, N., Eason, K.D. and Robey, D. *Managing Computer Impact.* Ablex, Norwood, NJ.

Eason, K.D. and Harker, S.D.P. (1991) Effective human factors contributions to the design process. In Shackel, B. and Richardson, S.J. (eds.) *Human Factors for Informatics Usability.* Cambridge University Press, Cambridge.

Eason, K.D. (1988) *Information Technology and Organisational Change,* Taylor and Francis, London.

Eason, K.D. (1982) The process of introducing information technology. *Behaviour and Information Technology* 1: 197-213.

Ehn, P. (1988) *Work-Oriented Design of Computer Artifacts.* Almquist & Wiksell, Arbetslivscentrum, Stockholm.

Galer, M., Harker, S.D.P. and Ziegler, J. (eds.) (1992) Methods and Tools in User-Centred Design for Information Technology, HUFIT. Elsevier, Amsterdam.

Harker, S.D.P. and Eason, K.D. (eds.) (in press) The Application of Information Technology. Taylor and Francis, London.

Harker, S.D.P. (1991) Requirements Specification and the Role of Prototyping in Current Practice. In Karat, J. (ed.) *Taking Software Design Seriously.* Academic Press, London, pp. 339-354.

Harker, S.D.P., Olphert, C.W. and Eason, K.D. (1990) The development of tools to assist in organisational requirements definition for information technology systems. In Diaper, D., et. al., (eds.) *Human-Computer Interaction - INTERACT '90.* Elsevier, Amsterdam.

Hedberg, B. (1980) Using computerised information systems to design better organisations and jobs. In Bjørn-Andersen, N. (ed.) *The Human Side of Information Processing.* North-Holland, Amsterdam.

Hirschheim, R.A. (1985) *Office Automation: A Social and Organisational Prospective.* John Wiley, Chichester.

Kinsman, F. (1987) *The Telecommuters.* John Wiley, Chichester.

Klein, L.K. and Eason, K.D. (1991) *Putting Social Science to Work.* Cambridge University Press, Cambridge.

Long, R.J. (1987) *New Office Information Technology*. Croom Helm, London.

McCosh, A.M. (1984) Factors common to the successful implementation of twelve decision support systems and how they differ from three failures. *Systems, Objectives, Solutions*, 4: 17-28.

Mitroff, I.I. (1980) Management myth information systems revisited: a strategic approach to asking nasty questions about systems design. In Bjørn-Andersen, N. (ed.), *The Human Side of Enterprise*. North-Holland, Amsterdam.

Mowshowitz, A. (1976) *The conquest of Will: Information Processing in Human Affairs*. Addison-Wesley, Reading.

Mumford, E. (1983) *Designing Secretaries*. Manchester Business School, Manchester.

Mumford, E. and Weir, M. (1979) *Computer Systems in Work Design: The ETHICS Method*. Associated Business Press, London.

Onsrud, H.J. and Pinto, J.K. (1992) Correlating Adoption Factors and Processes with GIS User Satisfaction in U.S. Local Governments. In Masser, I. and Onsrud, H.J. (eds.) *Use and Diffusion of Geographic Information Technologies*. Kluwer, Dordrecht, The Netherlands.

Pava, C. (1983) *Managing New Office Technology: An Organisational Strategy*. New York Free Press, New York.

Pye, R., Bates, J. and Heath, L. (1986) Profiting from Office Automation: Office Automation Pilots. Department of Trade and Industry, London.

Rogers, E.M. (1983) Diffusion of Innovation, Free Press, New York, Third Edition.

Rogers, E.M. (1992) The Diffusion of Innovations Model. In Masser, I. and Onsrud, H.J. (eds.) *Use and Diffusion of Geographic Information Technologies*. Kluwer, Dordrecht, The Netherlands.

Savage, C.M. (1990) *Fifth Generation Management*. Digital Press, U.S.A.

Tapscott, D. (1982) *Office Automation: A User Driven Method*. Plenum, New York.

Walton, R.E. (1989) *Up and Running*. Harvard Business School Press, Boston.

Wroe, B. (1986) *Contractors and computers: why systems succeed or fail*. PhD thesis, Loughborough University of Technology.

Zwart, P.R. and Coleman, D.J. (1992) Trade-Offs towards Real Time Spatial Information Networks. In Newton, P.W., Zwart, P.R. and Cavill, M. (eds.) *Networking Spatial Information Systems*. Bellhaven, London.

IMPLEMENTATION OF NEW INFORMATION TECHNOLOGY

PAUL S. GOODMAN
Carnegie Mellon University
Graduate School of Industrial Administration
Pittsburgh, PA 15213

ABSTRACT. This chapter examines factors which contribute to the successful implementation of new information technology. It is designed for people doing research on the implementation of new information technology, designing these systems, or implementing these information systems. The paper begins with a delineation of basic concepts, provides a brief overview of implementation literature, and suggests implications that might be drawn from that literature. It then goes on to examine seven critical issues in implementing new information technology; the major focus of this paper. Some of the issues discussed include: defining the appropriate social system for implementing new technology, defining the meaning of success, understanding the object and socially constructed characteristics of new information technology, and the roles of the socialization, commitment, reward, and learning processes on the implementation of new technology.

1. Introduction

This chapter focuses on understanding the factors which contribute to the successful implementation of information technology. It is designed for people doing research on the implementation of new information technologies (e.g., GIS Geographic Information Systems), designing these systems, or implementing these information systems.

The rate of technological innovations in organizations is expected to accelerate over the next decade (Reddy, 1990). These innovations will afford many new opportunities for improving the competitive positions of organizations, the quality of working life for the organizational participants, and the quality of life for society. The proliferation of information technology, in particular, will be found in all the working sectors of our life (e.g., public, private), as well as the in non-work sectors of our life (e.g., consumer purchasing, family management).

One critical issue is whether the technology will be successfully implemented and fully utilized over time.[1] It is one thing to successfully design new forms of information technology. It is another thing to insure that this technology will enhance individual decision making and/or improve organizational indicators such as productivity, quality, etc. In a recent review of this research literature, Goodman and Griffith (1991) report that there were many examples of lost opportunities where technology was introduced but never fully utilized. In the recent literature on the economics of information, one finds a similar picture (Barua et al. 1991; Loveman 1988). Firms invest millions of dollars in information technology to improve productivity, but the productivity effects are not discernable. The problem is clear. Whether you search in the research literature on the introduction of new

I. Masser and H. J. Onsrud (eds.), Diffusion and Use of Geographic Information Technologies, 45–57.

information technology, the economics literature on the impacts of information technology, or you search through more popular accounts of use of information technology, you find similar accounts--the objective benefits of the new technology are not being fully realized.

Given the expected proliferation of new information technology and the new ways this technology may substantially change human decision processes, it is important to understand more about the implementation process. The degree of successful implementation significantly conditions the flow of benefits from any new technology. If we can understand how to more successfully implement technology, we can then capture the strategic benefits of new information technology.

To understand the process of implementing new technology, this presentation is organized into four sections. First we present some basic concepts. Second, we provide a brief overview of the literature. The main section of the paper examines seven issues. These issues represent the basic dilemmas one must consider if you are designing an information technology, doing research on the implementation of a particular technology, or are responsible for managing the introduction of a new technology.

2. Basic Concepts

Implementation is defined as a process: (1) undertaken to translate a tool, technique, method, or other object into some form of utilization; and (2) bounded by the adoption decision and institutionalization (Tornatzky and Johnson 1982; Ettlie 1984; Lucas 1986).

The adoption decision (Dean 1987) refers to the processes by which a new piece of technology is selected for the organization. While the adoption decision clearly affects implementation, it is composed of different processes and is analytically distinct from implementation.

Institutionalization refers to the process by which a structure (e.g., technology) persists over time (Goodman and Dean 1982). The distinction about when implementation stops and institutionalization starts may be a bit arbitrary. Institutionalization assumes the new technology has been introduced and has reached some level of utilization. This process focuses more on the persistence of the utilization rate of the technology over time.

Technology is the other key concept. While there are many definitions of technology (Goodman 1986; Weick 1990), here technology refers to a system of components which act on or change an object from one state to another. The components include hardware, software, and programs to transform materials or information from one state to another. Implicit in this definition is the view of technology as a system of cause and effect relationships embedded in machines and methods (Sproull and Goodman 1990).

Our particular focus will be on information technology designed to enhance decision making and organizational performance. Examples of information technology in this class would include expert systems, electronic assistants, intelligent documents, and, of course, GIS technology and their applications.

Implementation success is another key concept for our analysis. In the literature a variety of different measures of success are used, such as use or utilization rates of the new technology and user satisfaction. In these studies, higher utilization rates and more satisfaction are equated with more effective utilization. However, focusing solely on these measures of effectiveness prevents one from thinking more carefully about what a successful implementation means. In this paper, implementation success is characterized as a multi-dimensional construct which incorporates:

(a) Multiple Constituencies. The introduction of any new technology in an organization will involve or impact a variety of individuals. Some of these include users, designers, co-workers, supervisors, people who maintain the technology, vendors, clients, senior management, and so on. These constituencies are likely to use different standards and referents to judge the effectiveness of the new technology, (Leonard-Barton 1987b).

(b) Different Time Frames. The time frame one uses to evaluate implementation success will bear heavily on the judgment of a technology's effectiveness. An evaluation done immediately after the introduction of a new technology will provide quite a different picture from one conducted a year after the introduction. Several studies provide evidence of the variability of success indicators over time (Sproull and Hofmeister 1986; Leonard-Barton 1987b).

(c) Different Levels of Analysis. Most measures of implementation success are at the user level focusing on dimensions such as individual utilization or satisfaction. Little attention has been placed on understanding the effectiveness of implementation at the organization level. Here we are interested in two questions: "to what extent has the new technology become imbedded in the organization?" and "to what extent has new technology impacted the effectiveness of the organization?"

As we move through the paper, this concept will be further delineated.

3. A Brief Overview of the Literature

What are the drivers of implementation success? What are the factors that account for some information technology being successfully implemented while others are not? Is there a systematic set of factors that explains implementation success across different forms of information technology and in different settings?

There is a growing literature exploring these questions. (See, for example, the work of Bikson et al. 1987; Ettlie 1984; Leonard-Barton 1987a and 1987b.)

One dominant approach in the literature focuses on classes of variables as predictors of implementation success. The typical classes of variables include organizational, technological, implementation activities, and individual characteristics. Each researcher, not surprisingly, pulls out different sets of variables to create his or her model of implementation. Some of the organizational variables proposed as drivers of implementation success include: the alignment between the organizational structure and the new information technology, the nature of job design, the selection strategy (McKersie and Walton 1989), availability of resources, and the reward system (Bikson et al. 1987; Leonard-Barton 1987). Some of the technological characteristics, such as transferability, divisibility (Leonard-Barton 1988), level of functionality and modifiability (Bikson et al. 1987), are related to implementation success. Some of the implementation variables include user involvement and leadership (Leonard-Barton and Deschamps 1988). Individual characteristics such as personal innovativeness and expertise (Leonard-Barton 1986) also have been related to implementation success.

Another approach to identifying drivers of implementation success focuses on processes. This approach argues that implementing any new technology involves a complicated set of rational, social, and political activities. There are many possible variables which might affect implementation success. So the theoretical strategy is to articulate a set of parsimonious processes which underlie the implementation of any new technology. By understanding how these processes function with different technologies and different contexts, we can better understand the drivers of implementation success. A recent paper by Goodman and Griffith (1991) identifies a set of theoretical processes and illustrates how they may be used to understand the implementation of new technology. Some of the processes discussed in this paper include socialization, commitment, reward allocation, and learning/adaption. In reviewing the literature, Goodman and Griffith (1991) identify a series of studies which support these processes as prediction of implementation success. For example, some studies relevant to socialization report that successful implementation requires continuous training (Goodman et al. 1985), and close timing between training and use (Leonard-Barton 1987a). The importance of vendor-firm training is found in research by Ettlie and Rubenstein (1980). Other research has supported the changing nature of rewards during the course of implementation and the role of different sources of reward in affecting implementation success (Leonard-Barton 1987a). Still other studies have examined the process of adapting and designing the technology over time in order to insure long run implementation success (Leonard-Barton 1987b).

The purpose of discussing this literature is to acknowledge that there is a growing body of studies on implementing new technology. It is also to state that we are accumulating evidence on drivers of successful implementation from both variable and process perspectives.

At the same time, there are clear limitations to this literature. First, there is no generally acceptable theory of implementation. Also, it is unlikely there will be one developed in the near future. Implementation is a complicated process which occurs in many organizational contexts. This complex set of interactions among different types of technologies set in different contexts makes developing an acceptable model of implementation both very difficult and very challenging.

Second, there is a set of conflicting findings in the literature. For example, there has been a lot of attention on using involvement practices to facilitate implementation. However, the empirical evidence (Locke and Schweiger 1978; Ives and Olson 1984; Bikson et al. 1987; Walton 1989) supporting involvement as a driver of implementation success is not clear

Lastly, the design of many of the empirical studies is not strong. Many are case studies, while others focus on a few variables and do not incorporate strong measurement or analysis procedures. There are very few multivariate, longitudinal studies of implementation success.

4. Critical Issues in Implementing Information Technology

Given what we know about implementation of new information technologies and about the current state of the literature, how do we move ahead in understanding the implementation process. In this section, we focus on seven critical issues. These issues represent dilemmas or questions to be resolved by the researcher interested in the implementation of new information technology or the designer or implementer of these new technologies.

4.1 DEFINING THE APPROPRIATE SOCIAL SYSTEM

Before designing or implementing any new information system, we need to define the appropriate social system. Focusing solely on the user of the information system is likely to lead to failure in implementing new technology. For example, we have examined implementations of new technology where the focus was primarily on the user of the system (Argote et al. 1983). The supervisor of the user and the support people were left out of the implementation process. This technology was never successfully utilized, in part, because the supervisors undermined the implementation process. They were threatened and reacted. The research done by Leonard-Barton (1987b) shows these multiple constituencies both define and determine implementation success.

There is a complex array of constituencies involved in any implementation of new technology. For example, in introducing a GIS technology and related applications in a city, one would have to consider all the operating departments, support groups (e.g., information systems) and the political apparatus. Across all these units are users, designers, supervisors, maintainers of the system, gate keepers, and senior management. In defining the social system for implementing this technology, we want to include all those who can affect the short and long run viability of the system.

The definition of the social system is the first step. It determines how we define implementation success and the drivers of implementation success.

4.2 DEFINING THE MEANING OF SUCCESS

In the conceptualization of implementation success, I suggested a multi-dimensional concept defined by different constituencies whose preferences and evaluation vary over time. In another paper (Goodman and Griffith 1991), I have presented four dimensions of implementation success: (1) the knowledge about how to operate, troubleshoot, and redesign the new technology, (2) the behavior or the extent to which different constituencies use, problem solve, or redesign the new technology, (3) the attitudes or the extent to which different constituencies express positive and negative feelings about the new technology, and (4) the normative/value consensus about the technology. Here I refer to whether there is consensus about performance strategies in using the technology and about the value of the technology. Initially, success may be measured by knowledge and behavior changes. Over time, we would expect to see the emergence of normative and value consensus. Without this consensus, the viability of the technology will be questionable. Currently, we do not have good theories and/or studies tracing the evaluation and interrelationship of these criteria over time.

The difficulty in establishing the meaning of implementation success is inherent in the concept of multiple constituencies. Because these constituencies represent different roles and different needs, we would expect differences in the interpretation of success at any time. Given our focus over time, we would expect that changing roles and needs would lead to different interpretations about the meaning of success. Implicit in this analysis is that the meaning of implementation success is inherently ambiguous. The challenge to the implementors of the new technology is that constructing the social meaning of success will be as important as managing other processes directly tied to implementation of the new technology. Detractors of the new technology can, of course, react to the inherent ambiguity of implementation success. They can focus attention on the negative aspects of the technology and implementation process as a way to derail the new technology.

How can the implementors of the new technology reduce some of the ambiguity of implementation success and create some common social meanings of success? One strategy would be to get a representative committee (users, maintainers, designers) from the defined social system to agree on some criteria and operational measures of that criteria. The related strategy would be to have this group collect and interpret this information. The basic concept, which is absent from the implementation literature, is that managing the meaning of success is critical to whether the new technology is successfully implemented.

4.3 IDIOGRAPHIC NATURE OF TECHNOLOGY AND THE DESCRIPTION PROCESS

Explicit in this presentation is the notion that there are not general models of implementation success which fit all types of technology. Rather, my position is that any particular information technology can be configured in a variety of ways and provide the same results. In addition, I argue that the meaning of any technology is dependent upon its organizational contexts. Organization contexts vary in terms of the types of people, reward systems, delivery systems, culture, and so on. Barley (1986) has pointed out that the same technology in different contexts can be socially constructed in different ways. This means that different constituencies assign different meanings to the functionality or consequences of the same technology. It is the interaction or adaptation among the technological system and organizational characteristics which really defines the unique aspects about the technological system. The basic argument is that we begin with a unique configuration of technology embedded in a unique organizational context. This technology has both objective and socially defined characteristics.

If one accepts this argument, then one must begin the implementation process with a complete and detailed description of the technology and its organizational context. What is surprising in the literature is that although there are many accounts of implementation success or failure, there are very few accounts of the technology under observation. The description must include an account of the objective characteristics of the technology (e.g., functionality, capacity). It must also include a description of the social construction of the technology from different constituency points of view. The meaning of the technology needs to be understood in the context of the task activities to which it is applied. New technology is designed around certain functional specifications. However, the meaning of functionality is grounded and evaluated in terms of ongoing job activities. So a description of these activities prior to and after the new technology would provide a way to understand the task-technology interaction.

The next task is to describe the technology in terms of its organizational environment. In this task we are particularly looking for alignments or misalignments between the technology and its organizational context. Dimensions for describing alignments include user knowledge skills and abilities, organization structure for decision making, organization reward systems, organizational strategy, existing technological delivery systems, and processes (Leonard-Barton 1988; Walton 1989).

What is the goal of this description? Its primary function is for diagnostic purposes. That is, it is the mechanism for identifying impediments to successful implementations.

Consider the following case: First, we know that people's social constructions of technology may determine how the technology gets used. Let's say we introduce a GIS technology which objectively provides many new benefits to a user compared to existing methods. For example, the technology might permit new forms of analysis through "as if scenarios" or new ways to proactively monitor critical issues such as incidents of crime. The critical issue is how the user (not the designer) defines the functionalities of the technology. We have seen cases in the industrial context where the user defined the technology as a substitute for past activities (human scheduling), and not in terms of the new functionalities which provided ways to redesign the production process. So a key task throughout the implementation process is understanding the different social constructions of the technology and identification of areas where the social constructions are in conflict with the intended objective functionality of the technology. We need to understand theoretically and empirically how these social constructions are formed.

Second, we know that alignment of the technology with the organization is critical. For example, we did a study on an information system designed to facilitate proactive monitoring of a production process. However, the organization primarily rewarded people for reactive behavior. There was no alignment between the proactive measurement system and an organization reward system to facilitate that behavior, and the system failed. GIS technology can be relatively complex in terms of hardware, software, and data base design. An important alignment issue is the general level of computer literacy among users and the level of skills and positions for those responsible for maintaining the system and applications. Lack of skills and/or resource commitments to maintain a system is one of the clearest predictors of system failure.

The basic concept is that rich descriptions of the technology and its organizational context generate diagnostic information on impediments to successful implementation. We need innovations in theory and methods to identify the facilitators and impediments in different contexts. Identifying the facilitators and impediments can affect the change strategy in introducing the technology, and suggest action plans for facilitating the implementation. It is important to note that the diagnostic process is continuous throughout the implementation. It is not simply to be initiated at the beginning of the implementation.

4.4 SOCIALIZATION: SELECTING AMONG TARGETS, KNOWLEDGE, MECHANISMS

Socialization is one of the critical processes underlying the successful implementation of new information technology. One critical question is to identify the targets for socialization. In much of the literature the focus has been on the user. In this discussion I have argued that defining the appropriate social system is critical for implementation. The social system definition is important because it identifies the targets for socialization. The process of socialization becomes complicated because there are different targets in the social system which require different types of knowledge through different socialization mechanisms. For example, the users in different functional areas, maintainers, supervisors, clients, and senior management require different types of socialization.

A related issue is defining the kinds of knowledge necessary for implementing new information technology. Most of the studies on implementation have not articulated the requisite knowledge, behaviors, and skills for the new technology. For example, in a study we did on vision systems used to improve dimensional quality in an automobile assembly plant (Goodman and Griffith 1991), the user had to have knowledge of how to operate the system, how to monitor data and do causal analyses, and how to maintain the system. In addition, the user needed knowledge pertaining to the relationship between this system and other systems on quality. At still a broader level, the user needed to have knowledge of the strategic role of this system. Throughout the three years we spent observing this implementation, there was never an articulation of these types of knowledge by the designer or implementors of this information technology. Indeed, the major focus of training was only on how to operate the technology. This technology, although it had clear objective value, was not successfully implemented.

The third issue concerns mechanisms for socialization. Much of the literature has focused on formal classroom methods as a way to learn. Only recently has there been attention to observation, trial, and error (Goodman and Miller 1990), simulation (Chew et al. 1991), peer communications, and other informal communications (Argote, et al. 1983) as modes of learning. In another paper (Goodman and Griffith 1991), I have argued that we need to move towards matching different types of knowledge requirements in information technology with different modes of socialization. That is, certain types of knowledge and skills require different types of learning mechanisms. For example, in the vision system study, the user needed to operate the system and analyze data from that system. The classroom represented an effective mode to provide knowledge about how to operate the system because the interface was relatively simple. Analytic knowledge and skill requirements, however, are quite different in that technology. Here, apprenticeship, simulation exercises, and/or trial and error opportunities represented a more effective way to acquire the requisite analytic skills. In some GIS technology with complex interfaces and system configurations, there is a real challenge in learning both how to operate the system and how to use applications. The issue is to fit the knowledge and skill requirements to the appropriate socialization mechanism. A traditional strategy is to use a classroom format in a specific time frame (e.g., two weeks). Another strategy is to train users and others in an apprenticeship format over a distributed time frame (e.g., once a week over 14 weeks). The latter strategy seems much more compelling, given the complexity of the task. The challenge in the socialization process is to develop a matrix of targets x knowledge requirements x learning mechanisms for the specific implementation under consideration.

4.5 PARADOX OF VALUE

The "paradox of value" is inherent in any implementation process. The basic argument revolves around focus of attention. In any implementation process the bias will be to focus on the benefits of the information system and understate the costs. In time, the actual benefits and costs will be discrepant with the promised or expected benefits. The theoretical argument and empirical evidence (Sproull and Hofmeister 1989) indicate the expectations will be primarily negative, that is, actual benefits are lower than expectations, and costs are higher. The size of the discrepancy is an important predictor of use and attitudes toward the new technology. The greater the negative discrepancy, the lower the use and more negative the attitudes.

If it is true that the paradox is inherent, a number of implications follow for the designer. First, acknowledgment of the paradox may lead to a more "realistic preview" of the technology and its benefits. That is, the more realistically we portray benefits and costs, the less the discrepancy between expectations and actual experiences. For example, let's assume there are very long lead times for introducing GIS technology and applications. Providing a realistic time frame for expecting benefits may reduce some of the problems generated by the "paradox of value." An interesting experiment might be to vary the "realistic preview" among users (in different departments or work areas), and measure the impacts on attitudes and use.

Second, developing strong initial commitment toward the technology can buffer the effects of the discrepancy and minimize the likelihood of lower utilization and the development of negative attitudes toward the new technology. That is, under conditions of strong commitment, users can tolerate negative discrepancies for longer periods of time. In the literature, participation of the user is the major way to develop commitment. However, given the mixed findings in the literature on the role of participation, we should not rely on mechanisms for participation (e.g., designer-user task forces, steering committees) as the sole way to create commitment. We should experiment with other mechanisms and assess their effectiveness. For example, utilizing the user as a trainer of the new technology and as a diffuser of the new technology represent some possible ways to induce commitment. Also, modifying the user's job so time is available to experiment with the new system, or modifying reward systems to reinforce system use are other ways to create commitment. Building commitment is an important process for buffering negative discrepancies until there is a better alignment between positive expectations and actual benefits.

4.6 CONGRUENT AND INTEGRATIVE REWARDS

The focus of this paper has been on discretionary forms of information technology, that is, information technology which the user can elect to use and which can enhance decision making and performance. Throughout our analysis, we pointed to many factors which affect utilization. One of the critical factors, which surprisingly has not received a great deal of attention in the literature, is the reward system. Rewards can flow from the technology, the organization, and informal sources. To the extent that these reward systems are congruent and reinforce the behaviors required by the new technology, the more likely the implementation process will be successful. For example, if computers are not readily available or the interface for a GIS system is too difficult, or the system is not easily modifiable, the costs in using the system will exceed any benefits, resulting in low use. If the information system is designed to facilitate planning, but the organization does not reward planning activities, the system will not be utilized.

The challenge, however, is not just the fit between the new information technology and the reward system. Throughout this analysis, I have indicated that there are multiple actors in the social system who can facilitate or inhibit the use of the new information systems. In some of my own work (Argote et al. 1983; Goodman and Griffith 1991), we have shown that differential reward systems have led to conflict between designers and users or users and maintainers, which ultimately contributed to the demise of technology. The task, then, is to design rewards which are congruent to the new technology and integrative across the different actors who impact on the information systems. The integrative rewards are designed to pull the different constituencies together to work toward the full utilization of the information systems.

4.7 LEARNING AND REDESIGN

We talk about implementation of technology as if the technology is fixed, with known structure and parameters, and the process of implementation as a rational means/end process. Actually, we have incomplete knowledge of many forms of new technology, and there are high levels of uncertainty about the technology and its consequences. Similarly, the process of implementation or change is often chaotic, disorganized, and not characterized by simple means/end hierarchies.

If we think about the new technology as characterized by high levels of uncertainty and the implementation process as affected by a series of rational political, social, and random forces, the role of learning or feedback and redesign must become an essential feature of any implementation. Basically, our point of view changes. The prevailing model has been, "what variables do I manipulate to insure the successful implementation of a new technology?" This model assumes a fairly rational and mechanistic environment. The learning model states that there is a high degree of uncertainty in the technology and the implementation process. It is hard to know *a priori* what variables to manipulate. Rather, the strategy is to put the technology in place, even if it is not fully constructed, and to design a learning environment which will generate information to redesign the technology and the organization environment. The challenge is to create a dynamic process of redesign. The learning or feedback and redesign can occur before the technology is actually introduced. Through a variety of techniques (e.g., simulation), one can anticipate some of the benefits and consequences of the technology, and generate new design options (Chew et al. 1991). Or, as the information technology is introduced into the organization, feedback systems can identify discrepancies between expectations and reality, and initiate redesign processes. The question for designers or users of GIS technology and applications is, "what formal mechanisms are in place (1) to measure use and user problems, (2) to feed back this information, and (3) to redesign the system or application?" The process of feedback and redesign is essential for implementation success over time. It should be introduced early in the implementation process. Most of the technological innovations we have observed over the past decade, which have not explicitly initiated a feedback and design process, have failed.

While research over the pat few years has pointed to the importance of learning and redesign, there has been much theoretical work on understanding the effectiveness of different redesign mechanisms in different organizational context.

5. Conclusion

We can expect to see a proliferation of new information technologies over the next decade and beyond. Many of these technologies have the capabilities of substantially improving our decision making abilities, performance, and quality of life. If history is any indicator, many of these new initiatives will unfortunately fail.

The challenge is building on what we know. There is a growing literature which is beginning to identify the classes of variables and processes that drive successful implementation of new technology. However, there are a lot of unanswered questions and need for new theory and methodological developments.

The seven issues are presented as a way to organize our thinking about implementation.

They can be used by the researcher examining the implementation of a GIS system or a person designing and implementing such a system.

6. Endnote

1. My interest is in the implementation process. We acknowledge there is a complicated set of decision processes in the determination of whether to adopt a new technology. We make no assumptions about whether information technologies are inherently good or bad. Some are, others are not. We begin with the assumption that the technology has been adopted, and the focus is on the process of implementing that technology.

7. References

Argote, L., P.S. Goodman and D. Schkade (1983) How workers react to a robot. *Sloan Management Review* 24: 31-41.

Barley, S.R. (1986) Technology as an occasion for structuring: Evidence from observations of CT scanners and the social order of radiology departments. *Administrative Science Quarterly* 31: 78-108.

Barua, A., C.H. Kriebel and T. Mukhopadhyay (1991) Information technologies and business value: An analytic and empirical investigation. Working Paper, Graduate School of Industrial Administration, Carnegie Mellon University.

Bikson, T.K., B.A. Gutek and D.A. Mankin (1987) Implementing computerized procedures in office settings: Influences and outcomes. (Report No. R-3077-NSF) The Rand Corp., Santa Monica, CA.

Chew, W.B., D. Leonard-Barton and R.E. Bohn (1991) Beating Murphy's law. *Sloan Management Review* 32(3): 5-16.

Dean, J.W. (1987) Building the future: The justification process for advanced manufacturing technology. In H. Pennings and A. Buitendam (eds.), *New Technology as Organizational Innovation*. Ballinger, Cambridge, MA, pp. 35-58.

Ettlie, J.E. (1984) Implementation strategy for manufacturing innovations. In M. Warner (ed.), *Micro-processors, Manpower and Society: A Comparative, Cross-national Approach*. St. Martin's Press, New York, pp. 31-48.

Ettlie, J.E. and A.H. Rubenstein (1980) Social learning theory and the implementation of production innovation. *Decision Sciences* 11: 648-668.

Goodman, P.S. (1986) Impact of task and technology on group performance. In P.S. Goodman (ed.), *Designing Effective Work Groups*, Jossey-Bass, San Francisco, CA, pp. 120-167.

Goodman, P.S. and J. Dean (1982) The process of institutionalization. In P.S. Goodman (ed.), *Change in Organizations*. Jossey-Bass, San Francisco, CA, pp. 226-279.

Goodman, P.S., J. Edwards, M. Seabright, L. Argote and E. Ravlin (1985) Issues in implementing advanced technology. Presented at the Academy of Management Meeting, San Diego, CA.

Goodman, P.S. and S.M. Miller (1990) Designing effective training through the technological life cycle. *National Productivity Review* 9: 169-177.

Goodman, P.S. and T.L. Griffith (1991) Process approach to implementation of new technology. *Journal of Engineering and Technology Management* 8: 261-285.

Ives, B. and M.H. Olson (1984) User involvement and MIS success: A review of research. *Management Science* 30: 586-603.

Leonard-Barton, D. (1986) The secondary adoption decision: Implementing a new technology within an organization. *Working Paper 9-787-006*, Harvard Business School.

Leonard-Barton, D. (1987a) Implementing structured software methodologies: A case of innovation in process technology. *Interfaces* 17: 6-17.

Leonard-Barton, D. (1987b) The case for integrative innovation: An expert system at Digital. *Sloan Management Review* Fall: 7-19.

Leonard-Barton, D. (1988) Implementation as mutual adaptation of technology and organization. *Research Policy* 17: 251-267.

Leonard-Barton, D. and I. Deschamps (1988) Managerial influence in the implementation of new technology. *Management Science* 34: 1252-1265.

Locke, E.A. and D.M. Schweiger (1979) Participation in decision making: One more look. In B.M. Staw (ed.), *Research in Organizational Behavior*, JAI Press, Greenwich, CT, pp. 265-310.

Loveman, G.W. (1988) An assessment of the productivity impact of information technologies. *Working Paper No. 90s-88-054*. Sloan School of Management, Massachusetts Institute of Technology.

Lucas, H.C., Jr. (1986) Information Systems Concepts for Management, McGraw-Hill, NY.

McKersie, R.B. and R.E. Walton (1989) Organizational change and implementation. Management in the 1990s *Working Paper*, Sloan School of Management, Massachusetts Institute of Technology.

Reddy, R. (1990) Technology and organizations--A technological perspective. In P.S. Goodman and L.S. Sproull (eds.), *Technology and Organizations*, Jossey-Bass, San Francisco, CA, pp. 232-253.

Sproull, L.S., and P.S. Goodman (1990) Technology and organizations: Integration and opportunities. In P.S. Goodman and L.S. Sproull (eds.), *Technology and Organizations*, Jossey-Bass, San Francisco, CA, pp. 254-265.

Sproull, L.S., and K.R. Hofmeister (1986) Thinking about implementation. *Journal of Management* 12: 43-60.

Tornatzky, L.G., and E.C. Johnson (1982) Research on implementation: Implications for evaluation practices and evaluation policy. *Evaluation and Program Planning* 5: 193-198.

Walton, R. (1989) Up and Running, Harvard Business School Press, Boston, MA.

Weick, K. (1990) Technology as equivoque: Sense-making in new technologies. In P.S. Goodman and L. Sproull (eds.), *Technology and Organizations*, Jossey-Bass, San Francisco, CA, pp. 1-44.

COMMENTARY ON "IMPLEMENTATION OF NEW INFORMATION TECHNOLOGY" BY PAUL GOODMAN

PETER A. BURROUGH
Faculty of Geographical Sciences
University of Utrecht
Utrecht, 3508 TC
The Netherlands

Paul Goodman's article examines the process of implementing and successfully embedding new information technology in an organization. He explains that the goal of implementation should be to introduce the new technology in such a way that it produces tangible benefits, not just for the organization as a whole, but also for the persons working in it and their clients and colleagues. When new technology has been introduced in such a way it is said to have become institutionalized (the word institutionalized is used here in a non-pejorative sense).

The main aim of the article is to examine why the implementation process does not always lead to successful institutionalization. Goodman identifies the main difficulties under seven, somewhat overlapping headings: appropriate social systems, the meaning of success, the ideographic nature of the technology and the organizational context in which it is to be used, socialization and skills, the paradox of value, rewards, and learning and redesign. The message of the chapter is that it is insufficient to concentrate on any one of these aspects by themselves, attention must be paid to all aspects, though as he points out, there is no general theory or model for success, and often each case must be examined on its merits.

Throughout the text there is one major thread, that above others, seems to be of major importance. More than profitability, or financial rewards to users or operators, this main thread is knowledge and communication. A new technology can only be implemented successfully if a) managers know how to use it strategically, b) operators know how to use it technically and c) they communicate effectively with each other. Effective communication requires the means for communication (telephones, electronic mail, meetings), the will to make the connection, and knowledge skills and understanding that permit exchange of thoughts at a similar level. Effective communication also stimulates commitment and the correct use of rewards, and above all opens the channels for effective redesign and learning that Goodman stresses is so important for successful institutionalization.

As we all are aware, learning is a long and tedious business. Most professional persons spend at least twenty years in some form of formalized education, and many must feel that this investment should set them up for a considerable period before they need to retrain thoroughly. As people progress in their careers they tend to specialize and if they become managers, they may become quite removed from the "shop floor". When confronted by a new information technology, such persons will surely accept it most readily if it latches on to an aspect of their own training, for then the investment in relearning is small. The new technology will also be

I. Masser and H. J. Onsrud (eds.), Diffusion and Use of Geographic Information Technologies, 59–60.

more readily accepted the easier it is to understand, and the easier it is to place in an existing organizational context. We can call this approach the "playing safe" model.

The "exciting risk" model is adopted by pioneers who are thrilled by new technology and what it can do, but who may not be able to place it in a strategic context nor fully to be able to communicate its advantages to others. For them, new ideas are paramount - we all follow trends and fashions, otherwise life would be too dull.

"Playing safe" and taking "exciting risks" are two extremes of behaviour that can occur in persons at all levels of a management-production team. Either extreme can lead to failure when new information technology is introduced. The "playing safe" approach is to wait and see, to defer decision-making until the learning process has caught up. Unfortunately, developments may be so rapid that catching up is very difficult or impossible. The "exciting risk" approach involves gambling that the new technology will very quickly produce worthwhile, extra benefits to compensate for the risks of premature adoption.

It is axiomatic that skills and knowledge are both essential for rapid appraisal and strategic use of new technology and for reducing the risks associated with implementation. This implies an urgent need for training and communication, both within an organization, and outside. For some, knowledge about what a given kind of technology does is more important than to know how it does it; the combination of both what and how is obviously stronger than either separately. Finally, the knowledge of how the application field is changing, and how technology is changing allows one to adapt to new opportunities.

In many ways the response of persons and organizations to new technology strongly parallels Darwin's evolutionary hypothesis. Although, according to Goodman, the rate of new information technologies being developed is increasing, not all these technologies will have equal chances of taking hold - some will be more attractive than others. New technologies can be seen as new ecological niches; successful implementation means that persons and organizations have adapted to take advantage of the new conditions. If the ecological niche is short-lived, then only pioneer species will develop, and they will quickly fade away. If the new ecological niche is strong, but existing species do not adapt, then they too will dwindle. If the ecological niche provides conditions that are too far removed from those that the existing organisms can use, then however strong it is, it will only be colonised slowly, if at all. The most successful species are those who can colonise and use the most ecological niches in such a way that they dominate them and can modify them to suit their own ends. As in ecology, symbiosis, predation, adaptation, dominance and communication all play a role in the introduction of new technologies.

Alternatively we can consider the new information technologies as being the new species that are looking for suitable ecological niches (implementation). Weak organisms (inappropriate technologies or specialist technologies) look for specialised niches. If the niches are damaged then they die out. Strong organisms (technologies) can modify their niches to provide symbiosis and other advantages.

If we consider technologies as organisms it is not surprising that though many may be introduced, many will fall by the wayside. There are many more organisms that have lived on the earth than are now living. In fact, we can expect that it is in the very nature of most organisms/technologies to die out sooner or later, which suggests that there will always be a great deal of chance involved in introducing any new technology.

THE APPLICATION OF RESEARCH AND DEVELOPMENT FROM THE INFORMATION SYSTEMS FIELD TO GIS IMPLEMENTATION IN LOCAL GOVERNMENT: SOME THEORIES ON SUCCESSFUL ADOPTION AND USE OF GIS TECHNOLOGY

WILLIAM E. HUXHOLD
Department of Urban Planning
University of Wisconsin-Milwaukee
P.O. Box 413
Milwaukee, WI 53201 U.S.A.

ABSTRACT. For years, GIS researchers have been seeking explanations for rates of slow diffusion rate of GIS technology in local government. This paper attempts to add to existing GIS research on this subject by developing theories about the successful adoption and use of GIS technology in local government from previous research and development in the broader information systems field and calls for a research agenda in GIS to test these theories in case studies of GIS implementations. Lacking any convincing arguments that geographic information systems are significantly different from information systems in general, this paper reviews two research projects conducted on the implementation and use of local government information systems in the 1970s and a recent development in the design of information systems in the 1980s to draw attention to available knowledge and tools from the IS profession that should be considered in developing a research agenda for GIS in local government. The results of this review indicate that certain organizational conditions must be met in order for integrated information systems to be successfully implemented and adopted and that these conditions appear to be similar to conditions found in the GIS literature on the importance of organizational issues in GIS implementation and use. Eight theories are developed about GIS implementation in local government from the experience of the author and others in the IS field. This paper suggests that future GIS research be directed towards the verification of these theories in order to gain a better understanding of local government GIS implementation.

1. The Dynamics of Design: From Processes to Data

The change from "process-oriented" systems design that was characteristic of the transaction-based systems implemented when computers were first introduced in local government (during the 1950's and 1960's), to "data-oriented" systems design brought about by the introduction of commercial data base management systems in the 1970's and 1980's, has changed the focus of information systems design. The focus has moved from task-oriented processing, which supports the existing procedures and responsibilities within the organizational structure of local government, to information-oriented processing, which supports the information needs of operations, management, and strategic planning of the organization, regardless of the organizational framework in which they are performed. Thus, data base management systems (which form the essence of geographic information

I. Masser and H. J. Onsrud (eds.), Diffusion and Use of Geographic Information Technologies, 61–75.

systems) allow designers to build information systems from a strategic, enterprise-wide perspective rather than from a departmental perspective of automating existing procedures. This enterprise-wide approach to systems design organizes data around the business of local government rather than the processes of individual offices, bureaus, and departments.

1.1 THE "CASE" FOR INFORMATION ENGINEERING

In recent years, the information systems (IS) industry has developed techniques to assist organizations in designing information systems with an enterprise-wide view of data and processes as opposed to the traditional departmental view of systems design. These techniques, referred to as "software engineering" or "information engineering" provide a structured methodology for planning, defining, designing, and constructing (i.e., developing) information systems that are directly linked to enterprise strategy and goals, satisfy enterprise-wide requirements, and provide for the most effective and efficient use of information in achieving the enterprise's mission (Arthur Young & Company, 1987). With a strong emphasis on data sharing, user involvement in the development process, and building logical models of business data and activities, information engineering is growing in popularity in the non-GIS sector of the information systems profession because it provides a structured and standardized approach to the traditional "top down" information systems design philosophy of the past.

The objectives of the information engineering approach to information systems development include:

* Responsive and accurate support to the information needs of senior managers by developing information systems that are of strategic importance to the enterprise.

* A focus of the IS team on relating information systems activities and products to the enterprise goals and critical success factors they support.

* Providing senior management with an increased under-standing of, and a greater ability to control, the enterprise's information systems resources.

* Assistance to the enterprise in gaining and maintaining a competitive advantage in the marketplace by identifying strategic uses of information technology.

* Decreasing the time required to bring new applications into productive use, and reduce the maintenance problems associated with keeping them cost effective and productive.

* Involving users more effectively in information systems development through the increased use of techniques such as joint application development (JAD) and prototyping.

* Improving the quality of information systems software by increasing the rigor of the methods used to create it, and by basing the system design on data and activity models of the underlying business.

Automated tools for using the information engineering approach to IS development have begun to appear on the market. Originally developed to improve the programming phase of development by using automated code generation, there are now a number of CASE (computer assisted software engineering) tools available that integrate the planning, analysis, and design phases of IS development with automated code generation to provide an automated tool for the complete structured system development life cycle process. These CASE tools generally have a hierarchy of four phases in the development process:

* Information systems planning, which produces a high-level model of the enterprise and its data. This model links information requirements to senior management's strategic planning and documents the major business areas of the enterprise by using techniques such as clustering and factor analysis of processes and entities.

* Business area analysis, which creates a logical information model for each business area that reflects the interrelationships of data, activities, and enterprise goals of the business area. A detailed data model and activity model for each business area is developed and compared to an assessment of the current business area environment for developing alternative design concepts. These design concepts are prioritized via an enterprise benefit analysis and a specific implementation approach is selected.

* System design, which transforms the requirements of the business areas into detailed application system specifications. Prototyping is used to involve the users in specific product design and the database design is completed. This phase also designs the conversion system, system test plans, acceptance criteria, and user and operator training programs.

* Construction, which creates programming code through an automated code generator. It also develops procedures for system operation.

To date, there are no known documented examples of the use of information engineering tools for the design of geographic information systems. (One known effort, however, is currently underway at the State of Wisconsin Department of Transportation, but as of this writing, the results of this effort have not been reported.) The private sector, however, has apparently been using commercial information engineering tools since as early as 1985, as reported in available sales material (Texas Instruments 1990). One such testimonial reported in this material comes from S.D. Scicon, a large contract software firm from the United Kingdom, which used CASE tools in a $13 million three-year phased program to automate an entire government agency: "...we're producing the government applications using 17 non-programmers and just three programmers" and "...everything's integrated together, so cause and effect ripples through the entire model. The further integration goes through the life cycle, the better." (p. 31)

1.2 WHY NOT INFORMATION ENGINEER GIS?

Perhaps it is the unique nature of local government that prevents the transfer of information engineering tools to GIS projects. Existing institutional frameworks that organize local government around government processes make it difficult for information management technology such as GIS to be successfully designed and implemented as an enterprise-wide resource. Local government organization structures tend to support the "process-oriented" systems because mandates are often used to establish sub-goals for individual offices or departments. This can result in a misalignment between individual and subunit goals with the goals of the organization as a whole because skills and motivation are focused on those sub-goals and organization-wide goals become secondary (Finch, Jones, and Litterer 1976, and Campbell 1991). Such was the case in Milwaukee, for example, where the City Engineer was appointed by the Mayor yet reported to the Public Works Director, a civil service position. In many U.S. counties, some departments report directly to elected officials such as a property appraiser, county clerk, or register of deeds. Often, state mandates and statutes can control the policies of a municipal tax assessor who may be elected, appointed by the mayor, or hired by a civil service commission. These varying and sometimes illogical lines of communication and control tend to reinforce the tendency to emphasize division or office goals, and to discourage a broader view that would rely on common goals to drive policy and action (Ventura, Huxhold, Brown, and Moyer 1992).

Brown and Friedley (1988) observe that the "parochial outlooks and limited objectives of each department and profession will interfere with a comprehensive, multiple use view of the organization's requirements" (p. 24). Such a condition suggests that a public organization such as a local government does not have an "enterprise strategy" which is a prerequisite for the information engineering approach. Conversely, it may be that local government has too many enterprise strategies (oftentimes competing strategies at that) because of the many external constraints imposed on its operation. Others (Ezigbalike, Coleman, Cooper, and McLaughlin 1988) contend that each user of the system has "individual and private expectations of the system, and the various expectations do not always coincide; the multidisciplinary nature of the (GIS) user community makes it difficult for their requirements to converge, and increases communications problems." (p. 284). Campbell (1991) finds that:

> `An organization is therefore conceptualised as a negotiated order in which coalitions are continuously being formed and reformed with each group trying to structure the available resources to serve their interests' (p. 3)

Geographic information systems, however, are not the only enterprise-wide systems that can be of value to local government. The successful use of enterprise-wide financial accounting systems (and, to some extent, personnel/payroll systems and material management systems) in local government has allowed multiple interests to gain benefits from their use. The fact that these other enterprise-wide information systems have been successfully adopted and used in local government suggests that it may not be the unique nature of local government that prevents the use of the enterprise-wide approach to systems development advocated by GIS observers. Thus, it may be that information engineering can be used to develop successfully adopted geographic information systems in local government.

Perhaps it is the unique nature of geographic information systems technology that prevents information engineering from being used to design GIS projects in local government. It has taken non-technical staff and managers almost a full human generation (about thirty years) to understand the potential benefits of computers in the workplace for basic services such as payroll, accounting, material management, etc. GIS technology, on the other hand, does not have the same amount of history and, thus, has a new learning curve to overcome. As a result, users may not be sufficiently educated in the technology to be able to specify requirements (Ezigbalike, Coleman, Cooper, and McLaughlin 1988). This line of reasoning, however, places the perverbial cart before the horse in that, while GIS technology may be considered "novel" when compared to the overall IS field (and no one as yet has shown convincing arguments that it is), information engineering begins with the enterprise strategy and business information needs which are the daily concerns of these non-technical users. GIS technology is a solution to those needs - but only if and when the needs are identified. It is much later in the development of information systems using information engineering when specific technologies are considered for implementation (the systems design phase is the third of the four IE phases). Thus, whatever unique nature geographic information has over other enterprise-wide information would not appear to affect the use of information engineering until later in the process when specific technologies are considered.

1.3 IMPLICATIONS OF INFORMATION ENGINEERING FOR GIS RESEARCH

The information systems industry has developed the capability to plan, define, design, and construct information systems that satisfy enterprise-wide requirements, are directly linked to enterprise strategy and goals, and provide for the efficient and effective use of information in achieving the enterprise's mission. The private sector has even developed commercial products for automating this development process and clients of the firms marketing these products have successfully applied those products to their information systems projects. This successful use of information engineering in the private sector causes one to consider the following theory about GIS in local government:

> *Information engineering methodologies can produce more effective geographic information systems in local government.

2. The Dynamics of Diffusion

The GIS literature reports a flurry of activity in local government to investigate the implementation of geographic information systems (GIS) technology; and it is not just the large urban jurisdictions who are jumping on the GIS bandwagon - towns like Granby, Vermont, with a population of 81 to cities like New York are either using or developing GIS capabilities. Unfortunately, there have been no studies documented that have attempted to determine just exactly what the rate of adoption is and, therefore, if there is any constraint on the diffusion of GIS in local government.

In 1987 there were 83,186 units of local government (county, municipal, township and town, school district, and special district) in the U.S., of which 22,242 were county and municipal governments (U.S. Bureau of the Census 1989). While GIS observers might expect that eventually all of these 83,186 units of local government will have GIS capabilities to some degree in the (most likely distant) future, it may be more realistic to consider that a majority of adoptions will occur in the 22,242 counties and municipalities. Whether the 83,186 figure or the 22,242 figure is used to represent a 100% saturation point for GIS in local government, it is clear that there is a large number of jurisdictions that are candidates for the diffusion of GIS technology.

As stated above, there are no published figures on how many local governments have implemented GIS technology to date, but one might surmise from the literature that it is nowhere near the 83,186 or the 22,242 possibilities. Unpublished research, however, has determined that the five largest GIS vendors had sold their systems to 506 local governments in the U.S. between 1985 and 1990 (Onsrud and Pinto 1992). Assuming a negligible adoption rate prior to 1985 and of systems sold by vendors not among the five largest, these numbers would translate to a 0.6% or 2.3% adoption rate for GIS in local government, depending upon which saturation figure is used. (Note: Of 256 survey responses received from these local governments in this research, the smallest local government reported a jurisdiction population of 100, lending support to the notion that it is not necessarily the size of a jurisdiction that determines its motivation to adopt GIS technology.)

What appears to be a low rate of adoption of GIS technology in local government (say 2.3%) may be explained by the results of two research and development projects that studied the larger field of local government information systems in the 1970s. Both studies involved in-depth reviews of information systems in a small number of local governments and their findings can help explain not only what difficulties exist in bringing new technology to local government, but also how the existing information systems environment in a local government can either create these difficulties or support diffusion. These projects are:

The USAC Program - A $26 million experiment to build integrated municipal information systems in six local governments and then transfer them to other local governments.

The URBIS Project - A case study of the 30-year history of information systems in seven local governments to determine what caused computing change in local government.

Briefly, the USAC Program (which failed in its objectives) demonstrated that a "supply-push strategy" that espoused the theory that there was an existing demand for integrated systems was wrong. It found that local governments have always developed information systems on their own, based upon their own perceived needs. Hence, a specific information management technology is not likely to be in universal demand (Kraemer and King 1979). The URBIS project demonstrated that the manner in which computer technology was controlled in local government had the greatest impact on how the technology was developed and used. Four patterns of control were identified, called "states of computing management". The project found that these states are useful for establishing a benchmark for assessing the likely consequences of a particular management action, such as adopting a new technology (Kraemer et al 1989)

2.1 THE USAC PROGRAM

In 1970, six medium-sized cities in the U.S. were awarded over $20 million by a consortium of ten federal agencies to develop integrated municipal information systems (IMIS) for their jurisdictions. This consortium, called the USAC Committee (for Urban Information

Systems Inter-Agency Committee), was formed to provide funds to local governments as an incentive for research and development in the application of information systems to urban management in anticipation of transferring these applications to many other local governments, building toward an intergovernmental information network of information sharing at all levels of government. The underlying USAC philosophy was that local governments were major links in the overall governmental system that warranted federal government assistance in improving local government planning, management, and operations. Kraemer and King (1979) write:

> "The USAC architects further believed that the most promising tool for achieving this improvement was increased use of computers and automated information systems: information systems could automate routine information processing tasks, thereby improving productivity in the information-dependent field of local government management; they could be used in analysis and redesign of municipal goals and activities; they could integrate data and data processing, creating a dynamic data base for use in planning, management, and integration of local operations. Ultimately, the designers of USAC felt that the use of computers could result in development of new sociotechnical systems offering increased local government efficiency and improved service delivery." (p.315)

The IMIS project was based upon seven major concepts which would define, for the first time, a complete data model and information flow requirements of an entire municipal government that would take advantage of the full potential of information systems technology for integrating data processing across an entire enterprise. These seven concepts include:

1. A **total municipal information system** was to be built, based upon four major functional areas: public safety, public finance, physical and economic development, and human resources development.

2. The system design was to be based upon **integrated data processing**, involving a single data base that could integrate functions both vertically among government functions and horizontally around the objects of people, property, money, and personnel.

3. Most of the data to be automated was to be based upon the **operational data files of municipal agencies** because planning and management data is considered to be a by-product of operational data.

4. The development process was to be conducted from the top down, from analysis, conceptualization, design, development, implementation, and evaluation, with **prototypes implemented incrementally**.

5. At all phases, **complete documentation** was to be produced for research and to facilitate transfer of the prototypes.

6. The prototypes of systems and subsystems were to be designed and implemented to **facilitate transfer to other cities.**

7. The projects in each city were to be conducted through a **consortium approach,** with the city controlling the research and development through a project director, a private systems development firm providing technical expertise and training, and a university providing technical advice and evaluation.

The USAC program was a failure. After 7 years and $26 million (the six cities added $6 million of their own), the USAC Committee disbanded following a report by a panel of experts who assessed the IMIS project and found that "virtually no total systems had been transferred from the USAC cities to other local governments, and that the USAC program had had little impact on encouraging intergovernmental cooperation in the exchange of data." (Kraemer and King 1979, p. 323).

The IMIS project is not without value, however. This experience found that, while an integrated municipal information system is feasible, it will not be successfully implemented unless the particular local government perceives that it will solve a particular problem or will address a particular need that is not currently being met:

> "...while the need for more effective information management technology is obvious in the general sense, local governments differ considerably from one to another, depending on size and locality. Thus a specific package of information management technology is not likely to be in universal demand" (p. 347)

The notion that a particular information systems technology itself (such as GIS) should spur diffusion is also questioned by Campbell (1991) who calls for research to:

> "...examine the role of geographic information and the needs of users in practice rather than assuming GIS will be used simply on the basis of theoretical opportunities such facilities offer." (p. 9)

2.1.1. *Implications of the USAC Experiment for GIS Research* The failure of the USAC project and subsequent analysis of its attempt to develop enterprise-wide information systems in local government can assist GIS researchers in their attempt to understand the diffusion and use of GIS technology in local government. Kraemer and King (1979) present four findings from the USAC project that, when applied to GIS technology, can produce the following theories about GIS in local government:

* A local government is not ready to adopt GIS technology because of the technology itself - it is ready to adopt GIS technology when it has identified a problem and sees GIS as a solution to the problem.

* The scope of a GIS project must be clearly understood and accepted by all potentially affected functional areas. Sufficient resources must be applied to a study of the feasibility of the application of the technology with detailed definitions of all information flows and the potential for changes in organizational structure and procedures.

* The goal of the GIS project must reflect the individual goals of each participant in the project. These goals must also align with the overall goals of the organization. Any changes to the scope of the project during its development must be communicated and adopted by all participants.

* A detailed, accurate estimate of costs is critical and much more valuable to the success of the project than the determination of potential benefits. If the system will solve a problem, the adoption decision will be based upon the question "Can we afford it?" rather than on any quantitative cost/benefit figures.

2.2 THE URBIS PROJECT

In 1973, the Public Policy Research Organization (PPRO) at the University of California, Irvine began the Urban Information Systems (URBIS) project to study local government computerization and identify which policies used for the management of information systems in local government most effectively produced benefits and avoided problems. The URBIS project surveyed over 400 U.S. cities with a population of 50,000 or more and included an intensive field study of forty of them. After that initial research, seven cities and counties were studied longitudinally to identify the forces affecting their "computer package" since their initial adoption of computers, and then determine the effect of those forces on the use of information systems in the agencies over the years. The results of these studies have been documented (Kraemer, King, Dunckle, and Lane 1989) and provide a challenge to GIS researchers in local government to determine if their findings also apply to GIS technology.

While the URBIS study included the entire local government information systems environment rather than the specific GIS environment, it would appear that the results of this study can be effectively applied to the GIS program in local government because the research was prompted by the researchers' interest in "the interplay between information technology and organizational form and function". They were interested in understanding why some organizations were successful at exploiting information systems technology after its implementation in the late 1960s, while others were not successful.

While the authors present hypotheses about the computing environment in local government (hypotheses that, by their own admission, require further testing) their findings may explain why today some organizations have been successful at implementing and using GIS technology, while others have not. For example, they state that their most important finding was a simple one: "management makes a difference". They found that the manner in which computer technology was controlled in local government through the decisions of managers had the greatest impact on how the technology was developed and used:

> "In all seven organizations, of all the factors we studied, the decisions of managers had by far the most significant influence on computing development and use." (p. xii)

Furthermore, they found clear patterns in the management of computers in the local governments and have classified them into four different "states of computing management". Each state (that combination of control and interests presiding over an organization's computer-related decisions and policies at any one time) produces a different "computing trajectory" which they define as the long-term growth of computing in an organization. These states of computing management and the different effects they produce, if proven through further testing (as the authors suggest), may provide a useful typology for analyzing the organizational placement of the GIS project and its potential for success or failure in local government. They are: the Skill State, the Service State, the Strategic State, and the Mix State.

The Skill State: "In the Skill State, IS management controls computerization and applies computing resources to the enhancement of technical quality in the organization. The computing acquisition and allocation system is directed by and for IS management; organizational interests are held secondary to the development of technically advanced systems. In the case studies, the Skill management state usually flourished in the presence of an IS manager committed to implementing a particular vision of computerization. In these instances, the Skill state ended when this manager departed. While users are provided with substantial computing capabilities under a Skill state manager, the technical sophistication of the systems created in a Skill state can exceed the ability of many users to employ them effectively.

`We consider the policy-making mode of a Skill state to be technocratic elitism. A technical elite, composed of technical specialists (programmers, analysts, IS managers) and key technical users (budget analysts, planners), dominates computing decisions through its monopoly on technical expertise. The elite controls computing growth and change to suit its technical interests and ambitions...

`In a Skill state, the computing trajectory is determined by the needs and desires of IS management. As a result, computing package growth is intensive in specific functional areas of interest to IS management. Application developments are oriented to maximize technical sophistication and appear mainly in selected departments. This growth represents the limited scope of technical interests being served. However, the IS organization is also expansionistic about computing development throughout the larger organization. While the expansion helps mollify users, it also stimulates additional demand. This, in turn, helps justify the acquisition of leading-edge technologies. Computing staff and budget allocations increase in a manner reflecting the resource investment needed to implement advanced computing, and additional external support from government agencies or vendors may be involved.

`Computing hardware acquisitions include advanced vendor offerings and may even precede general market availability in sites designated as product "showcases" by vendors. In a Skill state, hardware is primarily centralized to facilitate enhancements, and any hardware distributed to departments is operated by central computing staff. User charges may be nonexistent because of the experimental nature of systems under development." (pp.117-118)

If such a Skill state does exist, and one were to apply these findings to GIS in local government, one might conclude that a GIS project under the control of the data processing function in local government will be technically sophisticated, but expensive, complicated for many users, and not linked to the overall strategy of the organization.

The Service State: "In the Service state, departmental management controls computing and the operational interests of these departments are served. The computing acquisition and allocation system is directed by and for departmental management, and the development of functional applications has top priority.

`Organizational pluralism is the policy-making mode of the Service state. Organizational pluralism is a mode in which no person or single interest dominates computing decision making. Instead, decisions are made through the pluralistic establishment of compromises and coalitions among the varied interests competing for computing resources. This mode assumes that allocation policies meet the minimal needs of the dominate interests and that decision-making patterns change as the coalitions change...

`In a Service state, the computing trajectory is determined by the interests of department-level management. Computing package growth accelerates as computerization expands in and across departments. Computing staff and budget allocations show moderate but continuous increases commensurate with this expansion. Application developments, oriented to support the functions of user department operations, exhibit growth both within and across departments. This growth represents the variety of operational interests being served.

`Computing hardware acquisitions encompass a broad array of capacities and capabilities, based upon the myriad needs of users. Some hardware is distributed to departments, depending on the configuration required by a specific task. User charges also reflect a service orientation, with the organization encouraging computing use." (p.120)

If such a Service state does exist, and one were to apply these findings to the GIS case, one might conclude from the URBIS study that a GIS project under the control of a User Policy Committee, would include a variety of different vendor technologies, contain applications that are directed toward specific functions, decentralize technical responsibilities to the users, experience a continuous growth in costs, and incur directional changes as interdepartmental coalitions change.

The Strategic State: "Within the Strategic state, top management controls computerization and the broad interests of top management are served. The computing acquisition and allocation system is directed by and for top management; the general goals of the organization and the specific goals of top management receive top priority.

`The mode of policy-making in the Strategic state is generally characterized as managerial rationalism. Under managerial rationalism, critical decisions about computerization are made by the organization's elected and appointed leaders with the intent of serving the broader interests of the organization in carrying out its basic mandates. In the case of city governments, these basic mandates are to provide public services efficiently and effectively. This mode is often characterized as the ideal mode of decision-making behavior in the computing management literature... However, the case studies show that the agendas, ambitions, and interests of top managers also influence their decision.

`The computing trajectory of the Strategic state is determined by the needs and desires of top-level management. Computing package growth is limited to that required by top management practices and organizational operations that top management favors. Computing staff and budget allocations often show minimal incremental increases commensurate with this conservative orientation, although new application interests by top management can change this profile dramatically. Application developments, oriented to support managerial control over organizational functions, exhibit growth primarily in core administrative departments.

`In stable and conservative organizations, the Strategic state, computing hardware acquisitions are limited to those required to implement core applications. These applications are usually designed to run on central mainframe systems, or at least in concentrated clusters. The acquisition of smaller systems is often curtailed by policy, and the distribution of hardware is restricted in practice. User charges reflect the strategic orientation by recapturing most of the actual costs associated with computing and serve as a means of regulating user demand for access to computing." (pp. 121-122)

If such a Strategic state does exist, one may conclude that a GIS project controlled by the mayor, city manager, or top elected or appointed official, will reside on a centralized mainframe computer or otherwise be under central control, address uses that are interdepartmental and administrative in nature, and be conservative in technical sophistication and cost. Furthermore, these findings suggest that a GIS implemented in the Strategic state is most likely to succeed because of the organizational-wide perspecitve of GIS technology, provided that the political interests of top management do not interfere with the long-range goals of the project.

The Mix State: "In a sense, the Mix state is a default condition that exists in the absence of any of the three ideal state types. That is, the Mix state encompasses any set of conditions where the level of control and level of interest served do not directly correspond.

`Particular modes of policy-making are associated with each of the three previous management states because they show consistency between the hierarchy of power and the use of computing application. Their policies are likely to be consistent in their intent and implementation. Mix-state organizations may achieve a compromise in which several agendas are satisfied or one agenda of several actually dominates the change process. Under either of these conditions, computerization follows a specific computing trajectory, but the direction is entirely contingent on the specific computing context. However, since the Mix state represents inconsistency between the visions for computing and the power to effect the vision, organizations in the Mix state run the risk of generating policies that conflict or overlap. Because such organizations fail to resolve basic differences, the policies enacted may actually exacerbate problems in the computing package.

`Under Mix-state conditions, organizational resources often are not mobilized in concerted efforts, nor are they applied in a consistent manner. Consequently, an organization in a Mix state may fail to develop a computing trajectory as clear as those generated by organizations in the pure states.

`While the Mix state usually signifies an organization's inability to reconcile computing authority and intentions, the Mix state can also represent extended periods of transition when a city is functioning without clear management direction, or when the organizational role of computing has not been clearly defined. During such transition periods, the lack of computing direction is even more pronounced." (pp. 123-124)

If such a Mix state does exist, one may conclude from the URBIS study that GIS proponents in a local government that has a Mix state of computing management might postpone an attempt to adopt the technology until one of the three congruent states is present because in a Mix state there is no power to effect the vision of computing that GIS proponents have.

2.2.1. *Implications of the URBIS Project for GIS Research* The URBIS project findings have led Kraemer, King, et. al. to make a number of observations about computing in local government, some of which may assist researchers of the diffusion of GIS technology. First, they have observed that the key feature of computing in the Strategic state is the "growth of organization-wide systems and departmental systems that serve top management" (p. 262) Since successful GIS projects reported in the literature to date have been those which have been organization-wide and have served top management, perhaps successful GIS adoption and use may take place only in those organizations that are in a Strategic state. Second, they have observed that "a Mix state is more likely to occur than the congruent states" (p. 266). If most local governments are now in a Mix state, then perhaps there is no vision nor power to effect the vision of GIS technology in most organizations, thus accounting for a slow adoption rate. Finally, they have observed that "there appears to be a drift toward the Strategic state" (p. 263). Perhaps as more local governments drift toward the Strategic state, GIS adoption will increase. The challenge, then, for GIS researchers is to not only test the URBIS hypotheses with empirical data as Kraemer, King, Dunkle, and Lane suggest (p. 278), but also to determine if the degree of GIS technology adoption can be explained by the state of computing management that the local government is in at the time the concept is introduced.

Saarinen (1987), Brown and Friedley (1988), and Croswell (1991) all insist on an assessment of organizational risk during the planning phase of GIS implementation, suggesting that management may not be ready to accept GIS at the time the technology is proposed. This concept is supported in the broader information systems literature by Gibson, et al. (1984) and the results of the URBIS study have also led Kraemer et al to conclude that a determination of the state of computing management a local government is in at the time a new technology is introduced can predict not only the likelihood of adoption of the new technology, but also the potential for success of the project. This, then raises the question: "Do the URBIS project findings apply to GIS in local government?" If they do, then the following theories should be tested in order to gain a better understanding of the adoption and use of GIS technology in local government:

* The decisions by the managers who control and are affected by GIS technology determine the adoption and success of the project.

* A GIS project is most likely to be successfully adopted when a congruent state (Service, Skill, or Strategic) exists in local government. Furthermore, because of the organization-wide benefit of GIS technology, the Strategic state is more likely to produce a successful adoption and implementation of GIS technology because in that state, top management is in control of the IS decision-making process (as opposed to the IS Manager or a committee of user management who may not have an organization-wide perspective on the benefits of the technology).

* As more local governments move to a Strategic State of computing management, the GIS adoption rate will increase.

3. Summary

This chapter has presented eight theories about geographic information systems in local government that have been developed from research and development activities in the broader information systems field. Further, it suggests that, rather than developing new theories about the adoption of GIS technology in local government, GIS researchers should test these theories in a research environment to determine if they can explain the perceived slow acquisition rate and subsequent successful adoption of GIS in local government.

These theories are based upon the assumption that the development and use of geographic information systems do not differ significantly from the development and use of the broader field of information systems in general - that GIS is just another form of integrated information systems made possible by the development of data base management capabilities. This integrating nature of GIS and of data base management systems in general allows the integration of data across different functional units to facilitate their use as well as the integration of processes which can link logically related tasks that are normally divided by organizational or other boundaries in an organization.

The developments reported in this paper, however, indicate that this integration (whether geographic, financial, or human resources related) can be achieved only under the following conditions:

1. Existing complex relationships among all units of the organization must be analyzed from an enterprise-wide perspective;

2. The goals of all participants of the integration effort must be in alignment and the participants must perceive the effort as a solution to a specific problem or issue;

3. Top managers must not only be in control of the policies affecting the integration effort, but must also be affected by its results.

GIS researchers must test the theories presented in this chapter to, first determine if GIS is any different from the broader IS field, and then determine if the findings of the broader IS field can be used to provide an understanding of the adoption and implementation of GIS technology in local government.

4. References

Arthur Young & Company (1987) *The Arthur Young Practical Guide to Information Engineering*, John Wiley & Sons, Inc., New York.

Brown, P. and Friedley, D. (1988) Assessing Organizational Preparedness for a Comprehensive, Distributed LIS/GIS, *Proceedings from the 1988 Annual Conference of the Urban and Regional Information Systems Association*, 3: 22-36.

Campbell, H. (1991) Organisational Issues and the Utilisation of Geographic Information Systems, Regional Research Laboratory Initiative Discussion *Paper Number 9*, Department of Town and Regional Planning, University of Sheffield, Sheffield, UK.

Croswell, P. (1991) Obstacles to GIS Implementation and Guidelines to Increase the Opportunities for Success, *Journal of the Urban and Regional Information Systems Association* 3(1): 43-56, The University of Wisconsin Press, Madison, Wisconsin.

Evans, D. (1987) Distributed Data Processing in a Municipality: How to Make it a Win-Win Proposition, *Proceedings from the 1987 Annual Conference of the Urban and Regional Information Systems Association* 3: 157-168.

Ezigbalike, I., Coleman, D., Cooper, R., and McLaughlin, J. (1988) A Land Information System Development Methodology, *Proceedings from the 1988 Annual Conference of the Urban and Regional Information Systems Association* 3: 282-296.

Finch, F., Jones, H., and Litterer, J. (1976) *Managing for Organizational Effectiveness: An Experimental Approach*, McGraw-Hill, New York.

Gibson, C., Singer, C., Schindman, A., and Davenport, T. (1984) Strategies for Making an Information System Fit Your Organization, *Management Review* 73(1): 8-14.

Kraemer, K. and King, J. (1979) *A Requiem for USAC*, The Regents of the University of California, Irvine, California.

Kraemer, K., King, J., Dunkle, D., and Lane, J. (1989) *Managing Information Systems: Change and Control in Organizational Computing*, Jossey-Bass Inc., San Francisco, California.

Onsrud, H.J, and J.K. Pinto (in press) Evaluating Correlates of GIS Adoption Success and the Decision Process of GIS Acquisition. *Journal of the Urban and Regional Information Systems Association.*

Saarinen, A. (1987) Improving Information Systems Development Success under Different Organizational Conditions, *Proceedings from the 1987 Annual Conference of the Urban and Regional Information Systems Association* 2: 1-12.

Texas Instruments Incorporated (1990) *Information Engineering Facility IEF: A Totally Integrated CASE Environment*, Texas Instruments, Inc., Plano, Texas.

U.S. Bureau of the Census (1989) *Statistical Abstract of the United States, 1989*, U.S. Government Printing Office, Washington, D.C.

Ventura, S., Huxhold, W., Brown, P., and Moyer, D. (1992) Chapter 8: Organizational Issues, In Brown, P. and Moyer, D. (eds.), *Multipurpose Land Information Systems: THE GUIDEBOOK*, Federal Geodetic Control Committee, Rockville, Maryland.

IMPLICATIONS FROM MIS RESEARCH FOR THE STUDY OF GIS DIFFUSION: SOME INITIAL EVIDENCE

GARY C. MOORE
Faculty of Management
University of Calgary
Calgary, Alberta, Canada

ABSTRACT. Growth in the use of Geographical Information Systems (GIS) has been very steady over the past few years, and has attracted the interest of researchers and practitioners interested in understanding how to best manage its diffusion and use. Nevertheless, it would appear that studies of GIS implementation have been conducted independently of the more general area of study within Management Information Systems. It is argued in this chapter that because GIS and MIS share a common base, research into GIS ought to benefit from studies within MIS. To illustrate this argument, some results from the MIS literature with respect to technology diffusion are briefly presented. This is followed by the findings of a series of recent interviews with respect to GIS implementation in the oil and gas industry. These findings show support for a model of user acceptance of GIS based upon the Theory of Reasoned Action.

1. Introduction

1.1 BENEFITS OF THE DIFFUSION OF GIS

Recent studies have shown that growth in the use of Geographical Information Systems (GIS) has been very steady over the past number of years. In particular, as is evidenced by the reports in this book, GIS have penetrated a large number of government organizations, including local and city governments in various parts of the world. At the moment, however, the diffusion of GIS seems to be within a rather distinct use area - those individuals who work with *geographic* information, generally through the use of maps (Wiggins, 1992). Nevertheless, the potential boundaries of the use of GIS seem to be still quite far out from current use domains, with many new potential clients and applications in non-government sectors. Some of these potential uses are still based on geographical information, whereas many others could use non-geographical, but still *spatial*, information. One such application already developed is used for inventory control within a large warehouse (see McLaughlin & Post, 1993, for a further discussion of the growing spatial information market place). The potential growth for GIS can be shown by comparing this type of system to the more general area of personal computing within the Management Information Systems (MIS) literature.

While MIS have been defined in a variety of ways, the general view is that an MIS is the total package of information technology used to support all levels of activity within an

I. Masser and H. J. Onsrud (eds.), Diffusion and Use of Geographic Information Technologies, 77–94.

organisation. *Personal computing* in turn is generally defined as the hands-on use of information technology by an individual who is not an information systems professional. Thus, it can be argued that a GIS is a specific instantiation (or sub-component) of a more general information system, and its use fits within the general realm of personal computing. Furthermore, GIS (as defined in Burrough, 1986) and MIS (as defined in Davis & Olson, 1982) are each based upon the same five general components:

- a computer platform, consisting of appropriate hardware and operating systems,
- data repositories (such as files or databases),
- appropriate output devices (such as screens, plotters or printers),
- appropriate user interfaces, and
- users.

Other parallels between GIS and MIS also exist. For example, it has been argued that the *information engineering* approaches used within the MIS realm should be also applied to GIS (Huxhold, 1992). Also, decisions to implement many MIS applications must be made at the organisational level because of the size and scope of the application. Only after the organisation has acquired the technology can individual users take advantage of it. The same is true of GIS. The differences between MIS and GIS thus generally boil down to the fact that GIS explicitly use *spatial* data, an attribute which is not as obvious in general MIS applications. Nevertheless, in a series of interviews (discussed below) of individuals in the oil and gas industry, it was noted that most data within their organisations *could* have a spatial attribute. Users of general systems, however, tend not to think of their data in spatial terms and hence the spatial linkages among data are not being fully exploited.

Given that GIS and MIS are similarly configured information technologies (IT), using a GIS should provide the same wide ranging benefits as using a generic MIS. These benefits fall in three general areas. First, use of IT can *accelerate the speed* at which we carry out various activities, such as processing volumes of data. This is reflected by the *transaction processing system* component of MIS, which is generally the first area in which IT is applied within an MIS. Secondly, IT can *augment our capability* to carry out certain tasks, and thus actually change the substance of the task. Such tasks as advanced economic analyses, reflected in *decision support systems* within MIS fall into this category. Finally, IT can *automate certain tasks*, thus freeing users to carry out more value-added activities. Many *expert systems* fall into this category, as well as certain less exotic applications, such as process control systems.

In that GIS to date seem to be applied within a rather narrow range of their potential, significant benefits could accrue to organisations from the further diffusion of these systems. Such diffusion would enable users to exploit the spatial nature of their data in the three general benefit areas of acceleration, augmentation, and automation of work processes. Defining benefits of using GIS in particular is beyond the scope of this chapter. Instead, it is assumed that many additional benefits of using GIS could in fact be defined, and thus the question addressed here is how to best manage the diffusion of GIS.

1.2 MIS RESEARCH AS A BASIS FOR STUDYING GIS DIFFUSION

In that GIS and MIS share a common technological and application base, there should be considerable benefit to GIS researchers in building upon the experience gained by researchers studying the diffusion of MIS applications and the growth of personal computing. Indeed, researchers within GIS and MIS seem to also share many of the same concerns with respect to technology diffusion. Campbell (1991) drew links between use of IT in general and GIS in particular, and aligned many of the organisational issues identified in the MIS literature with those in GIS diffusion research. Zwart (1992) and Cartwright (1992) also outline issues at the individual level with respect to diffusion of GIS. Within the MIS domain, similar issues have also been addressed. This is due to the fact that most functionally sound MIS applications have failed to achieve all their desired benefits because of the failure to consider non-technical factors in implementation (Eason, 1992; Goodman & Griffith, 1991).

It would appear that GIS researchers are also facing some of the same methodological issues as MIS researchers. For example, MIS was plagued for years with the lack of a cumulative tradition and a "framework of the month" approach (Keen, 1980; Kwon & Zmud, 1987), and it is only in recent years that attempts have been made to couch MIS diffusion research within theoretical frameworks (Alexander et al., 1992; Moore, 1987; Brancheau & Wetherbe, 1990). Thus, it would behove GIS diffusion researchers to avoid such theoretical pitfalls, and to build on the cumulative traditions of the more general technology diffusion field. By building on such foundations, the research process should be more structured, directed, and productive. The findings from various case studies of GIS implementation, some of which are described in this book, should also be more easily integrated. To illustrate a portion of the theory being applied to IT diffusion, the next section will describe some of the work carried out to date in MIS diffusion research.

2. MIS Diffusion Research

As mentioned above, it is only in recent years that MIS researchers have turned to theoretical bases for studying diffusion issues. Since the mid-1980's, however, several streams of study outside MIS have been tapped. While some researchers have taken perhaps a more "classical" approach and used economic theory to study computing diffusion (e.g. Gurbaxani, 1990), most studies are factors based; that is, they examine particular variables for their impact on rates of diffusion or levels of acceptance of IT. While not all explicitly refer to the body of literature described by Rogers (1983), they can be related back to the perspective described by him. Rogers (1983) defined *diffusion* as:

> *the process by which an innovation is communicated through certain channels over time among members of a social system.*

Each one of the components of this definition has been the focus of a variety of studies. Some researchers, while not explicitly referring to diffusion, have studied the attributes of the technology (innovation) itself in order to understand how an individual would interact with it (e.g. Benbasat et al., 1991). Other researchers have examined the communications issues (e.g. Lind & Zmud, 1991; Kwon, 1990; Nilikanta & Scammel, 1990), still others the individual differences among adopters (e.g. Majchrzak & Cotton, 1988; Roskies et al., 1988), and finally some have examined various factors associated with different levels within the social system. Included among these factors would be the characteristics of one's job, work group, organisation, society and country. Nelson (1990) provides a good review of the MIS literature addressing the individual through organisational level variables. Studies of societal and national level variables are rarer and more recent (e.g. Gurbaxani et al., 1990; Mansfield, 1988). In general, much of the research is inconclusive, although some significant results have been found at the individual adopter level (e.g. Davis et al., 1989; Moore & Benbasat, 1991). Those who have reviewed this research suggest that some of the lack of results may be attributed to the early stages of this stream of research within MIS. Nevertheless, as the research accumulates, it is expected that it will start to show more conclusive results. Thus, it is suggested that researchers investigating the diffusion of GIS refer to this growing body of literature to help guide their efforts.

In all instances, assuming that overt coercion is not used, the diffusion of an innovation occurs through the collective, yet individually based, decisions of individual level adopters. It is perhaps for this reason that Rogers (1992) reports that the variable most studied in the diffusion literature is that of *innovativeness*, defined as:

> *the degree to which an adopting unit makes novel use of a product or process as compared to other members of the referent social system* (Moore, 1987).

The above definition shows that innovativeness can have a number of dimensions. These include the degree to which one is relatively *early in adopting* an innovation (Rogers, 1983), the degree to which one uses a previously adopted innovation in a *novel use domain* (Hirschman, 1980), and the degree to which the innovation actually *penetrates a particular use domain* (Moore, 1987).

Most innovation studies have concentrated only on the early adoption decision with respect to innovations, and thus have failed to consider post-adoption behaviour. These studies look primarily at the diffusion among adopters of the innovation, not the diffusion within a particular adopter's potential use domain. Nevertheless, to be truly successful (such that an organisation can reap maximum benefit out of its investment in IT, for example), use of the innovation should appropriately penetrate into a variety of use domains, and into as much as possible within a given use domain. For example, GIS may be used only to produce maps, and users may completely miss the spatial analysis capability of the system. Thus, in studying GIS diffusion, it is not sufficient to measure only the rate at which this innovation is diffusing among organisations, but once it has been acquired one must also consider the variety of functions to which it is being applied, and the degree to which it is being applied within those functional areas.

Because innovativeness is a behaviour, some researchers have also drawn upon the general literature in psychology and sociology fields as cornerstones for their work. One particular stream which has received attention in the MIS field is drawn from the consumer behaviour field and is based on Fishbein's and Azjen's (1975) *Theory of Reasoned Action* (Davis et al. 1989; Moore, 1987; Pavri, 1987). This theory has received considerable support in a variety of research domains (for a discussion see Petty & Cacioppo, 1981; Sheppard et al., 1988), and as it can be used to integrate an assortment of perspectives, the remainder of this paper will discuss its application to the diffusion of GIS.

3. Theory of Reasoned Action

The Theory of Reasoned Action (Fishbein & Ajzen, 1975) posits that an individual's actual *behaviour* is a function of his or her *intentions* (**I**) to perform that behaviour, and that one's intentions in turn are a function of:

1. one's own *attitude* (**A**) about performing the behaviour; and
2. one's *Subjective Norm* (**SN**) towards the behaviour, defined as compliance with one's perceptions of what others expect one to do.

An individual's actual behaviour (**B**) then can be predicted according to the following formula:

$$\mathbf{B} \cong \mathbf{I} = \mathbf{w_1(A)} + \mathbf{w_2(SN)}$$

where w_1 and w_2 are the relative weights given to each component by the individual. Behaviour does not always correspond directly to the behavioral intention in that several external factors may intervene between the formation of the intention and the carrying out of the actual behaviour (Ajzen & Fishbein, 1980).

One's *attitude* is formed from a weighted evaluation of what one believes the outcomes would be of actually performing the behaviour. For example, with respect to using GIS, one might believe an outcome to be "better data integration". The degree to which one believes that this will actually result is then weighted by the importance or relevance of this outcome to the individual. All such perceived potential outcomes are then combined with one another to form an overall attitude about performing the behaviour.

Moore (1987) argued that with respect to the adoption and use of an innovation, these beliefs about outcomes could be cast in terms of the perceived characteristics of innovations (Rogers, 1983; Tornatzky & Klein, 1982). In a study designed to develop an instrument to measure the perceptions of using an IT innovation, Moore & Benbasat (1992) showed that the following perceptions were positively correlated with attitude, and could accurately predict adopters versus non-adopters of an information technology:

Relative Advantage: the degree to which using an innovation is perceived as being better than using its precursor.

Compatibility: the degree to which using an innovation is perceived as being consistent with the existing values, and past experiences of potential adopters.

Complexity: the degree to which using an innovation is perceived as being difficult.

Trialability: the degree to which an innovation may be experimented with before adoption.

Image: the degree to which using an innovation is perceived to enhance one's image or status in one's social system.

Result Demonstrability: the degree to which the results of using the innovation are observable and communicable to others; and

Visibility: the degree to which the innovation is apparent to the sense of sight.

In a related study, Moore and Benbasat (1991) studied the innovativeness of individuals with respect to *personal computing*, which they defined as the direct, personal, hands-on use of IT. In their study, they found that after personal computing had been initially adopted, *relative advantage*, *compatibility*, and *ease of use* were significant predictors of the number of different applications an individual used (such as word processing, spreadsheets, database management systems and so forth), and the number of hours the individual used the computer.

The second major aspect to the Theory of Reasoned Action (TRA) is that of the *subjective norm*. Whereas one's attitude has an internal, personal focus, the subjective norm is externally oriented, reflecting the influence of others. It is formed from two basic components. One is the consideration of what one thinks others, or one's referents, expect one to do with respect to a particular behaviour (termed the *normative belief*). The second component is one's *motivation to comply* with each of those expectations. An individual thus forms a subjective norm for each referent, and an overall subjective norm for the set of referents. With respect to the use of computers, for example, one would consider the expectations of one's supervisor about computer usage, and when tempered by one's motivation to comply with that expectation, one would be more or less inclined to use computers. Moore and Benbasat (1991) examined the effects of the subjective norms for one's senior management, supervisors, peers, and subordinates within the organisation. The subjective norm was found to be strongly related to the decision to first adopt usage of computers, as well as to the number of different applications used and to the number of hours per week of usage.

The strongest subjective norms were for those of supervisors, followed closely by senior management, then peers and subordinates. These findings are of interest because they highlight the roles of all referent groups within organisations. It is almost an axiom that successful projects require senior management support, but this is often only for the initial acquisition phase. Full implementation success, where the technology permeates into all possible use domains, requires strong support from all referent groups. In spite of this, few implementation efforts or research studies have considered the effects of referents other than one's superiors.

4. Applying the Theory of Reasoned Action to GIS Diffusion

As discussed above, use of a GIS is achieved through virtually the same technology as other personal computing applications. More importantly, however, a GIS itself is an adaptable technology, in that it too can be applied to a variety of functions and can penetrate one's work to differing degrees. Thus use of GIS can be seen to be an extension of the Moore and Benbasat (1991) study, and therefore the lessons learned there should be directly applicable to GIS diffusion studies.

The successful implementation and eventual institutionalisation of a GIS is generally the result of several stages in the diffusion process. Because of the nature of the technology, it must first be acquired by an organisation, either at the departmental or corporate level. This first stage represents organisational level adoption, which is typically preceded by an initial round of assessment before a GIS is acquired. The second stage of decision making is at the individual level, after the organisation has acquired the technology. In this stage, the eventual end user of the GIS must initially decide whether to adopt GIS usage, and then decide to what extent it will be used. Throughout all stages, decisions are made about using or rejecting the technology. Furthermore, while the Theory of Reasoned Action was originally cast in terms of individual decision making, because organisational decisions to acquire technology are in fact made by individuals or groups of individuals, TRA should also apply for organisational level decisions. The implications of this are discussed below.

As an initial indication of the applicability of TRA to using GIS, interviews of eight individuals charged with managing the introduction of GIS into organisations were conducted. Six of these individuals were from firms who explored for and produced oil and natural gas; one individual worked for a city government; and the last represented a GIS vendor. Thirty-one actual and potential end-users were also interviewed within one of the oil exploration companies.

The interviews were of a semi-structured nature. First, those individuals involved in acquiring GIS for their organisations were asked to comment on what they thought the important factors were that influenced the degree to which GIS would diffuse within their organisations. This was followed by probing by the interviewer into areas of TRA not touched upon previously in the interview. The actual and potential end users of GIS were asked to speak on the reasons why they did or did not use the technology when it was available. The purpose of this paper is not to report a detailed analysis of these interviews; rather, it is to describe how the TRA model as described above fits the diffusion of GIS. Therefore, the results of the interviews will be used to illustrate, where appropriate, the model's applicability to the diffusion of GIS.

To understand the application area on which this study focused, the following provides a very brief description of the rationale behind the use of GIS in the oil and gas industry. First, use of spatially related data has a long tradition in the oil and gas business. The basic product in this industry occurs in nature, and thus from the very beginning, the product has a geographical reference point. It is therefore possible, as one moves down the value chain, to relate the operations of the firm back to these reference points. This, however, is not yet widely done except within a few departments in the firm. For example, those working in the geoscience areas have virtually always viewed their data through the use of maps, and most firms in this industry either have, or used to have, large mapping departments who produce maps for geologists and geophysicists. These maps are used for portraying what is known, for interpreting the data that are available, and for forecasting where the most likely locations of future discoveries will be. While the actual oil and gas reservoirs have no legal or jurisdictional boundaries themselves, the rights to explore for and exploit these reservoirs are allocated based on surface land holdings. These holdings in turn are typically based on land surveys. Thus, land departments evolved whose roles were to acquire rights to exploit the mineral base beneath a particular land area, and to acquire the rights to move about these surface lands. Mineral rights in the vast majority of instances are held by governments, while surface rights are held by the landowner. In any event, the land department also views its data through maps, but often draws upon completely different data bases than those in exploration. Finally, production departments, whose role is to get the proven product out of the ground and to market, have also used maps to guide their operations, but again typically use their own data bases. Thus, while all these departments are involved with either finding, gaining the rights to, or exploiting, oil and gas from specific areas, they have viewed their worlds through different lenses. This in turn has created problems for management in reconciling the firm's data, and it is the recognition of this problem which has spurred the move towards GIS type data management systems within the oil and gas industry.

4.1 ORGANISATIONAL LEVEL ADOPTION DECISION

As discussed in the previous section, organisational level decisions are still made by individuals or groups of individuals. In reaching a decision in group settings, the members of the group undergo generally the same persuasive forces as an individual in larger settings; that is, they consider their own attitudes and the expectations of others in reaching their own decision, which then acts as "input" into the group decision. Because of this, many of the same decision factors should apply at the eventual organisation decision level as at the individual level. One should, expect, however that some of the factors will have different weights or impacts on the decision. The discussion below highlights the similarities and differences that were discovered in examining the two levels of decision making with respect to GIS adoption.

4.1.1 *Organisational Level Subjective Norms.* All of the respondents indicated that a primary impetus for the acquisition of GIS was the existence of a champion within the organisation. The role of the champion is consistent with the general IS literature, which indicates that champions play key roles in IS implementation efforts (Curley & Gremillion, 1983; Beath, 1991; Beath & Ives, 1989; Howell & Higgins, 1990). Champions first serve to communicate expectations to organisation level decision makers that the technology should indeed be acquired. Next, champions should be able to inspire among the decision makers a high motivation to comply with the champion's expectations. Thus, one aspect of a successful champion is that he or she have enough political power to influence others (Howell & Higgins, 1990). In the interviews of organisational decision makers, the different influence levels was noted. In one organisation, a senior Vice-President was the GIS champion, and he seemed to have much more influence than departmental heads who championed GIS in other firms.

A second factor which was cited by all respondents was the role played by senior management. In one company, acquisition of GIS was delayed because one of the key executives thought that GIS were "just another IS issue" and that there had been "enough investment in IT without demonstrable returns". Even though others in the organisation lobbied for GIS and the executive himself had no veto power, this executive's expectation that the system not be acquired, and the high motivation to comply with his wish, served to delay GIS acquisition. The champions of the technology had to carry out a much more rigorous campaign to eventually bring the technology into the firm.

The positive effect of a senior executive's expectation was also cited. In one firm, when the Senior Vice-President in charge of exploration learned that the firm was "investigating" acquiring a GIS, he stated that "we cannot be that much different from other firms - if they have bought into the technology, then we ought to as well". This expectation spurred the IS manager into action, who was able to quote the executive's expectations, and a GIS was acquired in a significantly shorter time frame than initially anticipated.

The expectations of departments can also play a role in influencing the larger organisation. The Land Department of an oil and gas firm is charged with acquiring rights to drill on certain properties, negotiating leases, and arranging partnerships with other firms in exploring or developing lands. Their data is very geographically oriented, and is also shared across other departments in the firm. In one firm, the Land Department was initially unable to get other departments to work towards acquiring a GIS. To overcome this, it actually freed up much of its own budget to provide initial funding for the GIS, as well as staff time to help with initial implementation. This served to signal their expectations that other departments ought to contribute, and because of their pressure and commitment, the motivation to go along with their expectations was raised in the other departments.

A final example of the role of subjective norms at the organisational level concerns the parent-subsidiary relationship. In one instance, the respondent indicated his firm had bought a GIS because their parent company (which had 60% equity in the "subsidiary") had one. Even though the parent did not dictate that a compatible GIS be acquired, the subsidiary moved much more quickly to acquire it than they had originally planned. An interesting reciprocal arrangement occurred with yet another firm. In this instance, the subsidiary was the first to acquire the GIS, and when managers were making reports to the parent company, they apparently expressed surprise that the parent had not yet acquired the technology. Once again, the effect was to motivate the parent to move towards GIS acquisition.

4.1.2 *Organisational Level Perceptions of Adopting GIS.* As discussed above, the attitude of the decision maker towards adopting is the other key component in whether one adopts. Attitudes in turn are a function of one's perceptions of the outcomes of adopting. In the interviews that were conducted, three of the generic perceptions of adopting an innovation seemed to be more predominant, or influential, factors in the decision to acquire a GIS.

The first major factor was *compatibility*, or the degree to which use of the technology would fit with previous experience and values. In all interviews, the degree of experience in using spatial data was cited as a key factor in the acquisition decision. Departments whose data had spatial attributes but who did not operate in spatial terms were much slower to move to GIS than departments who explicitly recognised the spatial implications of their data. An example of the former type of department in an oil and gas firm is the Production Accounting Department, which keeps track of production from the firm's wells, and then accounts for the production in terms of paying royalties to partners and governments. Because the oil or gas *well* is the basic element in the system, and thus has a definite spatial reference, much of the data can be linked spatially. Accounting, however, is not used to organising its data in these terms. Thus, across all the oil and gas respondents, production accounting was cited as the most reluctant to contribute money and effort towards acquisition of new GIS, even though they share data with other departments which do operate in spatial terms, such as the Production or Land departments.

Even when different departments agree on acquiring GIS, other compatibility issues come into play. This is especially true in the orientation towards various data characteristics. The respondent from the city government cited differences between the engineering department and the planning department with respect to the desired accuracy of the data. Engineering typically desired that their data be accurate within a tolerance of three centimetres, whereas planning wanted the data generally accurate for a "city block".

The engineers' insistence on a high level of accuracy and very disaggregated level of detail, with the concomitant significant implications for database design and development, almost destroyed the cooperation among the other user departments and threatened the overall project. The high level of accuracy and detail was not compatible with the other departments' experience, and little value was attached to it. Other respondents also cited such problems with respect to data orientation across their different departments.

In one interesting example, the issues of compatibility were expressly deemed to be *not* important by the Chief Executive Officer (CEO). In this case, after the many organisational level benefits of acquiring the system were spelled out in an executive committee meeting, the CEO stated: "if we acquire the system, I hope we are not foolish enough to try to tailor the system to us - we should adapt to the system". In this instance, the system had certain data models and procedures pre-packaged with it, and tailoring the system to the firm's current operations would have incurred significant costs and effort. This "fit us to the package" perspective was subsequently communicated to all those involved in acquiring and implementing the new system.

Finally with respect to compatibility, although not a part of this small survey of users, Wegener (1992) makes the argument of how differences in language and style affect the diffusion of GIS internationally, especially to non-English speaking countries. Language is obviously another aspect of compatibility and could have major effects on the diffusion of GIS.

Relative advantage, or the degree to which using the technology is better than the previous ways of doing the work, was also cited by all respondents as a factor in acquiring GIS. All respondents saw GIS as a way of finally integrating their databases, and they also felt that the acquisition project would force departments to finally "clean up" their data so that the GIS databases could be populated. The move towards "corporate" data and away from "departmental" or private data was also seen as a benefit in terms of overall accessibility and workload. With data being entered "only once", some respondents indicated that their firms also expected to be able to reduce staff once they were in an integrated GIS environment. Some of this staff would leave from the data processing side of activities. With respect to system features, three firms expected to be able to almost eliminate their drafting/mapping departments and move to the automated mapping components of the GIS. Only one respondent explicitly mentioned the spatial analysis aspects of GIS as a motivating factor in acquiring the technology.

The third major factor in moving towards GIS was the *trialability* of the system, or the degree to which the firms could experiment with the technology, with minimal investment, prior to acquiring it. Most respondents indicated that the decisions to acquire a GIS were eased because their eventual vendors were able to refer them to other clients who had already implemented a GIS. As members of an informal "club", the other clients allowed the prospective purchasers to experiment with their installation. This trial of the technology included taking different individuals from various user groups over to the other sites. Having access to a working system was considered crucial in gathering the information necessary to make a "confident" decision to acquire GIS. The respondent whose parent company had already installed a GIS reported a similar process whereby the system was tested. In short, the ability to see and experiment with a production level GIS, not a vendor's mock-up was a critical element in the decision process.

Included within the aspect of trialability was the notion of a "satisfaction guarantee" which was written into some of the contracts. In these instances, vendors provided a "fixed cost" solution and contracted with the client to provide a specified level of performance. In contrast, other vendors provided only an initial "consulting" period, after which the client was responsible for paying for any additional assistance in implementing or refining the system. This served to make any potential clients more reluctant in acquiring the product, even though in some cases the product appeared to be superior on the surface to the other contenders. Two respondents indicated they chose the system which appeared to be "second best" because of the service guarantee provided by the vendor.

The issue of *ease of use*, or the degree to which the system is easy to learn and to use, did not seem to be a factor in the organisational decisions to acquire GIS. It was not cited as a major consideration by any of the respondents. This echoes the fact that even though GIS are generally seen to be difficult to use (Cartwright, 1992), many organisations are acquiring them. For organisational level decisions, if ease of use were a factor one would expect that GIS vendors would have been forced to improve their products for the end-users. Although this is the case in the general IT world, with "user friendly" being a catchword, it does not yet appear to be the case with GIS.

4.2 INDIVIDUAL LEVEL ADOPTION

As discussed above, the acquisition of a GIS by the organisation is only the first stage in a successful implementation. The system must eventually be used to its potential by end-users in order for the organisation to reap the full benefits of its investment in GIS. As mentioned in Section 1.2 , research has shown that most attempted systems implementations eventually either fail to realise all the intended benefit, or only realise a portion of them. The primary reason which has been found for this lack of success is that implementers have failed to consider both the organisational context into which the IT is being introduced as well as the needs of the eventual users within that context. Because of this, many intended users decide not to use the technology. The Theory of Reasoned Action can be used once again to help understand this situation. The discussion below draws upon interviews of 31 users and potential users of GIS within one organisation which has recently introduced the system. Again, the intent of this paper is not to rigorously present the results of the interviews, but to use them to illustrate the applicability of the theory.

4.2.1 *Individual Level Subjective Norms.* As reported by Moore and Benbasat (1991), the subjective norm had a very significant effect on the degree to which personal computing was adopted within organisations. Interviews of users and potential users of GIS also showed that the subjective norm has a significant impact on the acceptance of this particular technology.

Virtually all users of GIS reported that their managers and senior executive had high expectations that the technology be used. For example, several users from the Production Accounting department reported that when the GIS database was being investigated, their manager explicitly told them that the move to this technology was for the "corporate" good. He had to take this step because his department was relatively happy with the current technology, and many department members were resisting the change to a GIS orientation because the benefits would accrue mainly to other departments. The manager made it clear that he expected the new technology to be used, and he indicated also that individuals who did not "get on the team" might find themselves left behind, and perhaps even moved out of the group. The effects of these statements were to clearly delineate the manager's expectation, and to reinforce his subordinates' motivation to comply with his expectation.

The subjective norm also had an effect on how much the system was used after initial adoption. For example, respondents who were relatively heavy users of GIS reported that there were individuals within their company who actively promoted the system's use. These "champions" were not those who had brought GIS into the organisation, but rather users who saw the benefits of the system and became GIS advocates. The effectiveness of these champions varied, however, depending on the degree to which their audience identified with them. Thus, champions from outside one's department seemed to be less effective than those within the potential user's department. Similarly, individuals who seemed to be "techies" were less effective champions of GIS because they were perceived to promote technology somewhat unquestioningly.

One effect of the subjective norm which was anti-GIS effect was also noted. In this case, some non-users indicated that their department's best performer was not yet a user of the GIS. This fact influenced some individuals to reject usage of the GIS for the time being. Indeed, one respondent stated that "when [the high performer] starts using GIS, I'll start using GIS".

Finally, the effects of the expectations of subordinates was also noted. One manager stated rather emphatically that he had adopted GIS usage because his subordinates had moved rather quickly in that direction. The subordinates clearly indicated to the manager that he "should" use the system for spatial analysis activities, and while the manager is not yet "convinced of the merits" of GIS, he is now a user.

In general, it can be seen that the different subjective norms seem to have an effect on GIS usage. This supports the theory. Furthermore, it emphasises the importance of ensuring that expectations are clearly communicated, and that mechanisms be in place to increase respondents motivation to comply with these expectations.

4.2.2 *Individual Level Perceptions of Using GIS.* As outlined in Section 3., the perceptions that one has of using a technology contribute towards one's attitude towards using it, which in turn affects one's actual usage. With respect to GIS usage, the perceptions of *relative advantage* and *compatibility* were found to have significant effects. In fact, these two perceptions seemed to be definitely intertwined, such that those who were used to working with spatial data and thinking in spatial terms were not only more "comfortable"with the technology, but were also able to more quickly identify advantages of using GIS than those who had not explicitly worked with spatial data.

Among the advantages cited in using GIS were the ability to integrate data from different sources, to conduct spatial analysis more easily, and to produce one's own maps independently of the mapping department. The advantages to some, of course, may be threats to others. Within the mapping department, the threat is that jobs would either be eliminated or significantly downgraded. For others, moving to an integrated database environment meant giving up control over data entry and management as the file based systems were eliminated. For example, while one manager was outlining to his staff the organisation's decision to move to an integrated database environment, he was explicitly asked by a worried subordinate whether they "would have to use" another department's data. Clearly, in implementing GIS, those in charge of the process have to consider many organisational and individual level issues, and *be obviously prepared* to deal with concerns about implementation if support for the project is to be engendered.

Compatibility with the new environment is also an issue. For example, even the title

"GIS" is alien to many whose data is spatially related, but not necessarily in "geographical" terms. In interviewing the vendor of GIS, he cited the case of a GIS implementation designed to manage a large warehouse. The use of the "G" word initially raised questions as to the applicability of the system, and only after references to the system were changed to "Warehouse Management Systems" was the concept able to gain acceptance.

Trialability of a system is also important in influencing individuals to adopt GIS. For potential users within organisations, however, the issue is less the risk of spending money on an "unknown", but rather *accessibility* to the technology for trial. Users generally reported that once a work station was installed in a common area on their floor, they were much more able to use it, and to assess whether they should adopt the technology. Other aspects of trialability included the development of small task-oriented applications by GIS specialists that a less technical user could apply rather quickly. These smaller applications allowed users to try out the GIS with less investment in learning, and to decide whether to spend the extra time to learn more about the system. Finally, some users reported that they felt that a trial was not as necessary when they had the reports of colleagues who had used the system. This vicarious trial enabled them to make a more confident decision to adopt GIS usage than if they did not have these reports.

Ease of Use was generally cited in a negative sense in that few respondents found using the systems initially easy. Instead, those individuals who were heavier users of GIS also reported more user support being available to them than the light and non-users reported. In one department, support was provided by having a technologist physically co-located with the engineering staff who were using the GIS. This technologist helped the engineers "navigate" the system. In another instance, "boilerplate" script was written for some users to employ which eased their interaction with the system. Finally, the availability and quality of training were also cited by a number of respondents as influencing their use of the technology. The non-users all cited the *lack* of training as an issue, whereas many of the heavier users had been able to get sufficient training to make them feel comfortable tackling the system.

As can be seen, the *perceptions* of using GIS seem to have a significant effect on actual usage of the technology. Chief among these were relative advantage, compatibility, and trialability, and ease of use. This fits well with general research into the diffusion of innovations, as well with the Theory of Reasoned Action.

5. Summary and Conclusions

As outlined in Section 4, this initial study shows that the diffusion and use of GIS at both the organisational and individual levels can be researched and explained to a large degree using the Theory of Reasoned Action. Furthermore, diffusion of this technology fits well within research into the diffusion of innovations in general, and with the diffusion of information technology in particular. This study has shown that subjective norms and perceptions of using GIS do indeed influence use.

The diffusion of GIS appears to be in its early stages. While it has penetrated many governments, there appears to be a larger, and as yet generally untapped, market outside these agencies. It is recommended that those charged with managing diffusion of GIS, and those interested in research into this phenomenon, draw upon established lines of research as a basis for their work. The challenge is to identify those areas where the characteristics of GIS fit closely with those of other innovations, particularly IT innovations, as well as to identify those areas where GIS are different than other technologies. In the latter area, the aspect of using spatially oriented data that span departmental boundaries is an obvious difference from other technologies. How these differences will influence diffusion of GIS is a significant question that needs to be addressed.

Finally, one of the many challenges in researching the diffusion of GIS, and that of IT in general, is to address the antecedents of the subjective norms and individual attitudes. As discussed earlier, the Theory of Reasoned Action portrays both the subjective norm and attitude as dependent upon an individual's perceptions, which can be influenced by a number of factors. A significant research question is what, then, are these factors? Identifying and understanding them could help management and researchers to better manage the diffusion and use of the technology. These and other issues need attention to increase our understanding of technology diffusion, such that developers and managers can increase the success of technology implementation. The record to date has not been very good.

6. References

Ajzen, Icek, and Fishbein, Martin (1980) *Understanding Attitudes and Predicting Behavior*, Prentice-Hall, Inc., Englewood Cliffs, NJ.

Alexander, Mary B., Joyce Elam, and Carol Wasala (1992) Multiple Theoretical Perspectives for Studying the Assimilation of Emerging Information Technologies, *Proceedings of the Twenty-Fifth Hawaii International Conference on System Sciences (Volume IV: Information Systems: Collaboration Technology, Organizational Systems, and Technology)*, Kauai, Hawaii, January 7-10, 1992, 400-409.

Beath, Cynthia M. (1991) Supporting the Information Technology Champion, *MIS Quarterly* 15(3): 355-374.

Beath, Cynthia M. and Blake Ives (1989) The Information Technology Champion: Aiding and Abetting, Care and Feeding, In Paul Gray, W.R. King, E.R. Maclean, and H.J. Watson (eds.), *Management of Information Systems*, Dryden Press, Chicago, IL, 338-354.

Benbasat, Izak, Francis J. Lim, and Peter Todd (1990) The User-Computer Interface in Systems Design: A Review of Graphics, Iconic and Direct Manipulation Interfaces, and Interfaces for Computer-Supported Collaborative Work, in A.M. Jenkins (ed.) *Research Issues in Information Systems: An Agenda for the 1990's*, Wm. C. Brown Publishers.

Brancheau, James C, and Wetherbe, James C. (1990) The Adoption of Spreadsheet Software: Testing Innovation Diffusion Theory in the Context of End-User Computing, *Information Systems Research* 1(2): 115-143.

Burrough, Peter A. (1986) Principles of Geographical Information Systems for Land Resources Assessment, *Monographs on Soil and Survey No. 12*, Oxford Science Publications.

Campbell, Heather (1991) Organisational Issues and the Utilisation of Geographic Information Systems, Regional Research Laboratory Initiative *Discussion Paper Number 9*, Department of Town and Regional Planning, University of Sheffield, Sheffield, UK.

Cartwright, Timothy J. (1993) Geographic Information Technology as Appropriate Technology for Development, In Masser, I., and H.J. Onsrud, editors, *Diffusion and Use of Geographic Information Technologies*, Kluwer Academic Publishers, Dordrecht, The Netherlands.

Cooper, Randolph B., and Robert W. Zmud (1990) Information Technology Implementation Research: A Technological Diffusion Approach, *Management Science* 36(2): 123-139.

Curley, Kathleen F., and Lee L. Gremillion (1983) The Role of the Champion in DSS Implementation, *Information and Management* 6: 203-209.

Davis, Fred D., Richard P. Bagozzi, and Paul R. Warshaw (1989) User Acceptance of Computer Technology, *Management Science* 35(8): 982-1003.

Davis, Gordon B. and Margrethe Olson (1985) *Management Information Systems: Conceptual Foundations, Structure, and Development*, Second Edition, McGraw-Hill, New York.

Eason, Ken D. (1993) Gaining User and Organisational Acceptance for Advanced Information Systems, In Masser, I., and H.J. Onsrud, editors, *Diffusion and Use of Geographic Information Technologies*, Kluwer Academic Publishers, Dordrecht, The Netherlands.

Fishbein, Martin, and Icek Ajzen (1975) *Belief, Attitude, Intention, and Behavior: An Introduction to Theory and Research*, Addison-Wesley, Reading.

Goodman, Paul S., and Terri L. Griffith (1991) A Process Approach to the Implementation of New Technology, *Journal of Engineering and Technology Management* 8: 261-285.

Gurbaxani, Vijay (1990) Diffusion in Computing Networks: The Case of BITNET, *Communications of the ACM* 33(12): 65-75.

Gurbaxani, Vijay, J.L King, K.L. Kraemer, F.W. McFarlan, K.S. Raman, C.S. Yap (1990) Institutions and the International Diffusion of Information Technology, *Proceedings of the Twelfth Annual International Conference on Information Systems*, Copenhagen, December 1990, 87-98.

Hirschman, Elizabeth C. (1980) Innovativeness, Novelty Seeking, and Consumer Creativity, *Journal of Consumer Research* 7: 283-295.

Howell, Jane M., and Christopher A. Higgins (1990) Champions of Technological Innovation, *Administrative Science Quarterly* 35: 317-341.

Huxhold, William E. (1993) The Application of Research and Development from the Information Systems Field to GIS Implementation in Local Government: Some Theories on Successful Adoption and Use of GIS Technology, In Masser, I., and H.J. Onsrud, editors, *Diffusion and Use of Geographic Information Technologies.* Kluwer Academic Publishers, Dordrecht, The Netherlands.

Keen, Peter G.W. (1980) MIS Research: Reference Disciplines and a Cumulative Tradition, In Ephraim R. McLean, ed., *Proceedings of the First International Conference on Information Systems*, Philidelphia, PA, December 8-10, 1980, 9-18.

Kwon, Tae H. (1990) A Diffusion of Innovation Approach to MIS Infusion: Conceptualization, Methodology, and Management Strategies, *Proceedings of the Twelfth Annual International Conference on Information Systems,* Copenhagen, December 1990, 139-147.

Kwon, Tae H., and Zmud, Robert W. (1987) Unifying the Fragmented Models of Information Systems Implementation, In R.J. Boland and R.A. Hirschheim, editors, *Critical Issues in Information Systems Research,* Wiley and Sons, New York, 227-251.

Lind, Mary R. and Robert W. Zmud (1991) The Influence of a Convergence in Understanding Between Technology Providers and Users on Information Technology Innovativeness, *Organization Science* 2(2): 195-217.

Lind, Mary R., Robert W. Zmud, and William A. Fischer (1989) Microcomputer Adoption - The Impact of Organizational Size and Structure, *Information and Management* 16: 157-162.

Majchrzak, A., and J. Cotton (1988) A Longitudinal Study of Adjustment to Technological Change: From Mass to Computer-Automated Batch Production, *Journal of Occupational Psychology* 61: 43-66.

Mansfield, Edwin (1988) The Speed and Cost of Industrial Innovation in Japan and the United States: External versus Internal Technology, *Management Science* 34(10): 1157-1168.

Moore, Gary C. (1987) End User Computing and Office Automation: A Diffusion of Innovations Perspective, *Infor* 25(3): 214-235.

Moore, Gary C., and Izak Benbasat (1991) An Examination of the Adoption of Information Technology for End Users: A Diffusion of Innovations Perspective, Working Paper #90-11, Faculty of Management, University of Calgary, Calgary, AB.

Moore, Gary C., and Izak Benbasat (1992) Development of an Instrument to Measure the Perceived Characteristics of Adopting an Information Technology Innovation, *Information Systems Research* 2(3): 192-222.

Nelson, Debra L. (1990) Individual Adjustment to Information-Driven Information Technologies: A Critical Review, *MIS Quarterly* 14: 79-98.

Nilikanta, Sree, and Richard W. Scammel (1990) The Effect of Information Sources and Communication Channels on the Diffusion of Innovation in a Data Base Development Environment, *Management Science* 36(1): 24-40.

Pavri, Francis (1987) An Empirical Study of the Factors Contributing to Microcomputer Usage, unpublished Doctoral Dissertation, University of Western Ontario, London, ON.

Petty, Richard E., and Cacioppo, John T. (1981) *Attitudes and Persuasion: Classic and Contemporary Approaches.* Wm. C. Brown Company Publishers, Dubuque, Iowa.

Post, Gerry and John McLaughlin (1993) Developing the Spatial Information Marketplace: A Canadian Case Study, In Masser, I., and H.J. Onsrud, editors, *Diffusion and Use of Geographic Information Technologies,* Kluwer Academic Publishers, Dordrecht, The Netherlands.

Rogers, Everett M. (1983) *Diffusion of Innovations*, Third Edition, The Free Press, New York.

Rogers, Everett M. (1993) The Diffusion of Innovations Model, In Masser, I., and H.J. Onsrud, editors, *Diffusion and Use of Geographic Information Technologies,* Kluwer Academic Publishers, Dordrecht, The Netherlands.

Roskies, E., J.K. Liker, and D.B. Roitman (1988) Winners and Losers: Employee Perceptions of their Company's Technological Transformation, *Journal of Organizational Behaviour* 9: 123-137.

Sheppard, Blair H., Hartwick, Jon, and Warshaw, Paul R. (1988) The Theory of Reasoned Action: A Meta-Analysis of Past Research with Recommendations for Modifications and Future Research, *Journal of Consumer Research* 15: 325-343.

Tornatzky, Louis G., and Klein, Katherine J. (1982) Innovation Characteristics and Innovation Adoption Implementation: A Meta-Analysis of Findings, *IEEE Transactions on Engineering Management*, EM-29, 1: 28-45.

Wegener, Michael (1993) Universal GIS versus National Land Information Traditions: Software Imperialism or Endogenous Developments, In Masser, I., and H.J. Onsrud, editors, *Diffusion and Use of Geographic Information Technologies.* Kluwer Academic Publishers, Dordrecht, The Netherlands.

Wiggins, Lyna (1993) Diffusion and Use of Geographic Information Systems in American Planning Agencies, In Masser, I., and H.J. Onsrud, editors, *Diffusion and Use of Geographic Information Technologies*. Kluwer Academic Publishers, Dordrecht, The Netherlands.

Zwart, Peter (1993) On the Diffusion of GIS: A Pragmatic View, In Masser, I., and H.J. Onsrud, editors, *Diffusion and Use of Geographic Information Technologies*. Kluwer Academic Publishers, Dordrecht, The Netherlands.

SECTION 2: ASSESSING AND MODELING THE DIFFUSION OF GEOGRAPHIC INFORMATION INNOVATIONS

ERIK STUBKJAER
Institute of Development and Planning
University of Aalborg
Fibigerstraede 11
9220 Aalborg 0
Denmark

This section includes five papers which address the diffusion of geographic information innovations from varying experiential perspectives and research approaches. The paper by Ian Masser reports initial attempts at quantitative analysis of the diffusion of geographic information systems in Great Britain. He describes and analyzes the diffusion of GIS in local governments in Great Britain by making comparisons between evidence gathered throughout the country with the classic model of diffusion . The paper by Heather Campbell describes observed corporate and departmental GIS implementation activities in Great Britain and reports on the corporate goals toward which those activities seem to be directed. In the third paper, Lyna Wiggins reports on the extent of use of GIS in urban and regional planning agencies in the United States which she has been attempting to track over time. In the fourth paper, Harlan Onsrud and Jeffrey Pinto report the results of a survey of local government GIS users in the United States. Using regression and comparative analyses, their methodology attempts to identify factors and processes which lead to successful use of GIS. The fifth paper by Peter Zwart presents observations and reflections which call into question the adequacy of the "diffusion model" approach. He presents an alternative paradigm for visualizing and understanding the implementation of GIS within organizations. Finally, a commentary by Michael Wegener challenges the limits of the "embodied GIS" concept raised and discussed by Zwart.

In his oral presentation at the workshop, Ian Masser placed the work reported in his paper in the context of longer term research on local government planning activities. Among the objectives of his continuing work include determination of the adopters of the new innovation, the pattern of adoption, the perceived benefits, and the impact on existing administrative procedures. Thus, he primarily has focused on general characteristics of the diffusion process through society rather than on, for instance, diffusion processes and factors internal to an organization. The innovation Masser has studied is geographical information systems. This innovation was defined by him to "include... automated mapping/facilities management and land information systems (but to)... exclude... thematic mapping and computer aided design packages." Emphasis appeared to be primarily on software packages and the necessary hardware. The definition implies that only software packages which enable the editing of spatially referenced and topologically structured data are included in the definition. The

I. Masser and H. J. Onsrud (eds.), Diffusion and Use of Geographic Information Technologies, 95–97.

definition also implies that geographic digital data sets were included as part of the overall innovation studied and were not treated in themselves as separable distinct innovations.

The paper by Heather Campbell begins with a discussion of the appropriateness of a corporate approach to implementation of GIS in light of the increasing trend toward fragmentation of government services in Great Britain. This expands into a discussion of collaborative and alternative strategies for implementing GIS at the local government level. With this foundation, Campbell describes survey work carried out to investigate the impact of GIS on British local governments and to determine the strategies actually being employed. Finally, she draws on her survey findings to explore the main benefits and problems being encountered by British local governments. She concludes by proposing a three-fold typology for characterizing system implementations which she hypothesizes will be useful in further studies in isolating generalizable implementation factors and processes.

The paper by Lyna Wiggins addresses the diffusion of GIS technology throughout U.S. urban and regional planning agencies. If one were to consider only the findings of the Masser and Wiggins papers, a comparison between them seems to indicate that the different structures of local government in Great Britain and the United States account for the substantially different government applications of GIS in the two nations. The common core of planning tasks is supplemented in Great Britain by facilities management and other tasks related to utilities, while the core land planning tasks in the United States are often supplemented by pressing needs for improved land parcel mapping and tax assessment. The innovation addressed in the Masser & Campbell studies as well as the Wiggins study appears to be the same (i.e., GIS technological capability). However, the Wiggins results communicate more specific information on reported GIS applications, software brands, and other market information.

The strength of the Masser and Campbell studies lies in their articulation and evaluation of theory and in the establishment of groundwork for generating hypotheses. The series of studies by Lyna Wiggins similarly establish groundwork for future work, particularly for longitudinal studies.

The Onsrud/Pinto study reported in their paper begins by drawing on hypotheses already reported in the GIS and diffusion literature and tests those hypotheses using regression analysis and comparative analysis techniques which are commonly used in the organizational theory research community. Whereas the universe surveyed by Wiggins and Azad is U.S. local government planning agencies, the universe surveyed by Onsrud and Pinto is U.S. local governments regardless of application domain, who have already acquired GIS capabilities. The Onsrud/Pinto study provides initial findings for those attempting to address the complex problem of correlating organizational and behavioral factors with the ultimate use of GIS. In addition, its greatest contributions may be in its use and description of the analytical approaches employed.

The paper by Peter Zwart returns to theory, not in terms of the diffusion model, however, but as a philosophical critique of the perception and use of computer technology in western societies. While the four previous papers are essentially either descriptive or analytical, Peter Zwart introduces a normative component to the diffusion issue. The object of study is, again, GIS hardware and software, but Peter Zwart distinguishes between mainframe or workstation-based GIS systems and desk-top mapping systems. He argues that general purpose GIS provide or strive to provide functions to satisfy most conceivable query, mapping, and analytical needs - "to be all things to all needs" - while current desktop systems have less than full spatial analysis capabilities, simpler data structures, and considerably lower costs. The paper notes a market trend from a 'foreground stand-apart technological profile' to a desktop technology where the

GIS innovation mixes with the phone, the typewriter, and the calculator, and, gradually, merges with these inventions behind the computer's monitor. Also, it draws attention to the fact that GIS diffusion to date has resulted from the supply of appropriate software and data, rather than from user demand.

These observations are related to the conception of technology of the German philosopher Martin Heidegger (1889-1976). Using Heidegger's concepts Zwart suggests how to change today's GIS technology to make it 'ready-to-hand.' Zwart argues that in the future we should expect people to cease to recognize GIS as a separate technology and argues that the research and practitioner communities should strive towards such a result. This thought provoking statement transcends the engineering problem of providing a suitable user interface. However, as the review by Michael Wegener explains, the differences between an 'all but disappeared technology,' a 'transparent' technology, and a 'black box' conception of an information (GIS) tool need to be clarified.

The paper by Peter Zwart ultimately challenges the GIS research community. If GIS technology 'all but disappears' by being engulfed within larger technological systems, the identification of geographic innovations as objects of study, and the maintenance of these objects in the research focus long enough to accumulate scientific knowledge on them are at risk. This premise was heavily debated at the workshop, and illustrates, the workshop's critical reflection on the social impacts of GIS technology.

THE DIFFUSION OF GIS IN BRITISH LOCAL GOVERNMENT

IAN MASSER
Department of Town and Regional Planning
University of Sheffield
Sheffield, S10 2TN
United Kingdom

ABSTRACT Some of the findings of a comprehensive survey of GIS in British local government are analysed in the context of the hierarchical spatial diffusion model. The findings show that there was considerable interest in GIS in local government circles in early 1991. Although only about 16 percent of all authorities have acquired GIS facilities, nearly 70 per cent of them are considering purchasing them at some stage in the future. The findings of a chi squared and multivariate analysis of this data indicate that there is a strong hierarchical effect in the diffusion of GIS in Britain from larger to smaller authorities and that there is also a marked but less pronounced spatial dimension to the diffusion process which is associated with the north south divide.

1. Introduction

Like any other technological innovation the adoption of geographic information systems (GIS) can be analysed as a diffusion process in time and space (Onsrud and Pinto 1991). The time dimension of the diffusion process is usually represented as an S shaped curve which assumes that the adoption of GIS proceeds slowly at first, then accelerates rapidly as the innovation takes off, and finally tapers off as levels of adoption reach a saturation point (Rogers 1983, Rogers 1993). The most appropriate space time diffusion model for the analysis of GIS is the hierarchical diffusion model. This suggests that diffusions 'trickle down' from large places to smaller places as well as spreading out from central regions to those on the periphery (see, for example, Haggett et al 1977)

With these considerations in mind, this paper analyses some of the findings of a comprehensive survey of GIS in British local government. It is primarily concerned with the adoption (and non-adoption) of GIS by local authorities in England, Wales and Scotland. The first part of the paper provides an overview of the survey findings in both qualitative and quantitative terms. The second is devoted to a systematic statistical evaluation of these findings using chi squared and multivariate analysis. In the process the properties of a number of possible explanatory variables are evaluated and explored.

2. Overview Of GIS Adoption

A telephone survey of all 514 local authorities in Great Britain was undertaken in early 1991 to collect information about GIS adoption (Campbell and Masser 1991). The decision to

I. Masser and H. J. Onsrud (eds.), Diffusion and Use of Geographic Information Technologies, 99–115.

carry out a telephone survey was vindicated by the 100% response rate which compares favourably with the partial response rate achieved by postal questionnaire surveys. It also enabled much of the ambiguity that exists regarding the definition of GIS to be removed by discussion with the respondents and provided valuable subsidiary information relating to the experience of the individual local authorities. A broad definition of geographical information systems which included automated mapping/facilities management and land information systems that excluded thematic mapping and computer aided design packages was adopted in the research.

Table 1. Plans for GIS in local authorities in Great Britain

	Number	**Percent of Total**
Already have GIS facilities	85	16.5
Firms plans to acquire GIS within 1 year	44	8.6
Considering the possibility of acquiring GIS	227	44.2
No plan to introduce GIS	158	30.7
TOTAL	514	100.0

The extent of current interest in GIS in local government in Great Britain can be seen from Table 1. This shows that 356 out of the 514 local authorities (69.3 percent) surveyed were considering introducing GIS in one form or another. However, the introduction of GIS is at an early stage in most authorities. Only 85 authorities accounting for 16.5 percent of the total of 514 already had acquired a GIS, while a further 44 authorities (8.6 percent) had firm plans to purchase one in the 12 months following the survey. It is therefore likely that a quarter of all local authorities in Great Britain will have acquired a GIS by Spring 1992.

The extent to which GIS is regarded as an important tool for local government even in local authorities who had no plans to purchase a GIS can be seen from the comments made by respondents during their interviews. Only five authorities claimed that they had no need for a GIS in that the present system worked sufficiently well for their purposes. In contrast, lack of financial resources featured prominently in the reasons given by local authorities for not considering the purchase of a GIS. Many small authorities and at least one London Borough felt that GIS facilities were too expensive. Others drew attention to the shortage of funds because of poll tax capping while several small authorities felt that they were too small in size to justify an investment in GIS.

Organisational and technical issues also featured prominently in the decision not to consider GIS facilities. Some shire districts claimed that they were having difficulties convincing their managements of the value of a GIS. Others linked the introduction of GIS to a more corporate approach in their organisation which they foresaw could lead to serious operational difficulties. One metropolitan district pointed out that its plans for GIS had been thwarted by the introduction of competitive tendering at the departmental level.

As might be expected the level of computerisation in some authorities precluded the introduction of GIS facilities. Some authorities felt that they had neither the level of facilities nor the know how to successfully introduce GIS. Several others were waiting for the introduction of an authority wide information technology strategy before considering GIS. Other problems cited included the difficulty of operating split site systems and the problems of linking GIS to existing main frame facilities.

Most of those authorities who are still considering the possibilities of introducing GIS have already taken steps towards achieving this goal. Many of them have set up departmental or interdepartmental working parties to explore the issues involved in introducing GIS into their organisations. Some of them also reported that they were in the process of conducting feasibility studies. These include authorities of all types in all parts of Great Britain. The findings of these studies have not always been favourable to the introduction of GIS. A number of authorities reported that their plans for GIS had been scrapped as a result of the recommendations of their working parties. Again there was no distinct pattern as to the type of authority or geographical location.

Some authorities reported that they already had facilities which would assist in the introduction of GIS. A number of these saw their CAD facilities as a vehicle for this purpose. Others intend to extend existing packages such as ICL's PLANES to develop their GIS capabilities.

To overcome the high initial costs of introducing GIS some authorities had plans to bring in GIS in association with the introduction of land charges or building control systems in their organisation. One rural shire district also reported that it would be introducing GIS specifically to cope with the provisions of the 1990 Environmental Protection Act.

Some authorities are planning to take advantage of another organisation's system. These included most of the Merseyside metropolitan districts who have access to the GIS run by Merseyside Information Service.

As might be expected, the timescale for introduction varies considerably between authorities. A large group of authorities estimated that they would purchase a GIS within two or three years. A smaller group felt that they would wait for five years or more before introducing GIS. Some authorities, however, had no fixed plans for the acquisition of GIS but insisted that they were keeping abreast of developments in the technology. Others reported that they had specifically decided to wait and see how others got on before committing themselves.

From these comments it can be seen that there is a high level of awareness of GIS in local government in Britain at the present time even among authorities who have no plans to invest in such a system. However, there are marked variations between local authorities of different types and also between different regions within Great Britain with respect to the extent of both current and anticipated GIS availability.

Table 2. Plans for GIS by type of local authority

	Already have GIS Number	%	Total
Shire Districts-England & Wales	24	7.2	333
Scottish Districts	4	7.5	53
Total Shire Districts/	*28*	*7.3*	*386*
Metropolitan Districts	*22*	*31.9*	*69*
Shire Counties	32	68.1	47
Scottish Regions	3	33.3	9
Scottish Islands	0	0	3
Total County and Regional Authorities	*35*	*59.3*	*59*
TOTAL (italics)	*85*	*16.5*	*514*

Table 2 shows the number of authorities who have already acquired a GIS by local authority type. A distinction has been made in the analysis between local authorities in England and Wales and those in Scotland due to differences in the local government structure and responsibilities. Scotland has a simple two-tier structure of district and regional authorities with no comparable unitary authorities to the Metropolitan Districts south of the border. The Scottish Regions also have a broader range of functions than the Shire Counties of England and Wales including services such as water provision. The three Scottish Island Councils have been included under the regions in the analysis.

Around 59.3 percent of all county and regional level authorities in Great Britain already have GIS facilities. In contrast only 7.2 percent of Shire Districts and 7.5 percent of Scottish District authorities have GIS facilities. The proportion of Metropolitan Districts with GIS falls midway between the two extremes at 31.9 percent. It should be noted, however, that the number of authorities with GIS in each of these categories is broadly similar. Systems have already been adopted by 28 Shire Districts/Scottish Districts, 22 Metropolitan authorities and 35 Shire County/Scottish Regions.

Table 3 analyses the findings with respect to the regional distribution of GIS facilities. This shows that there is a marked North/South divide between the level of GIS acquisitions in Great Britain. Three quarters of all authorities who already have a GIS are located in the southern half of Great Britain and the overall proportion of authorities with GIS in the south is nearly double that of those in the north.

Table 3. Plans for GIS by region

	Already have GIS		
Regions	**Number**	**%**	**Total**
East Anglia	4	17.4	23
East Midlands	3	6.7	45
South East	33	23.1	143
South West	8	5.1	53
West Midlands	13	32.5	40
Total South	61	20.1	304
North	4	12.8	33
North West	5	12.8	39
Yorkshire and Humberside	2	7.1	28
Wales	6	13.3	45
Scotland	7	10.8	65
Total North	24	11.4	210
TOTAL	85	16.5	514

The national leaders in terms of acquisitions relative to the total number of authorities are the South-East and West-Midlands regions. In contrast, the lowest levels of GIS penetration are to be found in the East Midlands, Yorkshire and Humberside and Scotland. The case of the East Midlands is especially interesting as an example of a region which has more in common with the north of Britain than its fellows in the south.

Table 4. Percentage of authorities who already have GIS facilities by local authority type and region

Jurisdictions	**South**	**North**	**Great Britain**
Shire Districts and Scottish Districts	8.6	5.2	7.3
Metropolitan Districts	47.5	10.3	31.9
Shire Counties and Scottish Regions	71.0	46.4	59.3
All authorities	20.1	11.4	16.5

The main findings from this analysis are summarised in Table 4 with respect to authorities who already have a GIS. This shows that in all types of authority the proportion of authorities who already have GIS facilities in the south is nearly twice that of those in the north. The difference is most apparent with respect to Metropolitan Districts where 47.5

percent of all authorities fell into this category as against only 10.3 percent in the north. Although some allowance must be made for those authorities who make use of the facilities provided for them by agencies such as Merseyside Information Service, the contrast between north and south in this respect is striking.

3. Modelling The Diffusion Of GIS

Two main factors emerged from the previous analysis as key elements in influencing the adoption of GIS in local authorities in Great Britain. These are local authority type and location. With respect to the former there were clear differences between the levels of adoption in Shire Counties/Scottish Regions as against Shire Districts/Scottish Districts with Metropolitan Districts occupying an intermediate position between these two extremes. With respect to the latter there was also a clear difference between local authorities in the north as against those in the south of Great Britain.

Both these findings can be interpreted in terms of the diffusion model. The first can be seen as evidence of the trickle-down effect from larger to small authorities while the latter reflects the spatial diffusion of innovations from core to periphery.

To explore these matters in greater detail two types of statistical analysis were undertaken in relation to the survey data. The first of these made use of chi squared analysis to test the significance of the relationships contained in a number of two by two contingency tables containing the findings for individual local authorities. At this stage the opportunity was also taken to explore the performance of a number of additional explanatory variables in relation to the survey findings. The second type of statistical analysis involves a multivariate analysis to evaluate the combined impact of these explanatory variables on GIS adoption.

3.1 CHI SQUARED ANALYSIS

Table 5 contains the main findings of the chi squared analysis of the relationship between local authority size in mid 1990 and GIS adoption. A population of 150,000 was selected from experience and intuition as a surrogate threshold beyond which jurisdictions, for one reason or another, appear more readily able to acquire GIS capabilities. From Table 5 it can be seen that a population threshold of 150,000 is highly significant as an indicator of adoption levels at the aggregate level with a massive chi squared score of 79.05. This value reflects the fact that 59 out of the 148 authorities with populations greater than 150,000 had GIS as against only 26 out of the 366 authorities with populations less than 150,000.

These differences largely reflect local authority type and that the differences within groups of similar types of authority are generally less marked. In the case of the Shire Districts and Scottish Districts, for example, the chi squared value falls to just below the 5% significance level. The case of the Metropolitan Districts also presents an anomaly as the chi squared value not only falls well below the 5% significance mark but the pattern of the contingency table is also reversed, suggesting that in this case smaller authorities are more likely to have adopted GIS than larger ones. However it should be noted that only 5 districts fell into the former category.

Table 5. Relationships between local authority population and GIS adoption

Jurisdictions	Population	Adopted GIS	Not adopted GIS	Chi^2 Statistic
Shire District/Scottish Districts (n = 386)				
	+150,000	5	27	
	-150,000	23	331	
				3.63
Metropolitan Districts (n = 69)				
	+150,000	19	5	
	-150,000	3	2	
				1.96
All Districts (n = 455)				
	+150,000	24	72	
	-150,000	26	333	
				24.42*
Shire County/Scottish Regions (n = 59)				
	+150,000	35	17	
	-150,000	0	7	
				10.18*
All Local Authorities (n = 514)				
	+150,000	59	89	
	-150,000	26	340	
				79.05*

*In Tables 5 to 9, an * indicates a value which is significant at the 5 per cent level for a one tailed test with one degree of freedom.

The exception to this rule is the findings for the Shire Counties and Scottish Regions and islands which show a highly significant relationship between GIS adoption and population size. This highlights the lack of homogeneity in this group which is largely due to the inclusion of the three Scottish Island authorities with their very small populations in this category.

Table 6. Relationship between location and GIS adoption

Jurisdiction	Adopted GIS	Not adopted GIS	Chi^2 Statistic
Shire District/Scottish Districts (n = 386)			
South	20	213	
North	8	145	
			1.55
Metropolitan Districts (n = 69)			
South	19	21	
North	3	26	
			10.69*
All Districts (n = 455)			
South	39	234	
North	11	171	
			7.58*
Shire County/Scottish Regions (n = 59)			
South	22	9	
North	13	15	
			3.67
All Local Authorities (n = 514)			
South	61	243	
North	24	186	
			6.71*

Table 6 summarises the main features of the relationship between GIS adoption and the location of local authorities with respect to the north-south divide in Great Britain. From this it can be seen that the impact of the north-south divide (as defined in table 3) is particularly marked at the aggregate level with respect to all authorities and all districts where the chi squared scores are well over the 5% significance level. As in the case of population size these differences are less pronounced within some local authority types. For example, no significant differences are found in the findings for the Shire Districts and Scottish Districts and the chi squared score for Shire Counties and Scottish Regions also falls below the 5% significance level. However, the Metropolitan Districts stand out as a clear exception with a chi squared score of 10.69 reflecting the fact that only 3 out of 29 northern as against 19 out of 40 southern Metropolitan Districts have adopted GIS.

The findings of these two analyses generally support the hypotheses underlying the trickle down and core-periphery spatial diffusion models. However, they also point to a number of variations between local authorities in this respect. The impact of the north-south divide, for example, is particularly pronounced in the case of the Metropolitan Districts but not especially significant with respect to Shire Districts and Scottish Districts. On the other hand the impact of population size is more pronounced in terms of the Shire Districts and Scottish Districts than is the case for Metropolitan Districts. The findings for the Scottish Shire Counties and Scottish Regions highlight the heterogeneous nature of this group in terms of population size.

In addition to the size and location of variables discussed above the adoption of GIS in relation to two other explanatory variables was examined in the chi squared analysis. These variables reflect population change and political standpoint respectively.

3.1.1 *Population Change* Population changes are a useful indicator of the kind of problems facing a local authority and the extent to which these pressures stimulate innovative measures such as GIS adoption to deal with them. If this is the case adoption rates in authorities with growing populations should be higher than those in authorities with declining populations.

To test this hypothesis population change in British local authorities between 1981 and 1988 is considered in relation to GIS adoption in Table 7. The overall findings from this table suggest that in Great Britain the opposite is the case and that authorities with declining populations are more likely to have adopted GIS than those with growing populations. This is particularly evident at the district level where the aggregate findings for the Shire Districts and Scottish Districts give rise to a significant chi squared score of 13.15. The main explanation for these apparently contradictory findings is that the effects of population size on GIS adoption appear to be much more important than population change in Great Britain. At the district level GIS adoption is greatest in the larger urban authorities which tend to have declining populations and least in the smaller suburban and ex urban authorities which are experiencing population growth. It should also be noted that, although the findings for the Shire Counties and Scottish regions are not significant at the 5% level they are nevertheless interesting in that they tend to support rather than reject the original hypothesis with a clear majority of local authorities with growing populations adopting GIS as against the half and half split between authorities with declining populations in respect of non-adoption.

3.1.2 *Political Standpoint* Another possible explanatory variable is the impact of different political standpoints on GIS adoption. It could be argued, for example, that levels of GIS adoption in traditionally minded local authorities are likely to be lower than those in more progressive labour authorities who are more aware of the potential of technical innovation than their conservative or independent counterparts. Obviously such a generalisation based on political standpoint must be heavily qualified in practice as party political ideology cannot be regarded as in any way synonymous with a local authority's receptiveness to technical innovations. Nevertheless, it was felt that it would be useful to explore the impact of the largest political party in each local authority as a proxy for these qualities.

Table 7. Relationship between population change and GIS adoption

Jurisdiction	Population	Adopted GIS	Not adopted GIS	Chi^2 Statistic
Shire District/Scottish Districts (n = 386)				
	Increase	15	249	
	Decrease	13	109	
				3.07
Metropolitan Districts (n = 69)				
	Increase	4	12	
	Decrease	18	35	
				0.45
All Districts (n = 455)				
	Increase	19	261	
	Decrease	31	144	
				13.15*
Shire County/Scottish Regions (n = 59)				
	Increase	29	17	
	Decrease	6	7	
				1.20
All Local Authorities (n = 514)				
	Increase	48	278	
	Decrease	37	151	
				2.12

Table 8 Relationship between largest political party and GIS adoption

Jurisdiction	Political Party	Adopted GIS	Not adopted GIS	Chi^2 Statistic
Shire District/Scottish Districts (n = 341)				
	Conservative	13	206	
	Labour	14	108	
				3.30
Metropolitan Districts (n = 64)				
	Conservative	7	9	
	Labour	14	34	
				1.16
All Districts (n = 405)				
	Conservative	20	215	
	Labour	28	142	
				5.98*
Shire County/Scottish Regions (n = 53)				
	Conservative	20	12	
	Labour	12	9	
				0.15
All local authorities (n = 458)				
	Conservative	40	227	
	Labour	40	151	
				2.74

Note: Local authorities where parties other than the Conservative, Independent or Labour parties had the largest number of councillors are excluded from the analysis

Table 8 summarises the main findings of the impact of political standpoint on GIS adoption. From this it can be seen that there are important variations in political standpoint with respect to the different types of authority within Great Britain in May 1990. The Conservatives and their supporters tend to be the largest parties in the Shire Districts and Scottish Districts as well as in the Shire Counties and Scottish Regions. In contrast Labour is typically the largest party in the densely populated metropolitan districts including Greater London.

The findings for local authority type highlight a number of anomalies in GIS adoption when seen from a party political standpoint. In both the Shire Districts and Scottish Districts as well as in the Shire Counties and Scottish Regions where generally Conservatives and their supporters are in power, local authorities with a Labour majority are more likely than their Conservative counterparts to have adopted GIS, thereby supporting the hypothesis, even though the findings are not significant at a 5% level. However, in the Metropolitan Districts where Labour is generally in power, local authorities with Conservative majorities are more generally likely to have adopted GIS than their Labour counterparts, thereby refuting the hypothesis in this case although once again the findings are not significant at the 5% level.

When the findings for all districts are combined a chi squared score for 5.98 is obtained which is significant at the 5% level. However, this largely reflects the impacts of population size on GIS adoption and the tendency for the larger Metropolitan Districts to have Labour majorities. As in the case of population change, therefore, the association of large Metropolitan District populations, mainly in the north, with generally declining populations and Labour held seats may be regarded as the dominant factor (albeit arguably) at the district level in Great Britain in GIS adoption.

3.2 MULTIVARIATE ANALYSIS

The second stage of the statistical analysis explores the predictive value of the hierarchical diffusion model incorporating both trickle down and spatial diffusion elements and an extended version of this model which includes population change and political standpoint variables.

The most appropriate form of multivariate analysis for such purposes is some kind of multiple regression analysis. In this case, as the dependent variable (GIS adoption) has only two values (O for non adoption and 1 for adoption), it is necessary to use the logistic regression model rather than the conventional least squares model.

The main features of the two models can be summarised as follows:

Basic diffusion model:

Usegis = f (Lnpop, Nthsth)

Extended model:

Usegis = f (Lnpop, Nthsth, Popchange, Party)

where

Usegis — refers to the adoption or non adoption of GIS by the local authority. As noted above adoption = 1 and non adoption = 0.

Lnpop — is the natural logarithm of the estimated mid year local authority population in 1990.

Nthsth — is a dummy variable which refers to the location of the local authority relative to the north south divide as defined in Table 3. In this case a location in the south = 0 while one in the north = 1.

Popchange — is the change of the local authority's population measured by its 1988 population/1981 population.

Party — is the ratio of conservative and independent held seats in May 1990 to the combined total of conservative and independent and labour held seats in local authorities where one of these parties holds the largest number of seats.

In the case of GIS adoption the following signs might be expected for each of these independent variables. Given the strong association between population size and GIS adoption a positive sign is expected for population size. In the case of the north-south variable a negative sign is expected, given the adoption levels are lower in the north and local authorities in this region are given the value of 1 in the model.

Table 9. Findings for the basic diffusion model (t scores in brackets)

Jurisdictions	Constant	Ln Pop	N-S	Chi^2 Statistic
Shire districts/ Scottish districts (n=386)	-19.88* (-3.94)	1.52* (3.50)	-0.38 (-0.86)	2.47
Metropolitan districts (n=69)	-5.05 (-0.76)	0.40 (0.74)	-2.09* (-2.97)	4.88*
All districts (n=455)	-20.24* (-5.88)	1.58* (5.42)	-1.06* (-2.76)	34.69*
Shire Counties/ Scottish regions (n=59)	-16.21* (-2.32)	1.27* (2.45)	-0.23 (-0.36)	2.30
All local authorities (n=514)	-21.08* (-9.43)	1.65* (8.91)	-0.76* (-2.50)	108.70*

* t scores significant at the 5 per cent level for a two tailed test: chi squared values significant at the 5 per cent level for a one tailed test with one degree of freedom.

As noted in the chi squared analysis a positive relationship between GIS adoption and population change might be postulated in general terms but, because of the association between population size and adoption at the district level in particular a negative sign is more likely to be found in Britain. The party variable expresses the ratio of conservative and independent seats to the total number of conservative, independent and labour seats. Given the arguments presented in the previous section of this paper a negative sign might be expected for this independent variable.

The logistic regression model was estimated using maximum likelihood methods and the findings are summarised in tables 9 and 10. Table 9 shows that the basic diffusion model is a powerful predictor of GIS adoption particularly at the aggregate levels of all authorities and all districts. In the case of the former 47 out of the 85 instances of GIS were correctly predicted by this model and the chi squared value for the two by two contingency table of observed and predicted values is a massive 108.70.

The performance of the basic diffusion model with respect to the different local authority types is less spectacular than its performance at the aggregate level. Although reasonable chi squared values are obtained for both Shire Districts and Scottish Districts and Shire Counties and Scottish Regions, neither of them are found to be significant at the 5 per cent level. Only in the case of the metropolitan districts is a chi squared value obtained which is above the 5 per cent significance threshold.

The individual parameter estimates highlight some of the reasons for the varied performance of this model. Population size, as might be expected, proves to be a robust and powerful predictor in every case except at the Metropolitan District level. In contrast the highest t score for the north south divide variable is obtained at the Metropolitan District level. Although the estimated parameters for this variable have t scores which are significant at the 5 per cent level for all districts and all authorities, the t scores obtained for Shire Districts and Scottish Districts and for Shire Counties and Scottish Regions fall well below this level. In all cases, however, the estimates for the parameters show the expected signs : positive for population size and negative for the north south divide.

To a considerable extent the findings for the basic model confirm those of the chi squared analysis that were summarised in tables 5 and 6. Population size proves to be a good indicator of GIS adoption particularly at the aggregate levels and to a lesser extent at the Shire District/Scottish District and Shire County/Scottish Region levels. The North South divide is also a useful indicator at the aggregate levels and is particularly important at the Metropolitan District level.

Table 10 contains the main findings for the extended model. From this it can be seen that the inclusion of two more variables generally leads to some improvement in predictive power as measured by the chi squared value for the 2 x 2 contingency table of observed and predicted values. This is particularly the case with respect to Shire Districts/Scottish Districts and Metropolitan Districts.

As in the basic diffusion model the values of the estimated parameters for population size show that it is a powerful predictor of GIS adoption at the aggregate levels and also at the Shire District/Scottish District level. However, its performance at the Shire County/Scottish Region level in this model is weaker than in the basic diffusion model. The next most powerful predictor in terms of the t scores obtained for the extended model is the north south divide. Somewhat surprisingly this emerges as an important predictor at the Shire District/Scottish District level as well as the Metropolitan District and aggregate levels.

The performance of the two additional independent variables is generally patchy. Political standpoint emerges as a useful predictor at the Shire District/Scottish District and aggregate levels but has only one t score which is above the 5% significance threshold. Population change has no t scores above this threshold but scores moderately well at all levels.

Table 10. Findings for the extended model (T scores in brackets)

Jurisdictions Statistic	Constant	LnPop	N-S	Popchange	Party	Chi^2
Shire districts/ Scottish districts (n=341)	-14.24* (-2.38)	1.32* (2.81)	-1.09* (-2.06)	-2.30 (-1.24)	-1.39 (-1.83)	4.13*
Metropolitan districts (n=64)	14.25 (1.05)	0.69 (0.08)	-2.16* (-2.76)	-15.41 (-1.56)	0.07 (0.05)	12.00*
All districts (n=405)	-12.18* (-2.69)	1.26* (3.97)	-1.64* (-3.74)	-3.41 (-1.66)	-1.34* (-2.11)	43.26*
Shire Counties/ Scottish regions (n=53)	-39.81* (2.07)	1.32 (1.91)	0.60 (0.58)	22.31 (1.60)	-0.69 (-0.38)	3.50
All local authorities (n=458)	-16.58* (-5.76)	1.53* (8.04)	-1.18* (3.28)	-2.45 (-1.38)	-0.90 (1.58)	94.72*

* t scores significant at the 5 per cent level for a two tailed test : chi squared values significant at the 5 per cent level for a one tailed test with one degree of freedom.

Note: Local authorities where parties other than the Conservative, Independent or Labour parities had the largest number of councillors are excluded from the analysis.

In the extended model population size has the expected sign in every case. The expected sign for the north south divide is found in all but the Shire County/Scottish Region case. Population change also has the expected signs associated with the British context in all but the Shire County/Scottish Region case while party political standpoint has the expected signs at all but the Metropolitan District level.

4. Conclusions

The findings for the extended model support the conclusions reached in the chi squared analysis in overall terms. The inclusion of the two additional variables does not add greatly to the overall performance of the model at the aggregate level, and on the basis of the t scores obtained for these variables they must be regarded as much less reliable predictors than the two variables contained in the basic diffusion model.

What the findings of the analysis show is that population size is by far the most important predictor of GIS adoption in local government in Great Britain. Using this variable alone 45 out of the total of 85 cases of GIS adoption are correctly classified. The inclusion of the North South divide dummy variable increases this number to 47 or 55% of the total number. The extended model also allocates 55% of the cases of GIS adoption correctly. Although the explanatory power that is added by the North South divide is marginal in this case, this variable produces classifications which are significant in their own right at the aggregate levels.

The main conclusion to be drawn from this analysis is that there is a strong hierarchical effect in the diffusion of GIS in Britain from larger to smaller local authorities and there is also a marked but less pronounced spatial dimension to the diffusion process which is associated with the north south divide. By comparison with these indicators the contribution of population change and party political standpoint as predictors is minimal.

Nevertheless a number of important qualifications should be borne in mind when interpreting the findings of the analysis. It must be emphasised that the analysis is purely statistical in nature. It considers which local authorities have or have not adopted GIS in Britain. It does not consider what has been done with GIS once it has been adopted. This is the subject of the next chapter by Campbell in this book. It must also be pointed out that the analysis does not explore in any detail the motives for GIS adoption. Instead it makes use of a set of simple statistical indicators to highlight some general features of the broad diffusion process. The factors which contributed to the decision of particular authorities to adopt GIS will be explored in a series of case studies that are being carried out as part of the same project. It is anticipated that these will provide some useful insights into the political and technological context surrounding the decision to adopt GIS in particular authorities.

It must also be recognised that diffusion is a dynamic phenomenon and that the data collected represents only a cross section of this process. Although the findings of the survey suggest that seven out of ten local authorities in Great Britain at the present time are seriously considering introducing GIS there is no guarantee that this level represents the saturation point on the S shaped curve. Consequently it will be necessary to carefully monitor changes over the next few years to assess the long term impacts of GIS on British local government.

Despite thse qualifications the findings of the analysis indicate the extent to which spatial diffusion concepts can be used to explore the nature of the GIS adoption process. What is now needed is comparable studies of experience in other sectors in other countries. Such studies should also take advantage of the extensive literature that has come into being on this topic and look out for similar and contrasting patterns in this literature which will provide further theoretical insights into the factors underlying the adoption of GIS.

5. Acknowledgements

The research described in this paper was carried out as part of a project on 'The impact of GIS on local government in Great Britain' which is funded by the ESRC/NERC Geographic Information Handling initiative. The author would like to thank Dr. Robin Flowerdew for his valuable comments on an earlier draft of this paper.

6. References

Campbell, H. and I. Masser (1991) The impact of GIS on local government in Great Britain. *Proceedings of the Association for Geographic Information Conference*, Association for Geographic Information, London, 2.5.1-2.5.6.

Haggett, P., A. D. Cliff and A. Frey (1977) Locational models. Edward Arnold, London.

Onsrud, H.J. and J.K. Pinto (1991) Diffusion of geographic information innovations. *International Journal of Geographical Information Systems* 5(4): 447-467.

Rogers, E.M. (1983) Diffusion of innovations, Third edition, Free Press, New York.

Rogers, E. M. (1993) The Diffusion of Innovations Model, In Masser, I., and H.J. Onsrud, editors, *Diffusion and Use of Geographic Information Technologies*, Kluwer Academic Publishers, Dordrecht, The Netherlands.

GIS IMPLEMENTATION IN BRITISH LOCAL GOVERNMENT

HEATHER J. CAMPBELL
Department of Town and Regional Planning
University of Sheffield
Sheffield S10 2TN
United Kingdom

ABSTRACT. The process of implementation is critical to ensuring the effective utilisation of innovations in organisations. It is this process which determines whether the potential of the technology is realised. There is increasing recognition that personal, organisational and institutional factors are likely to have a profound influence on the implementation of geographic information systems (GIS). It is generally assumed that the costs of equipment and data preparation, combined with the capacity of GIS to integrate data sets from a wide variety of sources, makes the development of departmental systems inappropriate.

This chapter forms the second half of a discussion which was started in the previous chapter concerning the experiences of British local authorities with respect to GIS. It therefore builds on the preceding review of the diffusion of GIS in local government in Great Britain to explore the implementation strategies which have been adopted. The current debate concerning the merits of "corporate" as against "departmental" approaches to system implementation is reviewed. Given this framework the findings of a study which is currently investigating the implementation of GIS in British local Government is reviewed. The analysis indicates that the organisational and technical characteristics of the GIS being developed as well as the consequences of implementation vary considerably between individual systems. However, a number of broad trends have been identified and on this basis a three-fold typology of system implementation has been produced. These approaches are as follows: classically corporate, theoretically/pragmatically corporate and fiercely independent. Consideration will also be given to a suitable methodology for research in this field.

1. Introduction

The process of implementation is critical to realising the effective utilisation of innovations in organisations. The underlying objective of this process must be to ensure the utilisation and maintenance, or in the words of Yin (1979) the "routinisation", of new technology so it becomes part of the every day activities of staff. However, there has been a tendency for research to concentrate on the initial decision to adopt a new form of technology ignoring the on-going process of implementation which underpins the utilisation of innovations (Leonard-Barton 1988). Once the adoption decision has been taken, generally by senior management, individuals throughout the organisation make a number of personal decisions about the value of the new technique. Staff are frequently suspicious about the introduction of computer based systems, even in instances where a direct challenge to status and influence is not implied, as such changes create a sense of uncertainty (Mumford and Pettigrew 1975, Robey 1987). Given the limitations imposed by the organisational context and the skills and resources available, the way

I. Masser and H. J. Onsrud (eds.), Diffusion and Use of Geographic Information Technologies, 117–146.

in which an innovation is implemented has a significant affect on the response of potential users.

This chapter described the process of implementation with respect to the introduction of geographic information systems (GIS) into British local government. In many ways it forms the second half of a discussion which was started in the previous chapter. It therefore builds on the preceding review of the diffusion of GIS in local government to explore the implementation strategies which individual local authorities have adopted. Local government is one of the most important users of GIS with the range of potential applications extending from property registers and highways management to emergency and land use planning. Given the variety of demands to be placed on a GIS and the considerable level of interest, local government is likely to be a rewarding environment in which to study the process of implementation. Throughout the discussion implementation is regarded as an on-going process involving repeated cycles of development, learning and routine use. Each of these cycles may be regarded as building on previous knowledge while at the same time tackling new issues which may have arisen.

It is generally assumed that the ability of GIS to integrate data sets from a wide variety of sources, combined with the costs of equipment and data preparation make the development of isolated departmental systems inappropriate (Gault and Peutherer 1989, Grimshaw 1988, Coulson and Bromley 1990). The strategic and efficiency benefits associated with increased levels of data sharing between departments has led to widespread acceptance, especially by management consultants, of the importance of adopting a corporate approach as the basis for implementation. However, an increasing number of accounts indicate that in practice the realisation of these benefits may be more difficult than generally envisaged (Cane 1990, Craglia 1992, Openshaw et al. 1990, Winter 1991). Unfortunately, as yet little systematic analysis has been undertaken which evaluates the underlying assumptions concerning the process of implementing GIS and thereby provides a benchmark against which these isolated accounts may be assessed (Onsrud et al. 1991).

The discussion has been divided into four parts. Firstly, background is provided on the nature of British local government, in terms of the overall structure and the internal organisation of authorities. Secondly, the current debate concerning the merits of "corporate" as against "departmental" approaches to system implementation are examined. Given this framework, the third section, reviews the results derived from the experiences of British local government and develops a three-fold typology of the styles of implementation which have been adopted. The final section evaluates these findings and briefly considers a number of methodological issues.

2. British Local Government

2.1. THE STRUCTURE OF LOCAL GOVERNMENT IN BRITAIN

There is no common system of local government throughout Great Britain, rather England and Wales, Scotland and Northern Ireland have each developed distinctive structures. In the case of Northern Ireland local government has largely been taken under the control of central government, due to the special circumstances in the Province. As a result Northern Ireland has been omitted from the subsequent discussion.

A dual system of local government operates in England and Wales, with a differentiation made between metropolitan and non-metropolitan areas. For the metropolitan areas which comprise the six large conurbations and London, there are 36 metropolitan districts and 32

London boroughs, subsequently referred to as metropolitan districts. Since the abolition of the higher tier metropolitan counties in 1986 these multi-purpose unitary authorities have worked alongside a range of ad hoc bodies which were introduced at this time to coordinate metropolitan-wide activities such as passenger transport and the police and fire services. In contrast in the non-metropolitan areas there is a two tier structure consisting of 333 districts and 47 counties. These authorities are often referred to as shire districts and shire counties. The Scottish system in contrast makes no distinction between metropolitan and non-metropolitan areas, simply having a two tier structure of 53 districts and 9 regions. There are, however, three unitary authorities for the island communities which depart from this general structure.

In terms of the division of functions the basic level of authorities, excluding the unitary metropolitan districts (see above), largely take responsibility for local-scale services such as housing, public health, refuse collection and local land use planning. In the case of England and Wales the higher level county authorities have responsibility for the provision of services such as education, roads and transport, strategic land use planning, the emergency services and social services, while in Scotland the regions have a broader role including such activities as water provision. As a result of this division of responsibilities the counties and regions command much larger budgets than the shire and Scottish districts. It should also be noted that while there are some administrative differences between authorities in England and Wales, and Scotland the findings presented later in the chapter largely integrate the responses of individuals in the counties with the Scottish regions and the Scottish districts with the shire districts. Trends specific to Scotland are noted where appropriate.

The limited range of tasks undertaken by British local government distinguishes it form many other countries (Batley and Stoker 1991). For example provision of most health care facilities, energy related services and in England and Wales water provision are the responsibility of either unelected regional bodies or privatised companies. Furthermore there is no tradition of maintaining a cadastre in Great Britain. As a result cadastre or utilities type functions for which considerable scope has been seen for the utilisation of GIS capabilities are not administered by local authorities in Great Britain.

A second distinctive feature of the British local government system is the size of the basic unit of authorities. Table 1 indicates that the average populations of the Scottish, shire and metropolitan districts are respectively 95,100, 96,600 and 259,800. Norton (1991) in his comparative study notes that reorganisation in West Germany and Scandanavia resulted in the populations of the basic unit of local government rising to 8, 10, 12 or 20 thousand. It is clear that these authorities in no way approach the size of the much larger units which have evolved in Great Britain. Teesdale, the smallest shire district, had in 1990 a population of 24,800. The size of these authorities reflects concern at the time of reform in the 1970s with achieving economies of scale and efficiency in service provision (Redcliffe-Maud, 1974). In contrast to the basic unit of local government the population size of counties and regions is smaller than the highest level, sub-national government in many countries. This in many ways reflects the lack of regional government in Britain. (For a more detailed comparative analysis of local government see Batley and Stoker 1991).

The structure of local government in Great Britain is, however, being reviewed by central government. It is worth outlining a number of initiatives at this point as these are having a significant impact on the development of GIS and are therefore referred to in the subsequent discussion. It seems likely with the re-election of a Conservative government in April 1992 that the existing policy towards local government will be continued and perhaps reinforced.

Table 1. Average population of local authorities in Great Britain in 1990

Type Of Local Authority	**Average Population In 1990**
Shire districts	966,000
Metropolitan districts	259,800
Shire counties	684,000
Scottish districts	95,100
Scottish regions	559,000

At the end of 1991 the Government made provision for the setting up of a Local Government Commission to examine the structure of local government in Great Britain. This body has now been formed and whilst the outcome of the process appears to be unclear, there seems to be an underlying presumption on the part of ministers that the introduction of unitary authorities would lead to more efficient and accountable local government. In addition, central government are continuing to monitor closely the spending levels of local authorities and through measures such as community charge capping are restricting the amount of local taxation which individual authorities can levy. In relation to the implementation of GIS as well as the more general provision of services such proposals create considerable uncertainty and consequently a reluctance to invest in projects which are likely to take sometime to produce benefits. These issues will be considered later (see Section 4.2).

2.2. INTERNAL ORGANISATION OF BRITISH LOCAL GOVERNMENT

The reorganisation of local government in the mid 1970s resulted in the establishment of fewer but larger authorities. At the same time as the reform of the structure of local government, it was recognised that attention needed to be given to the internal arrangements for the management of service provision within authorities. This resulted in several studies being commissioned (see Maud Report (1967), Bain's Report for England and Wales (1971) and the Paterson Report for Scotland (1971)). The analysis and recommendations of these reports largely reflected contemporary thinking in the management literature. As a result it was proposed that authorities should take a more corporate approach. Authorities were encouraged to appoint chief executives who in conjunction with a small management board consisting of a number of chief officers would coordinate the delivery of services. These trends were mirrored at the political level by the proposed establishment of central policy and resources committees to identify the council's priorities and allocate resources. The overall aim was to devise a more centralised organisational structure through which a unified approach to service delivery could be developed.

During the mid-1970s, alongside the complete reorganisation of the structure local government a large proportion of authorities acted upon the guidance they were receiving from Bains and others. Consequently a corporate dimension was introduced into the internal management structure of most local authorities. However, in practice despite the presence of formal mechanisms for the centralised coordination of activities most authorities continue to maintain a strong departmental emphasis. In many ways this is fostered by the British legal tradition of administrative discretion which leads authorities to be mainly staffed by specialist professionals such as surveyors, architects, land use planners, social workers and housing managers. Individuals within these groupings tend to have their own professional allegiances and associated with this their own inter- professional rivalries. In some circumstances it may appear that these allegiances are stronger than their links to a particular local authority. For instance surveyors may perceive they have more in common with fellow surveyors outside the authority than other professionals within. In many ways this is the case, particularly in terms of educational experience, skills and most importantly a shared language. As a result any attempts at the centralisation of the management structure of local authorities must be set against this background. It is possible in many authorities to identify an on-going tension between the activities of the chief executive and the often conflicting priorities of individual departments.

The natural tendency of British local government towards fragmentation has in recent years been compounded by central government's legislation on compulsory competitive tendering. The 1986 Local Government Act compelled authorities to open up certain manual activities such as refuse collection and grounds maintenance to tenders from outside organisations. Given assurances concerning the quality of service provision, the organisation with the lowest tender would then be requested to undertake the task. It appears likely that this scheme will be extended into professional activities such as housing management and possibly architecture and computing (Department of the Environment 1992).

As a result of these proposals many authorities have reorganised their internal management structure into business units. These units, which generally represent a small sub-section of a department, have in many cases simultaneously enhanced their control over resources. The justification for this decentralisation of responsibilities is that these units will be better able to compete in an increasingly market orientated public sector than large unwieldy departments. If the implementation of the legislation is taken to its natural conclusion it is possible to envisage a local authority entirely consisting of separate enterprises whose individual objectives are enhanced profits rather than integrated and coordinated service provision.

2.3. SUMMARY

The large size and increasingly fragmented nature of local government in Britain are two important characteristics of the contextual background against which the implementation of GIS is currently taking place. Section 3 will draw on the discussion of the internal management structure of local authorities in reviewing the debate about the effectiveness of various approaches to the implementation of GIS.

3. Conceptual Framework

3.1. INTRODUCTION

The discussion has so far identified something of a paradox in that while local government is moving towards decentralised business units much of the literature is espousing the virtues of adopting a corporate approach to the implementation of GIS. This section will provide a framework within which the interrelationships between the different approaches to implementation and the context in which GIS are being introduced may be understood. For instance, is there necessarily a conflict between a corporate approach to implementation and current developments in local government? Can GIS act as a mechanism to counter the increasing trend towards fragmentation?

An important issue underlying this discussion is the nature of technology. It is clear that computer based technology is very much more than a combination of hardware and software. As a consequence of investigations based in local government in the United States the Public Policy Research Organisation of the University of California at Irvine have developed the concept of technology as a package (see, for example, Danziger et al, 1982, Danziger and Kraemer 1986, International City Management Association 1989). They suggest that any definition of technology should include not only equipment but also people, personal skills, operational practices and corporate expectations. It is then envisaged that this computer package is in turn embedded within the human and institutional context in which it is located, therefore raising issues concerning the nature of organisations.

A number of studies indicate that organisations are complex social and political environments, typified by competing and often conflicting priorities and motivations, (Danziger et al. 1982, Hirschheim 1985, Hirschheim et al. 1987, Pfeffer 1981). Furthermore research has shown that three sets of organisational factors have a critical impact on the effective implementation of computer based systems (Campbell 1990). These are the organisational context, the personalities of the individuals involved and the degree of organisational stability and environmental stability.

With these considerations in mind the critical underlying question for research examining the process of implementation must be to what extent technology shapes the context or the context the technology? This issue will be returned to at the end of the paper.

The subsequent discussion in this section examines what constitutes a corporate approach and explores the assumptions underlying the proposed effectiveness of a variety of approaches to GIS implementation.

3.2. WHAT IS A CORPORATE APPROACH TO GIS IMPLEMENTATION?

It has so far been assumed that a "corporate approach" implies inter-departmental cooperation with respect to the adoption, implementation and on-going maintenance of a GIS. However, it is important to examine this concept more closely as it is clear that in practice there is a continuum of more or less corporate styles of working (see Table 2). These range from local authorities where all departments are involved in system implementation to instances where the introduction of a GIS is the responsibility of a single department. The key coordinating department also varies between authorities, from the classical top-down approach exemplified by the proposals of the Bains Report (Bains 1972) on the management of local government where the chief executive's department performs this function, to a more bottom-up approach where the lead is taken by a service orientated department. It is even possible for these approaches to be combined. For instance, it is possible to envisage a case where several departments are developing a GIS with a department such as the chief executive's as the lead while at the same time a separate system is being introduced by a single department.

Table 2 indicates the range of managerial structures which local authorities may adopt. There is probably little disagreement that the first of these approaches is definitely corporate while the last is not. However, between these extremes there is a range of more or less corporate structures. In relation to the research a broad definition is adopted applying the term corporate to any situation where more than one department is involved. Given this wide definition it will be necessary to consider the extent to which different styles of corporate working, particularly in terms of variations in the lead department, influence the experiences of local authorities.

Table 2. Styles of corporate working with respect to GIS

Number Of Departments Involved	Lead Department
All	Chief Executive/Town Clerk
All	IT/Computer Services
All	Some other department
All	None
3 or more	Chief Executive/Town Clerk
3 or more	IT/Computer Services
3 or more	Some other department
3 or more	None
2	Chief Executive/Town Clerk
2	IT/Computer Services
2	Some other department
2	None
1	

3.3. ASSUMPTIONS UNDERLYING VARIOUS APPROACHES TO THE IMPLEMENTATION OF GIS

The critical question for those implementing GIS concerns the likely effectiveness of the variety of approaches described above. It is generally assumed in the literature that the advantages of increased levels of data integration combined with the propensity for duplication in instances where departments develop separate systems makes it most appropriate for authorities to introduce GIS as a corporate or even inter-governmental system (Coulson and Bromley 1990, Gault and Peutherer 1989, Grimshaw 1988). It is, however, important to examine in more detail the advantages and disadvantages of essentially corporate as against departmental approaches to the implementation of GIS.

There are two sets of reasons why departments might choose to collaborate over the development of a GIS. The first is based on the pragmatic decision that given the costs of acquiring equipment and data such as the Ordnance Survey's (the national mapping agency in Great Britain) digital maps it is necessary for several departments to share the costs. For instance, finance may be allocated from a combination of central and departmental budgets. However, these circumstances do not imply long term inter-departmental working as once the system has been purchased, it is possible that the implementation and maintenance of the technology will become the responsibility of individual departments. The second and more general justification for the adoption of a corporate approach concerns the strategic and efficiency benefits to be gained from the integration of formerly isolated data sets. This approach assumes long term cooperation between departments concerning not only the process of implementation but also the critical issues of the maintenance and utilisation of the information generated. It is these assumptions about the strategic and efficiency benefits associated with a corporate approach which this Section aims to examine in greater detail.

Table 3 attempts to identify the main advantages and disadvantages associated with corporate and departmental approaches to implementing GIS. The table indicates the logical benefits to be derived from inter-departmental cooperation in terms of enhanced provision of information, more informed decision making which it is supposed will lead to 'better' decisions and greater efficiency. Gault and Peutherer (1989) argue that given central government's general restraint on spending, the greater accountability implied by the introduction of the Poll Tax and the insistence on compulsory competitive tendering for services, there is a need for local authorities to adopt "more rational decision making procedures" (p.2). They suggest that to achieve this objective local government requires information processing facilities which will enable the integration of data sets, the analysis of the relationship between client needs and resources over time and the display of the resulting information in a manner which will highlight the spatial distribution of trends at a variety of scales. A corporate GIS is regarded as fulfilling these requirements and as a consequence the tool which will provide the basis for more rational decision making. Similarly Coulson and Bromley (1990) argue in favour of the adoption of a corporate GIS by local authorities. Their findings based on a survey undertaken in Swansea District Council indicate that while several departments utilise the same spatial data sets there is little coordination of information processing activities and as a consequence a significant duplication of effort (Bromley and Coulson 1989).

The identification of these benefits and the associated logic that such authority-wide gains would be lost if GIS were developed on a departmental basis has led to widespread acceptance of a corporate approach as the most appropriate strategy. Most discussions pay little attention to the form of corporate approach to be adopted, but generally imply some form of top-down structure involving all departments. However, underlying these analyses about the value of a corporate GIS are a number of important assumptions about the role of information in decision-making and the nature of organisations.

The linear relationship between the introduction of GIS, improved information processing facilities and better decisions suggests decision making to be a rational process. However, there are several studies which question this conceptualisation (see, for example, Campbell 1990, Cohen, March and Olsen 1972, Feldman and March 1981, Lindblom 1959, Lindblom and Cohen 1979, Pfeffer 1981, Weiss 1977). Campbell in a study of the use of geographic information in local authority planning departments demonstrates the tactical and political role of information in the process of policy making. The research findings indicate the frequent utilisation of information to rationalise and support a favoured policy option rather than as the determining factor in the selection of a particular course of action. Frequently factors such as political priorities or personal experiences have a greater bearing on the outcome of the decision making process than the available data. Furthermore, the research found that a large proportion of decision makers have a low level of basic information awareness due to such issues as the perceived irrelevance of information to their work, lack of confidence and competence in data handling or poor working relations with the technical specialists. Inappropriate assumptions

Table 3. The main advantages and disadvantages of corporate and departmental approaches to implementing GIS.

CORPORATE APPROACH		DEPARTMENTAL APPROACH	
Advantages	**Disadvantages**	**Advantages**	**Disadvantages**
Integration of data sets	Variations in priorities between departments	Independence	Departmental isolation
Increased data sharing	Differences between departs. in their ability to exploit GIS facilites	Control over: - priorities - the form and accessibility of information - technical specialists - equipment	Lack of support in terms of: - finance - technical specialists - training
Improved access to information			
More informed decision making	Differences in the level of awareness and spatial data handling skills between departments	Clear lines of responsibility	Absences of authority-wide benefits including lack of system and data compatibility
Increased efficiency due to reduced duplication leading to time, staff and cost savings	Inter-departmental disagreements over: -access to information - leadership - data standards - equipment - training		

concerning the role of geographic information in decision making and the skills and priorities of staff are likely to inhibit the effective utilisation of the data generated by a GIS. However, the key issue in terms of the problems local authorities will encounter during the implementation of a corporate GIS concerns the largely contextless nature of the underlying assumptions. As a result little consideration is given to the impact of the social and political processes present within organisations.

It was pointed out in Section 2 that there is a tendency towards departmentalism in British local government, consequently the implementation of a corporate information system has significant implications for traditional patterns of working. Table 3 indicates that their may be considerable practical problems in introducing change into an environment which is often typified by competition rather than consensus between departments. The priorities of individual departments are likely to vary considerably, from those concerned with the operational activities involved in road maintenance to the long term strategic decision making associated with urban and regional planning and the central coordinating tasks performed by a chief executive's department. In these circumstances not only are their requirements likely to be significantly different but also their ability to exploit the facilities offered by GIS technology. Furthermore, in contrast to Gault and Peutherer's (1989) argument it is likely compulsory competitive tendering and the development of business units will encourage staff to focus resources on activities which produce direct benefits to their section rather than the authority as a whole.

The adoption of a corporate approach and the resulting integration of formerly isolated data sets has important implications for the ownership and control of information. Departments often jealously guard their scope of activity and treat with suspicion any proposal which implies a change to these circumstances. The theoretical advantages of sharing information may be viewed less positively by departments and individuals if they feel it will lead to closer supervision of their activities or open up the decision making process to greater scrutiny. Such changes may lead to uncertainty about the likely stability of the existing balance of power in an authority. These 'boundary' issues are often articulated in disputes over the division of responsibility, leadership, data accessibility and information standards, as well as more covert forms of non-cooperation. Given the potential for discord there is a need to investigate more closely the balance between the costs and benefits of adopting a corporate approach to GIS implementation. Consideration will also need to be given to the merits of top-down strategies in which a non-service department takes the lead as against the more spontaneous coming together of a number of largely service orientated departments.

The discussion has so far demonstrated the disadvantages as well as the benefits associated with a corporate approach. It is, however, important similar attention is focused on the merits of individual departments developing GIS technology. Further to the loss of the authority-wide benefits highlighted by analyses based on rational assumptions, general local authority support in terms of specialist computing skills and training is likely to be limited. Nevertheless Table 2 indicates that there are a number of positive features associated with a departmental approach. The critical issue concerns the independence individual departments gain from such an approach and as a consequence the increased sensitivity to the needs of end users. In these circumstances departments are not dependent on technical specialists within a central computing section to interpret their requirements nor do they have to take their place within the overall priorities of the authority. Moreover, given the historical mistrust which often exists between departments there is no obligation on individual agencies to share their information resources.

The discussion of the advantages of a departmental approach to the introduction of GIS, however, assumes the same level of consensus exists at a departmental level as corporate approaches assume on an authority-wide scale. It is important to recognise that individual departments are themselves divided into a large number of separate sections, each with a separate area of responsibility and consisting of individuals whose underlying goals and motivations may be in conflict (Campbell 1990). The impact of social and political processes on the experience of implementing GIS must therefore be examined at a departmental as well as an authority-wide scale.

3.4. SUMMARY

This discussion has indicated the main advantages and disadvantages of corporate and departmental approaches to implementing GIS. It has demonstrated the contextless nature of arguments which stress the benefits of a corporate approach while also noting the potential strengths and weaknesses of individual departments developing their own systems. The importance of examining these issues in relation to the nature of the corporate approach adopted, has also been emphasised. However, in examining these issues consideration must also be given to the functions the GIS is intended to perform. For instance, the range of GIS applications extend from operational activities such as highways maintenance, to strategic concerns such as the analysis of socio-economic trends and managerial matters such as performance review. It is possible that the nature of the problems encountered may be related to the type of activities being undertaken. With these considerations in mind the next section will describe the sorts of approaches which have been adopted in British local government.

4. GIS Implementation In British Local Government

4.1. INTRODUCTION

Background to the issues discussed in this section is provided by the preceding review of GIS diffusion in British local government. The findings presented therefore draw on the same research project which is currently investigating the impact of GIS on British local government. The project has employed a combination of research methods including a telephone survey of all 514 local authorities in Britain and detailed case study investigations. The objective of this approach is to utilise the survey to provide an overview of the experiences of local authorities with the findings of this work supplemented by in-depth studies of current practice. Due to the stage of development of the project this paper concentrates on the results of the comprehensive survey. The aim is therefore to identify broad indicators of the issues raised by the conceptual framework.

A telephone based survey approach was adopted due to the large number of postal questionnaires which were being circulated to local authorities and the resultant concern about the size of response. The 100 percent response rate supports the adoption of this method, in addition to respondents giving valuable subsidiary information during the interviews. This method also removes the ambiguity which exists in some surveys over the respondents' perception of the definition of GIS, as the capabilities of the software and the precise nature of its use in the host authority can be related to the operational definition. The research has adopted a broad interpretation of GIS including automated mapping and facilities management type systems but excluding thematic mapping and computer aided design packages. One respondent generally the project manager was interviewed with respect to each separate system present within a particular authority. The survey is based on a period from February to June 1991.

The analysis of the survey findings identifies the technical and organisational characteristics of the systems being implemented and examines these in relation to the main benefits and problems respondents perceived to be associated with the introduction of GIS. The organisational features considered include the length of time since the GIS was purchased and the institutional arrangements which have been developed for GIS implementation. The latter is subsequently referred to as the 'approach'. Technical features such as the type of software and hardware being utilised is also examined. Given these findings a three-fold typology of the styles of implementation adopted by British local authorities has been developed.

It is important to note the unit of study in the subsequent analyses are systems rather than authorities. A system is regarded as a distinct piece or combination of software, which one or more departments within an authority are implementing. For instance, a situation where several departments are developing separate applications based on the same software is considered as one system.

4.2. ORGANISATIONAL CHARACTERISTICS

4.2.1. *Length Of Experience In Implementing GIS Technology*. Table 4 indicates that since the early 1980s there has been a gradual growth in the number of GIS purchased with nearly 40 percent of all systems acquired in 1990. As a result of the timing of the survey the figures for 1991 are incomplete, but there is some indication of a declining rate of take-up. It is probable that the recession, uncertainty over the future structure of local government and changes in the internal management of authorities have contributed to these circumstances.

Table 4. Length of experience in handling GIS technology

Year Of Purchase	Shire And Scottish Districts (%)	Metropolitan Districts (%)	Counties And Regions (%)	Total (%)
1991*	6.9	-	14.9	9.2
1990	34.5	63.6	31.9	39.8
1989	27.6	-	23.4	19.4
1988	13.8	18.2	19.1	17.3
1987	13.8	-	2.1	5.1
1986	-	9.1	6.4	5.1
Earlier than 1985	3.4	9.1	2.1	4.1
	(n=29)	(n=22)	(n=47)	(n=98)

*Survey undertaken between February and June 1991 therefore data incomplete for that year
Monitoring will be necessary to ascertain whether 1990 represents the peak period of system take-up. The suggestion of a levelling off in adoption is strongest in shire districts and counties. The pattern in the metropolitan districts fluctuates considerably from year to year with nearly two-thirds of all systems purchased in 1990.

Analysis of the length of time authorities have been implementing GIS suggests growing experience amongst a number of authorities. Taking June 1991 as the base date the average length of time since system take-up was slightly over two years. This increasing experience is important as some authorities are already beginning to replace their existing facilities. Amongst these are a small number which evaluated a micro based system prior to committing more substantial resources and a number where the original software proved unable to satisfy their requirements. This discussion focuses on the configuration of equipment currently in use (see Section 4.3).

4.2.2. *Approach to GIS Implementation* The findings of the survey provide some important results with respect to the basic organisational structure authorities have adopted as the basis for implementation. Table 5 demonstrates that within a number of authorities several systems are being developed simultaneously, as a total of 98 systems have been purchased by 85 authorities. Shire counties are the most likely to have more than one system, as there are 44 systems in 32 counties. The only other authority to have adopted more than one system is a large Scottish district. Implementation in some authorities with several systems is being handled by individual departments whereas in others there are both departmental and what the authorities themselves term, corporate systems. Evidence from the subsidiary information provided by respondents suggests the presence of several systems within one authority is more likely to reflect the desire of certain departments to have complete control over the form and speed of system development than the perceived inability of the GIS software they have bought to perform a variety of functions.

Table 5. Level of GIS adoption by authorities and systems

Local Authorities	**No. of Authorities Possessing GIS**	**Number of GIS**
Shire Districts	24	24
Metropolitan Districts	22	22
Shire Counties	32	44
Scottish Districts	4	5
Scottish Regions	3	3
TOTAL	85	98*

* The survey identified 4 further systems, 1 adopted by a joint board and 3 by metropolitan research and information units.

Looking in detail at the approach being adopted to the implementation of individual systems, Table 6 distinguishes between departmental GIS and those being introduced by several departments. The striking feature of these findings is the even split between departmental and corporate approaches. This contrasts with the prevailing orthodoxy discussed earlier, which suggests that strategic and efficiency benefits favour the adoption of a corporate approach. Given this broad trend there are distinct variations between different categories of authority. It is nearly twice as likely that a system in a county or region will have been implemented departmentally rather than corporately, with similar results found in the Scottish Districts. In contrast the introduction of 77 percent of systems in metropolitan districts involves several departments, while the figure for shire and Scottish districts is less at 58.6 percent. It might be expected that the development of departmental systems is linked to the presence of large resource rich departments such as those found in the counties, regions and metropolitan districts. However the extreme positions of these two categories of authority questions this assumption.

The division of responsibility for funding the development of a GIS presented in Table 7 confirms the significant role of departments. Around a third of all systems are funded

Table 6. Approach to GIS implementation

Local Authorities	Departmental Number	%	Corporate Number	%	Total
Shire and Scottish Districts	12	41.4	17	58.6	29
Metropolitan Districts	5	22.7	17	77.3	22
Counties and Regions	31	65.9	16	34.0	47
TOTAL	48	49.0	50	51.0	98

exclusively from central budgets with this figure rising to 45 percent in shire and Scottish districts. In contrast nearly 47 percent of systems in counties and regions are funded solely from departmental budgets. It therefore appears that the introduction of GIS into local authorities frequently requires a significant commitment of resources by departments, even in instances where a corporate approach has been adopted.

It was suggested in Section 3.2 that a corporate approach has two key dimensions. These are the number of departments involved and the location of prime responsibility for system development. In the subsequent discussion the term department refers to the units so defined by the respondents to the survey. However, it should be noted that authorities organise their services in a great variety of ways and consequently departments may vary in size and functions from large multi-purpose directorates responsible for instance for a range of technical services to specialist units such as parks and land use planning. It should also be reiterated that the responsibility for particular services varies between the separate categories of local authority,

Table 7. Sources of funding for GIS

Local Authorities	Shire And Scottish Districts (%)	Metropolitan Districts (%)	Counties And Regions (%)	Total (%)
Central	44.8	36.4	27.6	34.7
Departments	24.1	31.8	46.8	36.7
Central and Departments	24.1	31.8	19.1	23.6
Do not know	6.9	-	6.4	5.1
	(n=29)	(n=22)	(n=47)	(n=98)

therefore education is handled by county, region and metropolitan authorities, while all tiers of local government undertaken some sort of planning related tasks.

Table 8 indicates that where a corporate approach has been adopted system implementation generally involves no more than three or four departments. As a result 72 percent of corporate systems are being developed by only two, three or four departments. Furthermore only 12 percent are being introduced in line with the classical Bains type corporate approach, which would entail all the departments in an authority participating in the implementation of the GIS. Smaller authorities such as shire districts are the most likely to favour this framework.

Table 8. The number of departments involved in GIS implementation where a corporate approach has been adopted

Number Of Departments*	Shire And Scottish Districts (%)	Metropolitan Districts (%)	Counties And Regions (%)	Total (%)
All	23.5	5.9	6.2	12.0
7	5.9	-	-	2.0
6	-	5.9	12.5	6.0
5	11.8	5.9	6.2	8.0
4	29.4	17.6	43.7	30.0
3	11.8	47.0	18.7	26.0
2	17.6	17.6	12.5	16.0
	(n = 17)	(n = 17)	(n = 16)	(n = 50)

*Departments are defined by the respondent and vary in size and type between authorities

Table 9 shows that where a corporate approach has been adopted planning is most likely to be the lead department followed by information technology/computer services. The findings suggest that in instances where a top-down structure is favoured the lead is generally taken by the central computing department rather than the chief executive's department. However, only in the case of one in four corporate systems is the key coordinating role taken by one of the central departments. It is probable that within an authority, lead departments will have the strongest commitment to GIS. The findings presented in Table 9, therefore suggest interest is strongest amongst the technical and property related departments. Consequently the introduction of GIS appears to reflect a bottom-up demand for such facilities rather than a perceived corporate need by those at the centre.

The contribution of planning is most pronounced in the case of the shire and Scottish districts and the metropolitan districts where respectively 47 and 26.5 percent of systems in each of these local authority types are led by planning. These results reflect the considerable interest planning authorities have for geographically based information as well as their traditional responsibility for meeting the cartographic needs of authorities (see Chorley Report, Department of the Environment 1987). However, planning departments are far less significant in the case of systems being developed by county and regional authorities, with highways and information technology/computer services most likely to take the lead. There is also a higher proportion of systems for which responsibility is shared by all participating departments than

Table 9. Lead departments where a corporate approach has been adopted for GIS implementation

Lead Departments*	Shire And Scottish Districts (%)	Metropolitan Districts (%)	Counties And Regions (%)	Total (%)
Planning/ Development	47.0	26.5	6.2	27.0
IT/Computer Services+	17.6	17.6	25.0	20.0
Highways/Engineers	5.9	8.8	31.2	15.0
Combined Technical Services	-	26.5	6.2	11.0
Chief Executive	-	11.8	6.2	6.0
Property/Land Management	-	8.8	6.2	5.0
Legal and related services	11.8	-	-	4.0
Community Services	5.9	-	-	2.0
No Lead	11.8	-	18.7	10.0
	(n = 17)	(n = 17)	(n = 16)	(n = 50)

*The activities of departments have been interpreted as appropriate gives the information provided by respondents

+Information Technology/Computer Services includes sections in Treasurer's Departments

for the other categories of local authority.

These trends are mirrored by analyses which include all the departments involved in the development of a corporate system. The one group of departments which are more prominent are estates and architecture which while showing a reluctance to lead often seem keen to participate in such projects. As a result, three-quarters of the total number of departments participating in implementing corporate systems are accounted for by planning, property and highways management, engineering and computer services. However, active interest in GIS has not permeated all local authority activities. The results indicate that the often large community service type departments of housing, education and social services have tended not to become involved. This may be a reflection of the considerable administrative pressures faced by these departments due in part to frequent legislative changes by central government as well as the limited resources available for additional activities such as new data handling initiative.

Departmental systems are shown in Table 10 to be associated with very similar types of activity to the GIS being implemented corporately. Again technical services such as property management and planning are prominent with a particular tendency for departmental systems to be developed by the engineers and highways authorities. There is no evidence to suggest that departmental systems are being developed by a different category of activities from corporate GIS. The absence of the central computer service sections from this list reflects their corporate role within local government. Given the discussion in Section 3.3 and the comments of respondents it is probable that the decision of departments to act alone is in part an effort to side-step these centralising agencies and thereby avoid compromising their own departmental priorities. The extent to which planning authorities have developed departmental GIS is worthy of note given the smaller resource base of these departments compared for instance to highways, estates, education and housing.

Overall the findings of the survey indicate that there are very few examples of centrally led fully corporate arrangements for the implementation of GIS. In practice around half the systems are being developed by a single department while collaborative projects most often involve no more than three or four departments with a technical service type department taking the lead. There is also evidence to suggest that within the counties in particular there may be several separate systems being developed within one authority.

Table 10. Departments implementing their own GIS

Departments	**Shire And Scottish Districts** (%)	**Metropolitan Districts** (%)	**Counties And Regions** (%)	**Total** (%)
Highways/ Engineers	33.3	40.0	29.0	31.2
Planning/ Development	50.0	20.0	12.9	22.9
Property/Land Management	16.7	20.0	25.8	22.9
Combined Technical Services	-	20.0	12.9	10.4
Emergency Services	n/a	n/a	9.7	6.2
Community Services	-	-	3.2	2.1
Others	-	-	6.2	4.2
	(n = 12)	(n = 5)	(n = 31)	(n = 48)

4.3. TECHNICAL CHARACTERISTICS

4.3.1. Software Table 11 provides a breakdown of the software being introduced by British local authorities. It should be noted that systems with GIS capabilities which are essentially being used to perform other activities such as computer aided design have been omitted from this exercise. Additionally, it is the main system which has been included in the analysis in instances where packages have been purchased to provide specialist GIS facilities to supplement the core system.

Table 11. GIS software adopted by British local authorities

Software	**Shire And Scottish Districts** (%)	**Metropolitan Districts** (%)	**Counties And Regions** (%)	**Total** (%)
Arc/Info	3.4	9.1	40.4	22.4
Alper Records	24.1	9.1	6.4	12.2
GFIS	3.4	9.1	14.9	10.2
Hoskyns G-GP	3.4	22.7	2.1	7.1
Axis Amis	13.8	9.1	2.1	7.1
McDonnell Douglas GDS Maps	6.9	9.1	2.1	5.1
Coordinate	10.3	-	2.1	4.1
Wings	-	-	8.5	4.1
Other*	34.5	31.8	21.3	27.5
	(n=29)	(n=22)	(n=47)	(n=98)

A striking feature of the survey findings is the market share of around 22 percent held by Arc/Info (ESRI), with Alper Records (a package developed by a small British company) and IBM's GFIS representing a further 12 and 10 percent respectively. These three packages therefore account for around 45 percent of the systems adopted by local authorities with the remaining 55 percent accounted for by 17 packages and 4 home grown systems. The range of products purchased by authorities tends to indicate the overall immaturity of the market with as yet only tentative indications of specialisation on a few core products. Such a process may be encouraged by technical developments which are increasing the range of hardware on which particular systems can be run easily. Until recently a considerable number of authorities have purchased the GIS software developed by the manufacturer of the authority's main computer hardware.

A number of further issues are raised by examining software adoption in relation to time. The results indicate that the market share held by Arc/Info has grown from around 17 percent of all systems purchased in 1987 to 30 percent in 1990. At the same time the adoption of GFIS has declined from around 35 percent in 1988 to 2 percent in 1990. It is also clear that there has been a steady increase in the variety of systems being implemented by local authorities. For instance, in 1986 four different systems were purchased, in 1988 eight and in 1990, sixteen.

There also appears to be some relationship between the approach to system implementation and the GIS software adopted. Frequently, the large multi-purpose packages such as GFIS, GDS Maps (McDonnell Douglas) and Coordinate (Bull) are being introduced corporately. The pattern for the more limited automated mapping and facilities management systems is less clear, although there is a definite bias in the case of Alper Records and Wings (Wang) towards departmental implementation. The Arc/Info findings are particularly interesting, as contrary to expectations, there is no clear link between this large complex software and a corporate style of implementation. These results reflect the adoption of Arc/Info by shire counties which tend to have a bias towards departmental implementation. These findings suggest some association between software and approach but there are no grounds for assuming that the type of software determines the approach employed for GIS implementation. Such matters can only be properly assessed in the light of case study findings.

A number of more detailed issues are raised by examining the GIS software being implemented by local authority type (see Table 11). It is clear from this that while Arc/Info is dominant within the shire counties and the Scottish regions, accounting for around 40 percent of all systems, this pattern is not repeated throughout local government. In the metropolitan districts Hoskyns G-GP (a system developed by a small British company) represents nearly 23 percent of the total number of systems which have been introduced. This reflects the original development of this software by the Research and Intelligence section of the now abolished Greater London Council and the subsequent interest this provoked on the part of a number of the London boroughs. As a result, of the seven systems purchased five are in London. A similar presence in Scotland is enjoyed by Alper Records but the reasons behind this are less clear. The striking feature of the take-up of software in the shire districts is the broad range of systems and the absence of a clear market leader. In more general terms these findings indicate a greater demand on the part of the counties and regions for software which provides a variety of GIS facilities, while the other authorities tend to favour systems with more limited capabilities such as automated mapping and facilities management.

4.3.2. Hardware The findings presented in Table 12 indicate a number of important trends with respect to the main forms of computer hardware being used for GIS applications. Where systems have several means of access, for instance mainframe terminals and microcomputers, it is the highest level of capability, that is the mainframe component, which has been recorded in the analysis. Given the nature of the technology there is inevitably a significant link between the software adopted and the supporting hardware. For instance 77.3 and 91.7 percent of Arc/Info and Alper Records systems respectively are handled using workstations, while 90 percent of GFIS and 75 percent of Coordinate systems are on mainframes and all the Wings systems currently being developed utilise microcomputers. Around 50 percent of the 'other' systems being implemented are microcomputer based.

Overall the findings suggest workstations are the most common platform for GIS software with around 41 percent of all systems based on this hardware. The utilisation of mainframes and microcomputers are similar at 25.5 and 23.5 percent respectively. However, when the relationship between hardware and the time of adoption is examined it is clear that there has been a move away from mainframe computers. There is also evidence from discussions with respondents that some early GIS adopters have substituted workstations for mainframe hardware. This general trend, in turn, has wider organisational implications for the traditional relationship between the central computer services department of a local authority and user departments, as the latter are increasingly able to control their own computing resources.

The analysis of the hardware utilised by different types of local authority highlights a number of further issues. It is perhaps surprising that mainframes are the most frequently used hardware by metropolitan districts with the shire and Scottish districts also relying on this technology to a greater extent than the counties and regions. This may be due to some county systems being switched from mainframes to workstations, as the latter accounts for 46.8 percent of the total within this local authority category. The use of microcomputers, particularly by counties and regions, where it accounts for nearly 28 percent of all systems, raises a number of further issues concerning the relationship between hardware and whether the GIS is being developed by a single department or a coalition of users. The analysis indicates that a departmental system is likely to be supported by firstly microcomputers and secondly workstations. Conversely it is unlikely that mainframe hardware will be employed. Nearly 74 percent of all microcomputer based systems are being implemented by a single department while the corresponding figure for mainframe GIS is 16 percent.

Table 12. Hardware adopted by British local authorities to support GIS software

Hardware	**Shire And Scottish Districts** (%)	**Metropolitan Districts** (%)	**Counties And Regions** (%)	**Total** (%)
Workstation	41.4	27.3	46.8	40.8
Mainframe	27.6	31.8	21.3	25.5
Micro	20.7	18.2	27.6	23.5
Mini	6.9	18.2	4.2	8.2
Midi	3.4	4.5	-	2.0
	(n=29)	(n=22)	(n=47)	(n=98)

The findings of the survey therefore suggest that in terms of equipment there is a trend, especially in the counties and regions towards Arc/Info software and workstation environments. However, overall given the as yet low level of specialisation within the market there are considerable variations in the technical characteristics of the systems being implemented.

4.4. BENEFITS AND PROBLEMS ASSOCIATED WITH GIS IMPLEMENTATION

4.4.1. Introduction The results of the survey provide a general indication of the basic organisational and technical features of the GIS being implemented in British local government. The final part of this analysis draws on the survey findings to explore the main benefits and problems being encountered.

The survey questionnaire asked respondents with respect to both benefits and problems, to rank in order of importance three sets of factors. They were then requested to describe in more detail the group of benefits and problems ranked first. Provision was also made for issues not identified in the questionnaire to be raised. The results of this section of the survey are inevitably largely dependent on the perceptions of the particular individuals interviewed, however, it is possible to identify some general trends.

4.4.2. Benefits Table 13 shows that the main benefits which respondents perceived to be derived from the introduction of GIS were improved information processing facilities. This includes improved data integration, increased speed of data provision, better access to information and the availability of an increased range of analytical and display facilities. In other words this implies a greater range of tools with which to display and analyse information rather than more profound organisational goals. It is striking that these benefits were regarded as most important by 60 percent of respondents with 31 percent stressing better quality decisions including managerial, operational and strategic considerations and only around 6 percent linking the main benefits of GIS to achieving savings. Given the work on cost-benefit analysis which has been undertaken in relation to GIS the very low level of importance attached to the ability of such systems to achieve time, staff and financial reductions is particularly interesting. Respondents suggested that while financial justifications are important in gaining agreement from members and senior management to purchase a GIS, they perceived that in practice due to the cost of equipment and data preparation these savings were likely to be limited. Furthermore, there is no evidence to suggest a strong link between GIS and improved decision making. In instances where 'better' decisions were regarded as the prime benefit to be gained from GIS implementation, just over 38 percent linked these advances to routine operational decisions, while the figures for strategic and managerial activities were 28.8 and 25 percent respectively.

The general importance attached to the contribution of GIS to improving the information processing facilities available within local government and the limited role with respect to realising savings is mirrored by each category of authorities. The only additional issue is the greater emphasis the counties and regions place on the benefits GIS provides in terms of better decision making. This is associated with operational activities rather than strategic or managerial decision making. Overall, however, the general pattern is replicated by the experiences of each of the local authority groupings.

The key issue raised by these findings is the operational rather than strategic and managerial uses to which GIS in British local government is being put. In terms of the types of applications being developed the vast majority are concentrating on relatively simple operational systems, regardless of the modelling facilities which may be available within the software. The key application area for most local authorities has been to exploit the digital data available from the Ordnance Survey to develop automated mapping facilities. In addition, the other main areas of application include grounds and highways maintenance and estate management.

Table 13. The most important group of benefits associated with the implementation of GIS in local government

Benefits	**Shire And Scottish Districts** (%)	**Metropolitan Districts** (%)	**Counties And Regions** (%)	**Total** (%)
Improved information processing facilities	68.4	70.4	51.1	60.5
Better quality decisions	27.0	13.6	42.5	31.5
Savings	4.6	11.4	4.2	5.9
Other*	-	4.5	2.1	2.0
	(n=29)	(n=22)	(n=47)	(n=98)

* Survey undertaken between February and June 1991 therefore data incomplete for that year

The reasons for these trends are not entirely clear and will need further investigation through case studies. However, a few tentative comments can be made at this stage. There appears to be little support for the suggestion that strategic and efficiency type benefits are associated with corporate systems. More important is the association between the strategic use of GIS and the need for information sharing both within and between departments (see Masser and Campbell 1992). It is possible to question whether information sharing is likely to take place in any circumstances, but given the current move towards fragmentation (see Section 2.2) such trends appear highly unlikely in British local government.

4.4.3. Problems The findings shown in Table 14 demonstrate that a range of technical, data related and organisational difficulties are being experienced by those implementing GIS in local government. Most respondents emphasised that the introduction of GIS into their organisation had not proved a straightforward process. Only in four cases was it stated that no problems had been encountered. Overall there was a slightly greater emphasis on data related issues such as the lack of compatibility between existing data sets, problems of maintaining up-to-date data sets, the cost of data capture and the cost and availability of Ordnance Survey digital data, than the other sets of problems.

Given the general trends outlined above particular types of difficulties were being experienced by each category of local authority. For instance, technical problems were most pronounced in the shire and Scottish districts whereas the counties and regions were having fewer technical difficulties, but data related matters were proving more serious. It is probable that the decreasing level of technical problems encountered by the counties and regions through metropolitan districts to the shire and Scottish districts is a reflection of the greater resources and experience available within the larger authorities. It is noticeable that organisational difficulties are relatively constant regardless of local authority type.

Table 14. The most serious group of problems associated with the implementation of GIS in British local government

Problems	Shire And Scottish Districts (%)	Metropolitan Districts (%)	Counties And Regions (%)	Total (%)
Data related	21.8	31.8	42.5	34.0
Technical	47.7	27.3	17.0	28.4
Organisational	27.0	27.3	29.8	28.4
Other	-	4.5	6.4	4.1
No problems	3.4	9.1	2.1	4.1
System not sufficiently developed	-	-	2.1	1.0
	(n=29)	(n=22)	(n=47)	(n=98)

The additional comments of respondents give some more detailed insights into the difficulties being experienced by local authorities implementing GIS. Around 20 percent mentioned resource issues such as cost and a shortage of staff time as inhibiting the introduction of GIS, while a significant number were highly critical of the activities of vendors. In particular there was concern about the extravagant claims of vendors prior to software purchase, their lack of appreciation of the local government context, the limited scope for linkage between systems and inadequate post-sales support. The importance as well as difficulty of establishing the financial viability of vendors was also stressed. Criticism was also levelled at the Ordnance Survey. In particular dissatisfaction was focused on the incomplete digital map coverage for Britain which results in partial data sets for most local authority areas and the quality of some of the data produced in terms of the speed of up-date and edge matching of maps.

In addition to these external influences many of the comments of respondents related to the specific organisational circumstances in which GIS were being implemented. These included lack of previous technical experience in GIS, limited awareness amongst the technical specialists of user needs and the lack of appreciation by senior staff that the implementation of GIS involves the complex process of managing change. Several respondents were also concerned about the tension between developing a corporate system which tends to involve a long lead time and the imposition of standards, and the immediate service needs of departments. More specifically, given the computer storage space required for handling geographic information, some respondents were concerned about the introduction of direct charging for mainframe time. Such circumstances were often associated with an internal reorganisation into business units with a number of the less wealthy departments such as planning feeling they would have to withdraw from the mainframe based GIS projects being developed in their authority. In purely practical terms several of the smaller authorities were concerned about the shortage of skilled staff, particularly professional staff who have a knowledge of computing and GIS.

There is no clear pattern with respect to the relationship between the problems currently being encountered and the time of original adoption. However, a similar exercise examining the relationship between difficulties and whether a single department rather than a group were responsible for system development indicates more definite trends. The results suggest that technical and organisational problems are more likely to be encountered where several departments are involved with implementation, while data related issues are associated with departmental systems. The latter point tends to confirm the earlier findings in relation to the counties and regions as around 65 percent of the departmental systems are located in these authorities. It is also clear from these findings that while systems being developed by several departments are encountering more severe organisational problems, such difficulties are not completely resolved by departmental implementation with well over a fifth of all these systems experiencing organisational problems.

4.4.4 Summary The findings of the survey suggest that overall there is a large measure of consensus between respondents that improved information processing facilities are the main benefits to be derived from GIS. In contrast a wide range of technical, organisational and data related problems are being experienced. Furthermore, the nature of the factors influencing the results of the process of implementation are not entirely clear and will therefore require more detailed investigation.

4.5 TYPOLOGY OF GIS IMPLEMENTATION

The analysis of the survey findings indicates that the characteristics of the GIS being developed in local government and the consequences of implementation vary considerably between individual systems. However, given the distinctive nature of particular systems a number of broad trends have emerged. On the basis of these results a three-fold typology of system implementation has been developed. These are as follows:

1. Classically corporate - The traditional understanding of a "corporate approach" is reflected in this style of system implementation (see Section 3.2). Consequently a large number of departments, and possibly the whole authority, participate in the project with the lead taken by the computer services or planning department. Planning in this case is taking a central coordinating role probably linked to an existing responsibility for the provision of Ordnance Survey maps or socio-economic information, while the existing central status of computer services is more obvious. The results of this approach are reflected in a workstation or mainframe based system using software which is designed to provide limited automated mapping or facilities management capabilities. The choice of software highlights that even in this context GIS is not primarily expected to provide improvements in efficiency or the quality of decisions but rather to enhance the information processing facilities available within the authority. However, in attempting to implement these systems authorities are encountering considerable difficulties. These problems are often technical in nature and are not assisted by a shortage of skilled staff.

   A classically corporate style of approach has been adopted as the basis for implementing 15 percent of the systems found in British local government. This approach is most likely to be favoured by shire districts.

2. Theoretically/pragmatically corporate - This style characterises the approaches adopted to implementation, which while involving several departments do not accord with the classical definition of a corporate approach. As a result this approach is either typified by a 'bottom-up' demand for GIS facilities from service delivery departments or an attempt by the computer services department to achieve some measure of corporate working. On the basis of the survey findings it is difficult to distinguish between approaches which reflect a pragmatic decision amongst departments to pool resources to facilitate the purchase of equipment and attempts at a coordinated and wide-ranging structure for GIS implementation but which have as yet fallen short of this goal. Consequently such approaches are characterised by the involvement of only three to four departments, with the lead taken by computing, a technical service type department or in certain cases joint responsibility for the project. These systems are largely mainframe or workstation based, but there are no clear trends in terms of software. The main benefits of introducing a GIS are perceived to be enhanced information processing facilities, although systems in some county councils are expected to improve operational decision making. Given these objectives, a wide range of problems are being experienced, especially in relation to data and organisational issues.

   Theoretically/pragmatically corporate approaches are being favoured for the implementation of around 35 percent of systems. Such an approach can be found in all categories of local authority, but most especially the metropolitan districts and shire counties.

3. Fiercely independent - This approach to implementation is typified by the introduction and development of a GIS by a single department. The department is likely to be involved with technical service type activities such as highways and to have considerable experience of information handling, including the presence of in-house technical expertise. Most of these systems are based on a micro or workstation with a range of software being employed, although there is a slight tendency towards Arc/Info. It is expected that the GIS will provide improved information processing facilities, however, this goal is being inhibited by a number of data related issues. In addition organisational problems have not been fully overcome.

A fiercely independent style of implementation appears to typify the approach adopted in relation to around 50 percent of the systems which are being introduced. Such an approach is most often found in the counties and regions.

5. Evaluation

5.1. ISSUES RAISED BY THE EXPERIENCES OF BRITISH LOCAL AUTHORITIES

The results of the survey raise considerable doubts as to whether in practice local authorities are following the advice in much of the literature to adopt a corporate approach to GIS implementation. Such an approach is generally justified on the basis of the strategic and efficiency benefits which will result. However, the findings suggest that the majority of respondents perceive the main benefits to be improved information processing facilities rather than more fundamental administrative advances. The extent to which such conclusions by staff within local authorities are the result of either a detailed assessment of the potential of the technology or an admission that the organisational context makes such an approach unrealistic, requires more detailed investigation. Certainly the most striking feature of the findings is that around one in two of the systems present within local authorities are being introduced by a single department. Furthermore, in cases where several departments are involved in a GIS project, few of the approaches adopted follow the classically corporate structure of total organisational involvement and central control.

The typology of styles of approach to GIS implementation is very instructive in this connection. The typology suggests that a classically corporate approach is only favoured in small authorities, where there is likely to be a high degree of personal contact between staff in different departments, and where the GIS is being expected to provide a limited range of facilities management or automated mapping type functions. In the remaining cases the term corporate appears to either represent a theoretical position for a situation which in practice is far from corporate or a pragmatic move to ensure sufficient resources are pooled to embark on the implementation of a GIS project. It is clear from the problems being encountered that there is a certain fragility to these systems. It is interesting to note that while organisational difficulties were most severe in cases where systems were being introduced corporately such problems were not entirely eradicated by departmental implementation.

So to return to the question posed in Section 3.1, does technology shape the context or the context the technology? The range of approaches which have been adopted within British local authorities would tentatively suggest the latter to be the case. However, further research is undoubtedly required. It could be argued that the fragmented approach to GIS implementation in many contexts reflects a short term reaction to the legislative pressures being imposed by central government. In contrast such circumstances may simply mirror the innate characteristics of organisations in general and local government in particular (see, for example, Campbell 1990, 1991, Bjorn-Anderson et al. 1986, Hirschheim 1985, Masser and Campbell 1992, Markus 1983, Robey 1987). Consideration also needs to be given to the dynamics of GIS implementation. Do the problems encountered change over time and what relationship does this have to the organisational and technical characteristics of the system? Furthermore little is as yet known about the level of utilisation of GIS and the precise nature of the applications being developed. The simple action of adoption by no means implies utilisation. It is likely that the experiences of introducing GIS may be influenced by the type of functions the system is expected to fulfil. For instance effectively implementing a GIS to achieve operational tasks such as map reproduction may be more straightforward than as a strategic tool for policy makers or as an aid to managerial decision making. There is also a need to consider to what extent particular characteristics of either the organisational context or the implementation strategy itself influence the effective utilisation of GIS.

Most of the issues raised above cannot be investigated fully using survey based approaches and research is currently underway to supplement these findings with case study material. However, undertaking research in this field raises a considerable number of methodological issues. These will briefly be examined below.

5.2. METHODOLOGICAL ISSUES

Surveys, such as the one on which the findings of this paper have been based, can only provide broad indications of the issues involved in the implementation of GIS. The results of surveys of this type are themselves highly dependent on the method adopted and the sample size. Comparisons between the findings discussed above and the results of postal surveys which are necessarily selective indicate that the latter tend to overestimate GIS adoption by around five percent (see Woodsford 1991). However, if the critical issues described above are to be more clearly understood it is of great importance that attention is given to the development of suitable case study methodologies for this work (Campbell 1990, Kling 1987, Kling and Scacchi 1982, Onsrud et al. 1989, Pettigrew 1988, Rogers 1983, Van de Ven and Rogers 1988, Yin 1979).

Kling in a discussion looking at the general implementation of technology in organisations argues that it is important that case studies are structured in such a manner that the researcher gets behind the formal appearance of the activities of organisations and the individuals of which they consist. As a result he distinguishes between the discrete-entity approach which focuses on those directly involved with the technology and formal statements of intent, and the web model which attempts to examine the social and political network within which computer based systems are located. Consequently, in conducting a web type approach it is important to define the boundaries of the case study broadly and thereby interview a broad range of individuals including those not directly involved in the project such as senior management and potential users. The dynamic nature of system implementation also makes it important to undertake longitudinal studies which seek to identify the changing pattern of relationships and impact of processes over time.

6. Conclusion

The findings of the research presented in this paper clearly indicate the need to supplement existing basic information on the adoption of GIS with work examining the effectiveness of various approaches to implementation. It is premature before the findings of the case studies which are currently being undertaken as part of this project have been completed, to propose criteria or hypotheses to be tested. However, even given the limitations of the survey it is clear that without further research which examines the issues affecting the implementation of GIS in a range of organisational environments, we will lack the necessary understanding to avoid GIS becoming yet another failed technology and as a result considerable resources being wasted.

7. References

Bains Committee Report (1972) The new local authorities: management and structure, HMSO, London.

Batley, R. and G. Stoker (eds.) (1991) Local Government in Europe, Basingstoke, MacMillan.

Bjorn-Anderson, N., K. Eason and D. Robey (1986) Managing computer impact: an international study of management and organisations, Ablex, Norwood, New Jersey.

Bromley, R. and M. Coulson (1989) Geographical information systems and the work of a local authority: the case of Swansea City Council, Department of Geography, University College of Swansea.

Campbell, H. (1991) Organisational issues in managing geographic information, In I. Masser and M. Blakemore (eds.), *Handling Geographic Information*, Longman, London, 259-282.

Campbell, H. (1990) The use of geographical information in local authority planning departments, *Thesis submitted for PhD in Town and Regional Planning*, University of Sheffield.

Cane, S. (1990) Implementation of a corporate GIS in a large authority, *Proceedings of the First European Conference on Geographical Information Systems*, Amsterdam, 10-13 April, 158-166.

Cohen, M. D., J. G. March and J. P. Olsen (1972) A garbage can model of organisational choice, *Administrative Science Quarterly* 17(1): 1-25.

Coulson, M. and R. Bromley (1990) The assessment of user needs for corporate GIS: the example of Swansea Council, *Proceedings of the European Conference on Geographic Information Systems*, Amsterdam, April 10-13, EGIS Foundation, Utrecht, 209-217.

Craglia, M. (1992) Jumping in at the deep end: GIS in Italian local government, *Proceedings of the European Conference on Geographic Information Systems*, Munich, March 23-26, EGIS Foundation, Utrecht, 627-638.

Danziger, J.N., W.H. Dutton, R. Kling and K.L. Kraemer (1982) Computers and politics: high technology in American local governments, Columbia University Press, New York.

Danziger, J. N. and K. L. Kraemer (1986) People and Computers: the impacts of computing on end users in organisations, Columbia University Press, New York.

Department of the Environment (1992) Competing for quality in housing, HMSO, London.

Department of the Environment (1987) Handling Geographic Information. *Report of the Committee of Enquiry* chaired by Lord Chorley, HMSO, London.

Feldman, M. S. and J. G. March (1981) Information in organisations as signal and symbol, *Administrative Science Quarterly* 26: 171-186.

Gault, I. and D. Peutherer (1989) Developing GIS for local government in the UK, paper presented at European Regional Science Association Congress, Cambridge.

Grimshaw, D. J. (1988) The use of land and property information systems, *International Journal of Geographical Information Systems* 2(1): 57-65.

Hirschheim, R. A. (1985) Office automation: a social and organisational perspective, John Wiley, Chichester.

Hirschheim, R., H. Klein and M. Newman (1987) Information system development as social action: theory and practice, RDP 87/6, Oxford Institute of Information Management, University of Oxford, Oxford.

International City Management Association (1989) Computer use in municipalities: the state of the practice, 1988, International City Management Association, Washington.

Kling, R (1987) Defining the boundaries of computing across complex organisations. In R. J. Boland and R. A. Hirschheim, (eds.), *Critical Issues in Information Systems Research*, 307-362, John Wiley, Chichester.

Kling, R. and W. Scacchi (1982) The web of computing: computer technology as social organisation, *Advances in Computers* 21: 1-90.

Leonard-Barton, D. (1988) Implementation characteristics of organisational innovations: limits and opportunities for management strategies, *Communications Research* 15(5): 603-631.

Lindblom, C. E. (1959) The science of 'muddling through', *Public Administration Review* 19: 79-88.

Lindblom, C. E. and D. K. Cohen (1979) Usable knowledge: social science and social problem solving, Yale University Press, New Haven.

Markus, M. L. (1983) Power, politics and MIS implementation *Communications of the Association for Computing Machinery* 26(6): 430-444.

Masser, I. and H. Campbell (forthcoming) GIS is organisations: some conditions for their effective utilisation, In M. de Lepper, H. J. Scholten and R. M. Stern (eds.) *The added value of geographic information systems in public and environmental health*, Kluwer, Dordrecht.

Masser, I. and H. Campbell (1992) Information sharing - the impact of GIS on British Local Government, paper presented at NCGIA I9 workshop on Information Sharing, San Diego, 27-29 February.

Maud Report (1967) Local government administration in England and Wales, HMSO, London.

Mumford, E. and A. Pettigrew (1975) Implementing strategic decisions, Longman, London.

Onsrud, H. J., H. W. Calkins and N. J. Obermeyer (eds.) (1989) Use and value of geographic information initiative for specialist meeting: report and proceedings, *NCGIA Technical Paper 89-7, NCGIA*, University of California, Santa Barbara, California.

Onsrud, H. J. and J. K. Pinto (1991) Diffusion of geographic information innovations, *International Journal of Geographical Information Systems* 5(4): 447-467.

Openshaw, S., A. Cross, M. Charlton, C. Brunsdon and J. Lillie (1990) Lessons learnt from a post mortem of a failed GIS, *Proceedings of Association for Geographic Information Conference*, 22-24 October 1990, Brighton, 2.3.1-2.3.5.

Paterson Report (1973) The new Scottish local authorities: organisation and management structures, HMSO, London.

Pettigrew, A. M. (1988) Longitudinal field research on change: theory and practice, paper presented at the National Science Foundation Conference on Longitudinal Research Methods, Austin, Texas, 14-16 September.

Pfeffer, J. (1981) Power in organisations, Pitman, Massachusetts.

Robey, D. (1987) Implementation and organisational impacts of information systems, *Interfaces* 17(3): 72-84.

Rogers, E. M. (1983) Diffusion of innovations, Free Press, New York.

Redcliffe-Maud and B. Wood (1974) English local government reformed, Oxford University Press, London.

Van de Ven, A. H. and E. M. Rogers (1988) Innovations and organisations: critical perspectives, *Communication Research* 15(5): 632-651.

Weiss, C. H. (ed.) (1977) *Using social science research in public policy making*, Lexington Books, D. C. Heath and Co., Lexington, Massachusetts.

Winter, P. (1991) Selling a corporate GIS, *Proceedings of the Association for Geographic Information Conference*, Birmingham, 20-22 November, 2.6.1-2.6.5.

Woodsford, P. (1991) Geographic information systems and local government. Paper presented at Count Planning Officers' Society Workshop on GIS and the Environment, Birmingham, September 1991.

Yin, R. K. (1979) *Changing urban bureaucracies: how new practices became routinised*, Lexington Books, D. C. Heath and Co., Lexington, Massachusetts.

DIFFUSION AND USE OF GEOGRAPHIC INFORMATION SYSTEMS IN PUBLIC SECTOR AGENCIES IN THE UNITED STATES

LYNA L. WIGGINS
Department of Urban Studies and Planning
Massachusetts Institute of Technology
9-513, 105 Massachusetts Avenue
Cambridge, MA 02139

ABSTRACT. Geographic Information Systems (GIS) technology is diffusing rapidly through the government agencies of the United States. At all levels of government - municipal, regional, state, and federal - agencies are currently adopting, or express the intention of adopting, this technology. Using the results from two mailed surveys, in this paper we examine the adoption, intention to adopt, and some of the uses being made of these systems today. These two surveys build on previous empirical work completed by the author and her colleague, Steven P. French of Georgia Tech. The first survey (1991) was of members of the Urban and Regional Information Systems Association (URISA). The responses from URISA members in the government sector are reported here. Of particular interest among these results are the domination by a few GIS software vendors; the predominance of land parcel mapping and demographic mapping in reported applications; and the mapping of streets, political boundaries, and parcels as the most common geographic features. The second survey (1992) was of planners, assessors, and engineers in the cities and towns of Massachusetts. Of particular note is that one-third of Massachusetts towns, mainly the smallest places, still report no access to computer technology of any kind. Although 40 percent of the towns with computers report an intention to adopt GIS technology within the next 2-3 years, this percentage is down from the 57 percent reported from a similar survey completed in 1989. A deep economic recession in the region may account for this decrease in reported intention to adopt. Other questions about the condition and status of local tax assessment maps make it clear that completing a statewide cadastral layer based on existing local maps would be extremely difficult.

1. Introduction

Although the technology for Geographic Information Systems (GIS) has been around for twenty-five years, it is only in the past five years that we have seen the extremely rapid diffusion of GIS (Sinton 1992). Last year the GIS industry's revenues grew to around $3.5 billion, with a growth rate of around 33 percent for the year. Experts are forecasting a continued growth of around 25 percent per year over the next decade (The Economist 1992). Most of the growth in the GIS market to date has been in the defense, government and utility sectors, but there are indications that the largest new growth may appear in the business and commercial sectors.

Although it is clear that the industry and technology are changing rapidly, if we examine the history of the development of GIS we see a long tradition of involvement of the geography, urban and regional planning, and landscape architecture professions. In this paper, I will focus on the use and diffusion of GIS technology in American local and

I. Masser and H. J. Onsrud (eds.), Diffusion and Use of Geographic Information Technologies, 147–163.

regional governmental agencies. The research methodology used in this work is the mail survey. This methodology has cost and time advantages over phone and personal interviews in collecting information from a wide range of agencies. It has disadvantages over case study work in the depth of coverage of important issues in adoption. There are also potential response bias issues, discussed at greater length below. This paper will describe some of the results from a series of mail surveys conducted over the past four years (French and Wiggins 1989, French and Wiggins 1990, Garcia and Perryman 1991, Budic 1991). In this paper, the emphasis is more narrowly on aspects of the use and diffusion of GIS, although the surveys themselves covered a broader range of topics.

Because there are facets to survey research via mailed instruments that are important, the paper begins with a discussion of some of the issues that we have discovered and re-discovered in conducting this work. Next, we consider the use and diffusion results from two recent surveys of broader scope than our previous surveys. The first is a survey completed in the spring of 1991 of members of the Urban and Regional Systems Association (URISA) in the U.S. and Canada. The second is a survey of local governments in Massachusetts sponsored by the Massachusetts Geographic Information Committee (MGIC).

2. Lessons From Survey Research

2.1 RESPONSE RATES

In all of our work, we have relied on the advice of Donald Dillman, with respect to the procedures of survey instrument design, mailing and follow-up mailing procedures (Dillman 1978). It is worth noting how successful these procedures have been in obtaining remarkably high response rates. To start, we kept the survey instrument short (4 pages) and the questions well formatted and spaced on the pages. The brevity undoubtedly helped our response rates, but at the cost of excluding questions of interest. In brief, the procedure suggested by Dillman for mailed surveys consists of: (1) a first mailing, including a hand-signed cover letter and a stamped, self-addressed envelope; (2) a follow-up postcard, within approximately one week; (3) a second mailing to non-respondents, including a hand-signed cover letter and another stamped, self-addressed envelope. An example of the incremental increase from this 3-step procedure, for the California survey completed in 1988, was from 28 percent, to 45 percent, to a final response rate of 80 percent. We feel these procedures have led to the high response rates shown in Table 1.

The first nine surveys for the various states were conducted using absolutely identical procedures, those recommended by Dillman (1978). What factors may account for the variations in the response rates shown in Table 1? The respondents in each of these surveys were urban and regional planning professionals in planning agencies. In all cases, we addressed the mailing to the Planning Director, if there was one listed in the state directories of local governments. If there was not a Director listed, we sent the survey to the Chair of the Planning Board in the town. Yet the response rates vary from 64 percent (Vermont) to 90 percent (Virginia). Possible reasons for this variation include: (1) the presence or absence of mandated planning in the state (e.g., in many of the small towns in New England there was not a professional planner present, whereas in California even the smallest towns have state-mandated planning); and (2) the abundance of very small towns without any

TABLE 1. Response rates from mailed surveys (1988-1992)

Survey	Number mailed	Percentage response	Year of survey
California	501	80	1988
Connecticut	138	69	1989
Maine	52	71	1989
Massachusetts	359	57	1989
New Hampshire	27	81	1989
Rhode Island	18	67	1989
Vermont	36	64	1989
Maryland	63	89	1990
Virginia	132	90	1990
URISA	2587	62	1991
MGIC	353	48	1992

computing resources in New England. There is a potential response bias here, as we expect that those towns with no computing resources responded at lower rates than those with resources. There may also be a small longitudinal effect from 1988 to 1992, as both the saturation of microcomputers and awareness of GIS technology increased over this time period. The URISA membership survey was a much larger effort, and the second step postcard was omitted. Still, 62 percent is a highly respectable response rate from a professional organizational membership that is highly saturated with mail surveys and professional literature. Finally, the Massachusetts Geographic Information Committee (MGIC) survey was conducted with a single mailing, but was sent to three respondents in each town (planner, assessor, and engineer).

2.2 INTENTION QUESTION

In all of these surveys (except the URISA survey) we have included questions about the intention to and anticipation of GIS adoption by the respondents. The question was worded identically in all cases, and asked for the respondents' opinions of expected adoption of GIS within 2 to 3 years. From survey and market research, we know the danger of asking such questions about expectations. Yet we are fascinated by the question of the perceptions of GIS adoption in the near future by these planning agencies. We hope to follow up on each of these surveys at future dates in order to check the actual adoption rates with the earlier stated anticipation of adoption. To a limited extent, I have started to do this with Massachusetts.

The responses to the question of expected GIS adoption have ranged from approximately 40 percent (California in 1988) to 90 percent (Rhode Island in 1989). At the time of the surveys, the actual adoption rate was 13 percent of the agencies responding. The detailed breakdowns by state are presented in Wiggins and French (1991). For Massachusetts in 1989, the reported anticipation percentage was 57 percent and the actual adoption rate was 7 percent. In the 1992 survey, the actual adoption rate was 13 percent of the respondents (48 percent response rate). Although we did not ask a specific question about reasons for not adopting, the deep recession during the past 3 years in Massachusetts has done little to encourage GIS adoption by local governments. The negative economic conditions have heavily impacted the construction industry, and there has been little residential, office or industrial construction during this period. In our earlier surveys of California, we noted the association between computer and GIS adoption and urban growth rates (French and Wiggins 1989). We suspect that rapid growth, and particularly growth control measures, encourage GIS adoption, and we plan to explore this relationship further. Since GIS and computer technology have both improved rapidly over this period, and prices have decreased, it is tempting to account for the low adoption rates by the state's economic problems. We see that the expressed adoption intentions of three years ago in Massachusetts are largely unfulfilled in 1992.

3. URISA Survey (1991)

This mail survey was conducted in the spring of 1991, and was sent to all 2587 current URISA members. URISA is a professional organization, now 30 years old, dedicated to the improvement of information systems for urban and regional planning and development. The membership has grown in both number and diversity over the past five years. The URISA membership is highly computer-literate, and generally knowledgeable about GIS technology. The majority of the URISA members work in the government sector, although the private sector membership (particularly from the hardware and software vendor community) has been rapidly increasing. The survey instrument, designed to inform the URISA board about members' views of the organization, also included some of the same questions from the earlier surveys of local planning agencies.

To make the results of this survey more comparable to the earlier state surveys, here I have selected only those cases from the government sector. Our earlier surveys were restricted to planning professionals in city, county and regional planning agencies. The URISA survey responses presented here are not restricted to planners, but all of the reported responses are from URISA members reporting government employment. Responses from federal and state governmental agencies were broken out separately here, since the acquisition and use of GIS by these levels of government differs from the local governments of our early surveys. Table 2 shows the population break down for the municipal, county and regional respondents from the URISA survey. We note the probable response bias in this survey towards larger cities and regions. It is clear from our earlier surveys that GIS adoption rates are much higher in larger places, and this is reflected in the URISA membership.

3.1 SOFTWARE USAGE

In the earlier surveys of planning agencies in 9 states, we found extremely consistent results on software usage questions (French and Wiggins, 1989; French and Wiggins, 1990; Wiggins and French, 1991). We discovered that the "Big 3" packages of word processing, spreadsheets and database managers were dominant over all other applications, with word processing achieving an overall 91 percent saturation in planning agencies with access to computers. Spreadsheets were reported to be in use by 67 percent of the agencies, and database managers by 56 percent. Reported use of other packages were much lower -

TABLE 2. URISA survey: Jurisdiction populations

Population	Municipal government respondents (%)	County government respondents (%)	Regional government respondents (%)
10,000 or less	0.03	0.01	0.00
10,001 to 50,000	0.16	0.05	0.03
50,001 to 250,000	0.41	0.33	0.22
250,000 to 500,000	0.17	0.24	0.23
500,001 or more	0.24	0.37	0.52
n =	327	220	88

business graphics (23 percent), statistical packages (18 percent), project management (14 percent), telecommunications (14 percent), GIS (13 percent) and CAD (8 percent). The overall percentages for URISA members were, as expected, much higher in all categories. This is not surprising, as URISA tends to attract professionals interested in information systems and computers.

Software usage by URISA government members is given in Table 3. Note the still clear domination of the "Big 3", but also the high percentages for GIS adoption among this group. Other reasonable statistics in this table include the high usage among regional agencies of graphics and statistical packages. Regional planning agencies tend to have higher requirements for such applications in their production of economic and population forecasts.

TABLE 3. URISA survey: Use of software

Software packages	Municipal gov. (%)	County gov. (%)	Regional gov. (%)	State/ Province gov. (%)	Federal gov. (%)
Word Processing	0.95	0.94	1.00	0.94	0.94
Spreadsheets	0.92	0.91	0.92	0.85	0.74
Database Management	0.90	0.87	0.90	0.89	0.85
GIS	0.78	0.78	0.89	0.78	0.83
CAD	0.64	0.48	0.45	0.42	0.30
Graphics	0.58	0.46	0.68	0.43	0.43
Telecommunications	0.53	0.49	0.63	0.65	0.68
Statistics	0.50	0.44	0.75	0.51	0.53
Project Management	0.48	0.44	0.39	0.49	0.53
n =	330	223	93	144	53

3.2 HARDWARE USAGE

Our earlier survey results showed a strong dominance among planning agencies of hardware consisting of standalone microcomputers. It is easy to see the striking difference among the URISA members, who have high overall adoption of GIS technology. In Table 4, we see that this group's dominant hardware configuration in 1991 was a mix of micros and larger systems. Also note the second largest percentages are for systems of networked micros. Given the horse power and storage requirements of GIS implementations, such configurations are currently appropriate and necessary.

TABLE 4. URISA survey: Use of hardware

Hardware configurations	Municipal gov. (%)	County gov. (%)	Regional gov. (%)	State/ Province gov. (%)	Federal gov. (%)
Standalone Microcomputers	0.02	0.04	0.10	0.02	0.12
Networked Microcomputers	0.36	0.31	0.43	0.38	0.33
Mini or Mainframe linked to PCs or terminals	0.13	0.20	0.09	0.08	0.16
Mix of Micros and Larger Systems	0.49	0.46	0.38	0.52	0.39
n =	309	204	89	132	49

3.3 GIS SOFTWARE USAGE

Table 5 includes the top 12 GIS packages reported in use by the government respondents to the URISA membership survey. The packages reported in use by these agencies vary from single-user desktop mapping systems, to more complex multi-user, high-level function GIS. Reviews and comparisons of most of these packages may be found in Van de Mark (1991), Wiggins and French (1988) and Levine and Landis (1988). Table 5 includes multiple responses., and the cell values are total agencies reporting use of each package. A striking picture is the dominance of Arc/Info software among this group. Recent market surveys show Arc/Info and Intergraph with the largest worldwide market shares. Intergraph dominates in the transportation and utility sectors. Overall, URISA has less utility members than the related AM/FM (Automated Mapping/ Facility Management) organization, but still, Arc/Info dominates Intergraph more strongly than expected, given results from other GIS market surveys. However, we again note that this table includes government employee responses, a group more oriented toward land-oriented planning applications, and many of

TABLE 5. URISA survey: Reported use of GIS packages

GIS Package	Municipal gov.	County gov.	Regional gov.	State/ Province gov.	Federal gov.	Total
Arc/Info	95	90	36	65	28	314
Intergraph	45	11	3	11	5	75
Geo/SQL	13	1	8	7		29
MapInfo	10	4	11	2	2	29
Atlas*GIS	9	6	7	3		25
GDS	7	9	3	1		20
Erdas		1	4	7	6	18
SPANS	1	3	3	2	9	18
TransCAD	7	2	4			13
GFIS	6	4	1			13
Autocad	4	1	3	1	2	11
Geovision	3	3	2	1	1	10

these users need both the parcel-level analysis and land suitability applications that are Arc/Info's strong suit.

3.4 APPLICATIONS OF GIS

Probably the most significant results from the URISA survey are in the areas of reported GIS applications. Since, as a group, the URISA membership provides a good snapshot of active government GIS users in the U.S., their reported activities are of great interest in studying current GIS use. Table 6 provides a look at the applications reported by 711 URISA members from five government sectors: municipal, county, regional, state/province, and federal. Notice the high response percentages of land parcel mapping applications, even at the regional and state level. We were surprised at these high rates of cadastral mapping . Although it is clear that many municipal GIS are designed around a base map consisting of parcels of land, the reported use of such parcel maps by agencies at higher levels warrants a more detailed look.

Other responses of interest include the state and federal percentages for land suitability analysis, natural resource inventory applications, and environmental impact assessment. We note that the highest reported percentage for these state and federal agencies is natural resource inventory applications. Combining this with a large number of responses on environmental impact assessment and land suitability analysis, we may infer that the use of GIS by these agencies as dominated by large area analysis, site selection and land use suitability analysis, and plan preparation.

Demographic mapping applications by municipal government are reported by many URISA members, perhaps reflecting the availability of the 1990 Census and the use of GIS in election redistricting during the period of this survey. Notice also that the largest percentage category for the regional agencies is also census mapping. This reflects the role

TABLE 6 URISA survey: Applications of GIS

GIS application	Municipal gov. (%)	County gov. (%)	Reg. gov. (%)	State/ Prov. gov. (%)	Fed. gov. (%)	Total number reporting application
Land Parcel Mapping	0.62	0.67	0.48	0.50	0.28	411
Demographic Mapping	0.68	0.66	0.40	0.27	0.12	385
General Plan Preparation	0.40	0.40	0.55	0.33	0.26	283
Census Mapping	0.35	0.45	0.67	0.28	0.19	281
Address Matching	0.44	0.43	0.53	0.20	0.14	278
Facility Management/ Utilities	0.53	0.24	0.30	0.30	0.21	261
Zoning Review	0.43	0.44	0.23	0.11	0.02	236
Event Mapping	0.37	0.27	0.23	0.24	0.16	209
Land Suitability Analysis	0.18	0.33	0.44	0.36	0.26	202
Natural Resource Inventory	0.08	0.29	0.39	0.55	0.49	196
Vacant Land Inventor	0.31	0.30	0.34	0.11	0.12	190
Monitoring Growth	0.22	0.30	0.41	0.17	0.09	177
Environmental Impact Assessment	0.22	0.30	0.41	0.17	0.40	147
Transportation Modeling	0.16	0.18	0.35	0.24	0.19	145
Capital Improvement Planning	0.29	0.17	0.16	0.12	0.09	144
Permit Tracking	0.22	0.22	0.15	0.17	0.09	140
Other	0.12	0.12	0.19	0.18	0.37	110
n =	279	187	88	114	43	711

of the regional agencies in providing population analysis and forecasts. We were surprised by the low percentages, relative to land parcel mapping, of address matching applications. Address matching applications allow GIS users to create "pin maps" for data files containing street addresses. For example, if an environmental agency has a database of underground storage tanks, including fields for the street address of each tank, an address matching application would locate (approximately) the location of each tank according to the street link on which it is located, and would place a symbol ("pin") at this location. Another example, for a historical preservation agency, would be a map displaying in symbols of various colors, the historical buildings of different architectural types in a municipality. Red dots might represent Victorians, blue dots might represent Arc Deco buildings, etc. Again, the tabular database would need to contain the type of building and its street address so it could be matched to an approximate location on a street centerline map. Since these applications are readily accessible using the added functionality now available in low-end mapping packages for desktop computers, as well as in higher-end packages, we had suspected that they might be the most frequently reported application of GIS users.

3.5 FEATURES MAPPED

We have consistently included a question on features being mapped on GIS in our survey series. The results for this question, also broken down by the five government sectors, are given in Table 7. The three highest feature categories are streets and highways (perhaps reflecting the use of TIGER files), political boundaries (perhaps reflecting the wide availability of these coverages from a variety of sources not requiring digitizing on site), and parcels (a reflection of the parcel mapping applications discussed above). We particularly note the importance of the TIGER files to many start-up, lower-end GIS operations. The TIGER (Topologically Integrated Geographic Encoding and Referencing) line files were developed jointly by the U.S. Census Bureau and the U.S. Geological survey for the 1990 Census. The TIGER files replaced earlier road centerline files, and increased the geographic coverage to include the entire nation (Bureau of the Census 1990a, 1990b). The TIGER files contain cartographic information about the road network, and a coding scheme to relate the Census data to this base. Data included for each street segment (from intersection to intersection) include street names, and address ranges for the left and right sides of the street segment (for metropolitan areas). Also included are jurisdiction, block and tract boundaries. This enabled the Census to match a particular address to its census block and tract. It is these files that allow the address matching applications to be performed, as described in the previous section. Some observers have suggested that the TIGER files may be the most important product of the 1990 Census in their role a providing a nationwide digital map (National Resource Council 1990). The roles of various levels of government in improving the accuracy of this map and its associated data, and in maintaining it over time, is currently being debated.

Other frequently mapped features include land use and zoning, natural geographic features, census tract boundaries and utility infrastructure. Reasonable differences across the five groups appear for such map features as flood plains and wetlands, where the state and federal agencies are taking a lead mapping role. Similarly, the local and county governments are more frequently mapping parcels, land use and zoning, and utility infrastructure.

TABLE 7. URISA survey: Features mapped

Features	Munici-pal gov. (%)	County gov. (%)	Region-al gov. (%)	State/ Prov.gov. (%)	Fed. gov. (%)	Total re-ported
Streets /Highways	0.79	0.84	0.76	0.65	0/67	549
Political Boundaries	0.58	0.72	0.74	0.64	0.72	466
Parcels	0.74	0.71	0.47	0.47	0.28	447
Land Use / Zoning	0.57	0.58	0.59	0.48	0.33	389
Natural Geographic Feature	0.35	0.50	0.51	0.64	0.77	343
Census Tracts	0.42	0.54	0.58	0.33	0.19	315
Utility Infrastructure	0.57	0.28	0.30	0.34	0.21	287
Flood Plains	0.25	0.40	0.36	0.65	0.70	280
Topography	0.29	0.33	0.31	0.50	0.65	253
Soils Information	0.10	0.37	0.35	0.46	0.28	192
Wetlands	0.10	0.29	0.33	0.50	0.49	190
Hazardous Waste Sites	0.12	0.13	0.16	0.25	0.28	112
Wildlife Habitats	0.03	0.10	0.14	0.42	0.35	102
Other	0.16	0.17	0.14	0.01	0.00	90
n =	279	187	88	114	43	711

4. MGIC Survey (1992)

The Massachusetts Geographic Information Committee (MGIC) is a voluntary organization of GIS users in Massachusetts state and local government, and the GIS vendor and consulting community. The MGIC provides advice to the Office of Management Information Systems of the state on GIS policy. Subcommittees include the State Mapping Advisory Committee (SMAC) which advises the United States Geological Survey on mapping priorities for the state, a subcommittee on GIS data standards, and a subcommittee formed to collect information on mapped information in the state. This last subcommittee developed a survey instrument mailed in January of 1992. Included on this survey instrument are several of the same questions discussed above. Also included are a series of detailed questions on tax assessing maps. The goals of the survey were to: (1) update the 1989 survey results on GIS adoption by local governments in Massachusetts; (2) provide a data base of existing digital coverages and their geographic extents with a goal of increased data sharing potential; and (3) to evaluate the accuracy, scales and age of tax assessment maps in the state. As noted above, the results reported here are summarized from a single mailing. To gain coverage of all cities and towns in the state in these diverse areas, the survey was mailed to planning, engineer/public works, and tax assessment professionals. This has led to an interesting analysis problem, where responses from multiple sources in the same town disagree with each other as to map availability and scales. We have made phone calls where possible to resolve conflicting information. Where various respondents still disagreed, we have removed the case.

4.1 SOFTWARE USAGE

Table 8 illustrates the software usage reported by 168 towns in Massachusetts. We note that here we included the software usage only for those towns reporting having computer capabilities. The pattern of the "Big 3" applications is clearly obvious, and we note the 10 percent reported adoption of GIS. Otherwise, the rank order of the software applications is identical to that reported in 1988, and similar to the results we found in all 9 states in the earlier surveys.

TABLE 8. MGIC survey: Software usage

Software Package Used	**Percentage of towns (with computers) reporting usage**
Word Processing	97.1
Spreadsheets	97.1
Data Base Management	75.0
Statistics	31.7
CAD	21.2
Telecommunications	15.4
GIS	13.5
Graphics	12.5
Project Management	11.6

n = 104

4.2 HARDWARE USAGE

We note, in Table 9, that 33 percent of the towns in Massachusetts are still reporting no computer usage. These are the small towns and villages characteristic of New England. Overall, the percentages of towns planners reporting no computer usage was 32 percent for the New England states in 1989. For those Massachusetts towns and cities reporting computer usage, the most common configuration continues to be standalone microcomputers, but with some noticeable additional increase compared to the 1989 responses of both a mix of systems and of networked microcomputers.

TABLE 9. MGIC survey: Use of Hardware

Computer Configuration	Number of towns	Percent of towns
No computers	55	0.33
Standalone Micros	41	0.25
Networked Micros	20	0.12
Mini or Mainframe	23	0.14
Mix of Micros and Larger Systems	18	0.11
n =	168	1.00

4.3 INTENTION QUESTION

Table 10 provides a summary of the responses on current and planned usage of permit tracking, land-parcel databases, and GIS. The percentage reporting current usage of GIS increased from 7 percent of the sample in 1989 to 13 percent in 1992. Through participation in statewide GIS activities, the author is able to personally verify the recent adoption by several towns. We note that the question about intention to adopt GIS is still showing a relatively high 40 percent. However, this is down from the 56 percent reported in 1989. We remind the reader of our earlier description of the economic recession experienced by this state during this period, and suggest this as a possible reason for the decreased optimism about GIS adoption in the near future.

TABLE 10. MGIC survey: Applications

Currently Use	Number	Percent
Permit Tracking	41	0.24
Land-parcel Database	57	0.34
GIS	16	0.10
Plan to Use (next 2-3 years)	**Number**	**Percent**
Permit Tracking	40	0.24
Land-parcel Database	43	0.26
GIS	68	0.40
n =	169	1.00

4.4 TAX ASSESSMENT MAPS

To meet a goal of increased spatial data sharing between levels of government and also between cities and towns, a more detailed look at available paper maps for parcels is required. Massachusetts is among many states in the U.S. currently evaluating ways to better share geographic information between levels of government. Such data sharing is viewed as cost effective, minimizing the duplication of multiple agencies mapping the same area and features. It is also viewed as enabling information from a variety of sources to be usefully combined for planning and decision making. For example, an environmental agency might invest considerable resources in accurately mapping the wetlands in a state. Local agencies might then use this map layer for planning analysis at the local level, combining this layer with other layers mapped by the local government, such as parcels.

Although state-of-the-art GIS systems allow the user to change projection systems, and to "edge match" and "rubber sheet" maps so they will align more adequately, there are still difficulties in combining maps of widely different scale and accuracy. If a state wishes to develop a statewide cadastral (property boundaries) map, then the varying quality of the tax assessment maps becomes a major issue. Some observers of the development of state and regional-level GIS have suggested that a cadastral layer is not necessary for many of the planning analyses conducted over a land area as large as a state. They argue that the scale and accuracy of maps of major natural and manmade features at 1:24000 are adequate for many functions. Others disagree, and suggest that since the parcel is the unit of taxation, land use control, and land use change, that a cadastral layer is central to state land use planning. If a small scale, accurate cadastral layer is available, then other layers can be registered to detailed base.

Short of starting from scratch with a consistent standard statewide, several states have considered compiling a state-wide cadastral coverage from the existing maps from their municipalities. In the colonial states like Massachusetts, the original land grants tend to be of irregular shape, as opposed to the square sections later used in the western states, and old tax maps in the eastern states are often not based on accurate ground control surveys. Often these old maps are simply not worth digitizing and correcting for a GIS. Instead, whole new photogrammetric mapping is required, with Global Positioning Systems (GPS) used to establish ground control points (Raymond 1990).

In the MGIC survey, we included two pages of detailed questions about the tax assessment maps of the towns and cities. In Table 11 we show the reported scales from 131 towns and cities in Massachusetts. We note that the scales vary all over the board, from 1:240 to 1:9600 base maps. The most common scale, by far, is the 1:1200 base map.

TABLE 11. MGIC survey: Parcel map scales

Scale	Number of towns
1:240	1
1:360	1
1:480	14
1:500	1
1:600	8
1:720	3
1:960	10
1:1200	58
1:2400	23
1:3600	1
1:4800	8
1:6000	1
1:7200	1
1:9600	1

Table 12 shows the reported coordinate systems used on the tax assessment maps. With 147 towns and cities reporting, 51 percent of the respondents either didn't know what the coordinate system was for their maps, or reported that the maps did not show a coordinate system. From the author's personal experience in this region, many of the old tax maps were completed without ground control, and although the map sheets usually have scale bars, coordinate systems are difficult to determine. Fifteen percent of the towns reported that longitude-latitude was used, and 25 percent reported a state plane coordinate system.

TABLE 12. MGIC survey: Coordinate systems

Coordinate System	Number of towns	Percent
Longitude-latitude	22	0.15
State Plane	37	0.25
Other	13	0.09
None Reported	75	0.51
n =	147	1.00

Also crucial to the concept of a state-wide cadastral layer, is the attribute (tabular) data about parcels. Table 13 shows the responses from 147 cities and towns on the current media of this tabular data. We note that 29 percent of the respondents reported having this data available only on paper, while 67 percent have the data available in digital form.

TABLE 13. MGIC survey: Tabular tax data

Media	Number of towns	Percent
Paper Only	42	0.29
Digital	14	0.10
Paper and Digital	84	0.57
Not Reported	7	0.05
n =	147	1.01

It is clear from these preliminary results that establishing a state-wide cadastral layer will be difficult at best. Existing map bases are of varying ages, use different coordinate systems (or none), and are at widely varying scales. Digitizing from existing parcel maps is probably not a feasible solution for a state-wide cadastral layer in Massachusetts. Yet the cost of new orthophotography, new digitizing, and accurate establishment of ownership boundaries from deeds is extremely high for an entire state. A possibility of establishing state standards to be followed (voluntarily) by towns and cities as they move into GIS is currently under discussion in several forums in Massachusetts. As these tax assessment maps are updated, such standards might in fact provide for a consistent cadastral layer over the long term.

5. Conclusions

Survey research is valuable for obtaining snapshots of GIS use and adoption from a large number of planning agencies over large geographic areas. It is possible to get high response rates from mailed surveys if the process is carefully conducted. Because of the media, and the necessity of short forms for high response rates, such surveys can not easily substitute for the more detailed look at case studies of planning agencies adoption, implementation, and use (Onsrud, Pinto and Azad, 1992). The software and hardware usage patterns over our 11 surveys, with respondents from all over the U.S., have shown remarkably consistent patterns. Questions about the intention to adopt GIS technology have also shown the great optimism out in the field. Whether these intentions to adopt are fulfilled in the next several years should be the subject of a continuing set of longitudinal surveys.

6. Acknowledgments

The author wishes to acknowledge her long-term collaboration with Professor Steven French, Director, Program on Urban Planning, Georgia Institute of Technology, in this work. We thank the Lincoln Institute of Land Policy (Cambridge, MA), the National Capital Planning Commission (Washington, D.C.), the Urban and Regional Planning Association (Washington , D.C.), and the Massachusetts Geographic Information Committee (Boston, MA) for their partial funding of the costs of this research. We thank the volunteers from the New England chapter of the Urban and Regional Information Systems Association for their assistance in the mailing of the URISA survey, and to Dick Cohane of Boston Edison for his assistance in the mailing of the MGIC survey. We also acknowledge the help of the following graduate students in the survey instrument design, mailing and data entry: Michael Henderson (Stanford University), Kelly Heffernon (Cal Poly, San Luis Obispo), Jagadish Pokharel (MIT), Tom Van Buren (MIT), and Peter Vaz (MIT).

7. References

The Economist (1992) The Delight of Digital Maps. *The Economist*, March 21, 69-70.

Bureau of the Census (1990a) TIGER: The Coast-to-Coast Digital Map Data Base. Washington, D.C., Bureau of the Census.

Bureau of the Census (1990b) TIGER/Line Technical Documentation. Washington, D.C., Bureau of the Census.

Budic, Z.D. (1991) GIS in Local Government. Survey work for the states of North Carolina, South Carolina, Georgia and Florida, in progress at the Department of City and Regional Planning, University of North Carolina at Chapel Hill.

Dillman, D.A. (1978) *Mail and Telephone Surveys: The Total Design Method.* John Wiley and Sons, New York.

French, S.P. and L.L. Wiggins (1989) Computer Adoption and Use in California Planning Agencies: Implications for Education. *Journal of Planning Education and Research* 8(2): 97-108.

French, S.P. and L.L. Wiggins (1990) California Planning Agency Experiences with Automated Mapping and Geographic Information Systems. *Environment and Planning B* 17(4): 441-450.

Garcia, M.W. and M. Perryman (1991) Computer Adoption and Use by Local Government Planning Agencies in Arizona and New Mexico. Paper presented at the Second International Conference on Computers in Urban Planning and Urban Management, Oxford, England, July 6-8, 1991.

Levine, J. and J. Landis (1988) Microcomputer-Based Geographic Information Systems for Planning. In A Planners Review of PC Software and Technology, edited by R.E. Klosterman, Planning Advisory Service Report No. 414/415, American Planning Association, Chicago, 45-56.

National Research Council (1990) Spatial Data Needs: The Future of the National Mapping Program. A project of the Mapping Science Committee, Board on Earth Sciences and Resources, Commission on Physical Sciences, Mathematics, and Resources, and the National Research Council, Washington, D.C., National Academy Press.

Onsrud, H.J., Pinto, J.K., and Azad, B. (1992) Case study research methods for geographic information systems. *Journal of the Urban and Regional Information Systems Association* 4(1): 32-44.

Raymond, E.W. (1990) Old Tax Maps: Are They Suitable for GIS. *Remote Sensing and Database Development News* 1:6-8.

Sinton, D.F. (1992) Reflections on 25 Years of GIS. Special Insert, *GIS World* 5(1).

Van Demark, P.H. (1991) Head-to-Head: A Comparison of Four MS-DOS Desktop Geographic Information Systems. *Journal of the Urban and Regional Information Systems Association* 3(1): 102-124.

Wiggins, L.L. and S.P. French (1988) Thematic Mapping Packages for Microcomputers. In *A Planners Review of PC Software and Technology*, edited by R.E. Klosterman, Planning Advisory Service Report No. 414/415, American Planning Association, Chicago, 23-34.

Wiggins, L.L. and S.P. French (1991) Comparative Analysis of Geographic Information Systems Use in American Planning Agencies. Paper presented at the Second International Conference on Computers in Urban Planning and Urban Management, Oxford, England, July 6-8, 1991.

CORRELATING ADOPTION FACTORS AND ADOPTER CHARACTERISTICS WITH SUCCESSFUL USE OF GEOGRAPHIC INFORMATION SYSTEMS[1]

JEFFREY K. PINTO
College of Business
University of Maine
Orono, ME 04469 U.S.A.
and
HARLAN J. ONSRUD
Department of Surveying Engineering
University of Maine

ABSTRACT. The intent of this research was to gather initial information on specific interpersonal, organizational, and institutional variables that affect an organization's decisions to acquire, implement, and use GIS. We selected United States local governments as an appropriate social system of interest for this initial investigation. One goal was to explore whether factors determined as critical to the adoption of innovations in other fields were also deemed significant by adopters in the field of GIS in U. S. local governments. Another was to examine whether factors important in the acquisition of GIS are also predictive of the success of use of GIS within an organization over time. A further goal was to investigate correlations between characteristics of users and their satisfaction with GIS. A final goal was to demonstrate the application of quantitative methods for investigating GIS adoption variables.

I. Introduction

"Diffusion" refers to the process of communicating an innovation to and among the population of potential users who might choose to adopt or reject it (Zaltman et. al. 1973). Numerous studies have addressed the adoption and diffusion of technological innovations in other fields. (Rogers 1983) However, few formal social science investigations of geographic information technologies have been reported. (Wellar 1988, Onsrud and Pinto 1991) Although investigations in the GIS domain should be pursued along several fronts, the intent of this study was to evaluate GIS adoption issues in the context of a single but broad institutional environment, namely local governments, and to focus on the relationship between acquisition and utilization issues. We began with these two areas of inquiry in order to provide a link to foundation adoption principles developed in technology transfer studies in other fields and to provide knowledge in the GIS applications domain from which other studies might extend.

The research project described by this article extends from foundation discussions set forth in Onsrud and Pinto (1991) and Onsrud and Pinto (1993). The background and overview sections which follow extract freely and extensively from those previous articles. The article then describes a research investigation of variables in GIS adoption success within

I. Masser and H. J. Onsrud (eds.), Diffusion and Use of Geographic Information Technologies, 165–194.

the context of U. S. local governments, analyzes the research results using a range of statistical methods, evaluates correlates between user characteristics and user satisfaction with GIS, and draws some conclusions.

2. Adoption Factors and their Correlations with Use Success

2.1 BACKGROUND

In this first section we explore whether factors determined as important in the acquisition of GIS are also predictive of the success of use of GIS within an organization over time. We have argued previously (Onsrud and Pinto 1991) that an understanding of the GIS diffusion process can aid in allowing those who could benefit from an innovation, such as a new technology, to begin accruing those benefits earlier. By identifying critical social factors and processes in the acquisition, implementation, and utilization of a technology, it is expected that decision making responses of individuals, groups, and organizations may be predicted and therefore also may be accommodated or redirected through prescriptive strategies. By identifying critical human and technical factors within classes of potential users, diffusion studies also have the potential for directing the design efforts of system developers to those system characteristics and improvements most valued by end users.

Under a content model approach, data are collected in an attempt to identify those variables which are significant to the adoption process for the particular class of potential users being considered. The variables we focused on in this particular study were drawn primarily from the observations and conclusions of Rogers (1983) in his review of the diffusion of innovations literature base and which were additionally indicated to us through a small number of GIS tracer studies as being potentially important to GIS users. One might argue that different subsets of past studies from the organizational theory and business management literature might have been additionally or alternatively drawn upon; such as more specifically focused implementation, use optimization, or technology assessment studies. However, we argue that the diffusion perspective is a rational perspective from which to begin. There are limits to the amount of material that feasibly may be addressed in any single research effort and the diffusion of innovations approach provides a broad but manageable research tradition upon which one might build. Factors and processes indicated in the implementation, use optimization, assessment, and GIS experience literature bases then immediately arise as additional areas of inquiry for contemporaneous or future research.

Factors that might be particularly germane to GIS adoption as opposed to other information system or innovation adoptions but were not addressed by our study should not distract from the results. For instance, one might argue or hypothesize that legislation and other regulatory mandates have a far greater effect on the acquisition and utilization of GIS by local governments than they would on other innovations or information systems. Similar arguments can be made for other potential factors. Regardless, our intent was not to test all possible factors and effects of consequence to the GIS adoption process but only a subset drawn primarily from the diffusion and technology adoption literature.

Paul Goodman argues in this volume (Goodman, 1993) that where there is a high degree of uncertainty in a technology and a high degree of uncertainty in the implementation process, it is unlikely that some combination of values of variables will predict successful adoption. He argues that one should not expect to know, a priori, what variables to manipulate in which way to effect adoption or optimal use of an innovation. We have considerable empathy for this position and the conclusions of our work lend some credence to that position. However, we also argue that no organization, such as a specific U. S. local government, stands completely unique to all other similar organizations. Similarly, we argue that GIS as an innovation has at least some characteristics which remain relatively consistent wherever implemented and used. Thus, commonalities and general principles based on those commonalities are possible and probable.

This research attempts to isolate some of those generalizable principles. However, we have also argued in the past that there is no distinct set of factors and processes which is wholly generalizable across the complete range of GIS environments. All we may ultimately strive for is a model which incorporates factors and processes which account for some maximum percentage of the total causes of success (e.g. 40% or 90%) and the remainder are idiosyncratic to the environment in which the innovation is being adopted. At some point, one must look to subjective causes on a case by case basis to improve the utilization of the GIS in an organization and to increase its value to the organization. However, the search for generalizable principles which may aid the adoption process and increase the organizational utilization of GIS has only begun. As of yet, we have little idea where the limits on finding useful generalizable principles might lie (i.e. Is it 40% or 90%?) (Onsrud and Pinto 1993). Thus, we see the research approach described below as complementing rather than conflicting with the "dynamic process of redesign" approach advocated by Goodman for optimizing use of an innovation.

2.2 RESEARCH DESCRIPTION

We initiated a two-stage study. The first stage of the research was use of "tracer" case studies as one of the steps in developing a content model of implementation. Tracer case studies consist of selecting a small sample of organizations in which a successful diffusion or acquisition of the technology occurred. As a result of these case studies and a previous literature review, a listing of critical determinants of adoption (content factors) was developed. Data and personnel from the tracer cases were also consulted at numerous other times in preparing the questionnaire and in interpreting results. Because the experiences and insights of these selected organizations are drawn upon throughout the entire research process as a means of helping to verify the validity of processes and results, they are referred to as "tracer" studies.

The second phase of the research study consisted of a larger-scale survey research format. The major GIS vendors were asked to place us in contact with local government organizations to which their GIS systems had been sold. The questionnaire sent to these organizations was constructed to elicit data adequate for in-depth statistical testing. The questions were designed to enable us to determine the relative importance of a wide variety of content factors in the successful introduction and adoption of GIS.

The research reported in the remainder of this section adheres to the same conceptual framework and research workplan as reported in Onsrud and Pinto (1993) However, the sample analyzed is different. As such, the research results reported here should be compared and contrasted with that previous work. In addition, to avoid repetition, please refer to that article for a discussion of background concepts and a detailed description of the workplan. A summary of the research plan stages is contained in Table 1, extracted from that article.

Table 1. Research stages

Step	Activity	Reason
1) Getting Started	Definition of research question	Focuses efforts
2) Select Unit of Analysis	Specify population	Sharpens external validity
3) Identify Variables	Possible a priori constructs	Provides theoretical grounding
	Literature and construct research	Preliminary instrument development Identification of content variable list
4) Refine Variables	Select tracer cases and conduct series of interviews	Refines research constructs Tests for instrument flow, readability, and logic
5) Develop Survey Instrument	Refine questionnaire	Based on case interviews, modify questions and/or research constructs
6) Conduct Final Interviews	Engage in "round two" of interviews with GIS managers	Final check for revised questionnaire to ensure clarity and completeness of constructs
7) Prepare Mail Survey	Identify relevant sample population	Determine scope of research study and appropriate target population
	Generate random sample	Based on population determine appropriate size of representative sample
8) Mail survey	Three wave mailings to identified target sample	Follow accepted social science mail survey methodology
10) Analyze & Interpret Data	Analysis to establish linear (variance) relationship	Determine predictors of successful diffusion (content model)
	Frequency analysis and structural equation modeling	Determine process model of successful diffusion

2.3 RESULTS AND STATISTICAL ANALYSIS

To the best of our knowledge, a total of 506 questionnaires were mailed out by vendors to their local government clients. A total of 256 usable responses were returned, for a response rate of over 50%. Because the questionnaire was fifteen pages long and we lacked the ability to phone or otherwise directly contact potential respondents, we felt the return rate was very good. T-tests of the mean responses for each of the research variables for early mailing returns versus late mailing returns showed no response bias. We initially requested that vendors send the questionnaire to their U. S. local government clients. However, some sent the form to at least some of their foreign clients. Therefore, for the analysis which follows, we used only the 208 responses from U. S. local governments. The distribution of respondent organizations by State is shown in Table 2. The distribution by jurisdiction type along with median population, median area, and median budget is shown in Table 3.

Table 2. Responses by State

State	Response Sample	State	Response Sample
Alabama	6	New York	11
Alaska	9	North Carolina	20
Arizona	8	Ohio	1
California	30	Oregon	6
Colorado	4	Pennsylvania	5
Connecticut	2	South Carolina	2
Florida	26	Tennessee	4
Georgia	3	Texas	8
Hawaii	2	Utah	4
Illinois	4	Vermont	2
Iowa	1	Virginia	4
Kansas	4	Washington	11
Kentucky	1	Wisconsin	3
Maryland	3		
Massachusetts	3		
Michigan	2		
Minnesota	6		
Nebraska	2		
Nevada	5		
New Jersey	1		
New Mexico	5		

The questionnaire required participants to think of the installation and use of their GIS within their operations. The acquisition, implementation and use of the system was to be their frame of reference while completing the questionnaire. Participants were presented with a set of factors that could be related to the adoption of GIS in their organization and were asked to what degree each of these issues helped or hindered the adoption process.

All of the respondents to the survey were "successful acquirers" of GIS, although not necessarily "successful users" of GIS. Of the 34 factors indicated in the literature and

Table 3. Characteristics of sample and responses

Universe: All U.S. local government organizations who had actually acquired GIS capabilities from GIS vendors prior to the time of the study (Fall 1990)

Sample: Entire Universe (Sales of GIS capabilities to U.S. local governments by other than the top five vendors were determined to be negligible)

Responses:

1. n = 208

2. Responses by Type of Organization

Organization Type	Number	Percentage
Town	4	1.9
City	60	28.8
Municipality	19	5.1
County	96	46.2
Multi-district	6	2.9
Other	22	10.6
Missing	1	.5

3. Population of Area Covered by Organization

   Mean = 738,480
   S.D. = 1,443,424
   Range = 400 to 10,000,000
   Median = 240,000

4. Area in Square Miles Covered by the Organization

   Mean = 6,321
   S.D. = 39,224
   Range = 1 to 516,000
   Median = 520

5. Approximate Annual Budget of the Organization

   Mean = 254,211,594
   S.D. = 1,040,756,927
   Range = 400 to 1,100,000,000
   Median = 3,700,000

through our tracer studies as possibly important to the adoption process, twelve of the factors were rated on an average by these successful acquirers as being highly important to their adoption decisions (i.e. average rating above 3.66) as shown in Table 4. For all factors, responses were received on both the "not important" and "important" sides of the neutral scale of 3.0.

Table 4. Factors included in the questionnaire

1. Ability to talk with previous adopters
2. Accessibility to GIS vendors
3. Accessibility to informal educational opportunities
4. Availability of learning materials
5. Availability of formal educational opportunities
6. Visibility of uses and benefits*
7. Availability of existing data*
8. Ease of transferring data*
9. Data accuracy*
10. Ease of generating results*
11. Extent of existing computer usage skills
12. Availability of skilled GIS people
13. Compatibility of GIS with existing computer system
14. Ease of making pilot study
15. Cost of hardware and software*
16. Cost of data entry
17. Cost to retrain staff
18. Close proximity to other GIS users
19. Advantages of GIS over existing processes*
20. Ability to serve several units of the organization*
21. Ability to serve those outside organization
22. Adapting GIS to meet your specific needs*
23. Ability to revert in case of GIS failure
24. Availability of informal methods to cover costs
25. Existence of informal communication channels
26. Existence of formal communication channels
27. Adverse ramifications to advocates if GIS fails
28. Adverse ramifications to organization if GIS fails
29. Benefits to advocates if GIS is a success
30. GIS consistency with corporate goals*
31. Ability to expand types of uses*
32. History of successful computer systems in organization
33. History of unsuccessful computer systems in organization
34. Importance of GIS champion*

* Factors rated (on average) by respondents in the upper third of the scale in importance to their adoption decisions (i.e. above 3.66 where 1 = not important and 5 = very important)

Two major statistical procedures were employed in the first segment of the study. First, factor analysis was used to condense the compiled set of 34 content factors down to more general and meaningful dimensions. Second, stepwise regression was employed to determine the efficacy of these content factors in predicting GIS implementation success.

2.3.1 *Factor Analysis.* Factor analysis is a generic name given to a class of multivariate statistical methods whose purpose is data reduction and summarization. In general, it deals with the problem of analyzing the interrelationships among a larger set of variables, (e.g., questionnaire scale items) and then explaining these variables in terms of their common underlying dimensions (factors) (Norusis, 1985). The goal of factor analysis is to find a way of condensing the information contained in a large number of original scale variables into a smaller set of dimensions (factors) with a minimum loss of information. In other words, the technique allows researchers to search for and define the fundamental constructs assumed to underlie the original variables (Hair, Anderson and Tatham, 1987).

In effect, factor analysis enables us to "clean up" the data by simplifying the amount of information within the scales. In the present example, the large number of originally determined content factors (34) hypothesized a priori to influence GIS diffusion, made it first appropriate to determine whether some of this wide variety of factors were conceptually and statistically similar in that they were assessing the same general dimension or issues. To illustrate, it is likely that the factors relative to 1) cost of hardware and software, 2) cost of data entry, and 3) cost to retrain staff may, in fact, be sub-dimensions of a more general factor related to implementation cost. In such circumstances, it would be more appropriate (and far simpler) to test for the overall effects of implementation cost on adoption success.

As a result of the factor analysis, we were able to take the original questionnaire variables, made up of a total of 34 items, and condense them into a set of ten independent factor groupings. Table 5 shows the results of the factor analysis on the set of independent variables. Ten separate and statistically significant factors emerged (i.e., each had Eigen values greater than one). Each of the groupings was named according to characteristics common to all the factors or the dominant characteristic within the grouping. The ten factor solution accounted for 61% of the total variance. In order of loading, these factors have been labeled 1) Access to Learning, 2) Ease of Use, 3) Effects of Use, 4) Cost, 5) Utility, 6) Benefits to Extended Users, 7) Communication Channels, 8) History of Success and Failure, 9) Fallback Options, and 10) Proximity to Other Users.

A second factor analysis was performed on the dependent measures of utilization success (e.g., also "adoption success" since by our definition utilization success achieves adoption success). The measurement of adopters' perceptions of success or failure of innovations has been a major focus of recent implementation problem studies (e.g. Ives, Olsen and Baroudi 1983, Brancheau and Wetherbe 1988, Rivard and Huff 1988, Davis 1989, Galleta and Lederer 1989, Tan and Lo 1990, Igbaria and Nachman 1990). In our study, individuals directly involved with the GIS answered a small number of straight forward questions on their perceptions of the usefulness and value of the GIS to the organization. Such questions have been the predominate method of gauging user and organization satisfaction with technological innovations in past studies. That is, measures of "user satisfaction" were used as surrogates for direct measures of use success or failure. Although used extensively in past studies, use of this surrogate as a measure of success may not be construed as a comprehensive measure (See, for instance, the discussion by Goodman in this volume.)

Table 6 shows the results of a factor analysis on the aggregate success measure. Previous literature and pre-test interviews have suggested that GIS utilization success should be measured through consideration of issues related to both use of the GIS and perceived value of the GIS (Gillespie 1991, Didier 1990, Onsrud, Calkins and Obermeyer 1990). The results offer statistical support for a two-factor solution, confirming the need to assess utilization success in terms of both system use and perceived value.

Table 5. Factor loadings table for independent variables
(N = 208)

	1	2	3	4	5	6	7	8	9	10
Access to Learning										
Accessible to venders	.58									
Informal educ. opps.	.81									
Informal lrning mtrls.	.77									
Formal educ. opps.	.72									
Visibility of benes.	.34									.31
Ease of Use										
Availablty of existing data		.65								
Ease of data transfer		.75								
Avail. skilled people		.39						.31		
GIS compat. w/ system.		.53								
Effects of Use										
Adverse ram. to indivs.			.81							
Adverse ram. to org.			.81							
Benefits to indivs.			.79							
Cost										
Ease of pilot study				.42						
Cost of Hardware and Software				.82						-.50
Cost of data entry				.84	.31					
Cost to retrain staff				.71						
Utility										
Data accuracy					.39					-.42
Advant. over curr. sys.					.44					
Consist. w/ org. goals					.61				.30	
Potent. to expand uses					.58					-.32
Benefits to Extended Users										
Useful to mult. units						.66	.33			
Useful to customers			.40			.73				
Adapt GIS for org needs		.38				.61				
Existance of champion						.48				
Communication Channels										
Formal commun. channels							.75			
Informal comm. channels							.72			
History of Past Successes and Failures										
Existing computer skills								.70		
History of successes								.67		
History of failures								.74		
Fallback Options										
Fallback if GIS fails									.53	
Informal cost covering									.78	
Proximity to Other Users										
Prox. w/ other users										.74
Eigen Values	5.5	2.4	2.2	2.1	1.8	1.5	1.4	1.2	1.2	1.1
% Var. Explained	16.2	7.1	6.6	6.2	5.4	4.6	4.3	3.6	3.4	3.3

Total Var Explained = 61.0%

Table 6. Factor loadings table for dependent measures of adoption success (N = 208)

	Loadings 1	Loadings 2
Use		
GIS accepted and used	.81	
GIS performs as intended	.75	
GIS made our job easier	.59	.43
GIS allows us to do more	.50	.44
Satisfied w/ GIS intro.	.78	
Perceived Value		
Value of GIS in daily ops.	.48	.63
Value in middle dec-mking		.78
Value in upper dec-mking		.83
Value in providing pub. access		.73
Eigen Values	4.7	1.2
% Var. Explained	52.1	11.3
Total var. explained = 63.4%		

2.3.2 *Regression Analysis.* Following the use of factor analysis to condense the total number of independent variables into ten more meaningful dimensions, a stepwise multiple regression analysis was performed on the data set. Multiple regression analysis is a statistical technique that allows us to analyze the relationship between a dependent variable (utilization success) and one or more independent (predictor) variables. The objective of multiple regression is to use the several independent variables whose values are known to predict the single dependent value that we wish to determine (Hair, Anderson and Tatham, 1987).

Stepwise multiple regression automatically calculates the independent variables that individually offers the most predictive power and "loads" this variable first into the linear regression equation. This technique is repeated until it is determined that all significant predictor variables have been included in the regression equation. As a result, stepwise regression allows the researcher to determine the minimal set of independent variables that have the most predictive power.

The correlation matrix of all research variables is given in Table 7. Table 8 shows the results of the first set of stepwise regression analyses on the data set (N=208). Three separate regression analyses were performed, using different dependent measures of utilization success: (1) level of system use, (2) perceived value of the GIS, and (3) an aggregate success measure incorporating both use and value measures.

The results indicate that when an *aggregate measure of success* is employed, the stepwise analysis identified four variables that were statistically significant: Benefits to Extended Users, Ease of Use, History of Success and Failure, and Cost (i.e., highly utilized systems tended to rank the importance of these variables high and less used systems tended to rank the importance of these variables low). These four variables had a cumulative coefficient of determination value (r^2) of .28. In other words, these four variables were able to account for approximately 28% of the total causes of aggregate system success.

Table 7. Correlation matrix for adoption data

Construct	1	2	3	4	5	6	7	8	9	10	11
1. Access to Lrning											
2. Ease of Use	.31										
3. Cost	.27	.36									
4. Bnfts to Extd Usrs	.13	.17	.19								
5. Utility	.14	.35	.48	.20							
6. Effects of Use	.07	.23	.21	.13	.44						
7. Com Channels	.23	.27	.19	.30	.31	.22					
8. Fallback Options	.13	.41	.25	.26	.27	.15	.23				
9. Hstry of S & F	.18	.25	.28	.29	.20	.18	.17	.21			
11. Proximity	.16	.13	.15	.19	.09	.09	.25	.11	.14		
12. System Use	.15	.42	.16	.25	.20	.14	.20	.11	.08	-.02	
13. System Value	.16	.24	.15	.29	.14	.15	.19	.08	.15	.03	.52

All coefficients above .18 have $p < .01$

All coefficients above .14 have $p < .05$

Reliabilities (Cronbach alpha) on diagonal

Using *perceived value* of the GIS as the dependent variable yielded five predictors: Benefits to Extended Users, History of Success and Failure, Utility, Costs, and Communication Channels, with a cumulative r^2 of .31. Finally, *system use* had a significant linear predictive relationship with Benefits to Extended Users, and Utility with a cumulative r^2 of .14.

A second set of regression analyses were performed on the data set following an assumption on the part of the researchers. The original regression was run on the total data set which included several respondents who had only recently purchased and installed a GIS. There is, however, a strong possibility that recent buyers may not yet be in the position to adequately assess the value of the system or the level of usage within their organizations. As a result, the second set of regression runs were based on a reduced data set in which all participants who had installed a GIS within the last 12 months were eliminated. This resulted in a sample size of 151.

The results of this second set of stepwise regression runs is shown in Table 9. These findings show strong parallels to those determined through the full data set. For the *aggregate measure of success*, Benefits to Extended Users, Utility, Costs, and History of Success and Failure were significant predictors, with a cumulative r^2 of .29. Using *perceived value* of the system as the dependent variable yielded five predictors, Benefits to Extended Users, History of Success and Failure, Utility, Cost, and Communication Channels, with a cumulative r^2 of .28. Finally, *system use* had a significant linear predictive relationship with only Benefits to Extended Users, for a cumulative r^2 of .11.

Table 8. Stepwise regressions : total U.S. local government sample (N = 208)

(1) Dependent Variable - System Use

Variable Loading By Step	adj. r^2	adj. cum. r^2	Beta	T-Value	Equation F
1) Benfts to Extended Users	.09	.09	.48	3.64	13.25***
2) Utility	.05	.14	.23	1.78	8.06***

(2) Dependent Variable - Perceived Value

Variable Loading By Step	adj. r^2	adj. cum. r^2	Beta	T-Value	Equation F
1) Benfts to Extended Users	.17	.17	.68	4.71	22.16***
2) Hisory of Success & Fail	.08	.25	.36	3.93	16.03***
3) Utility	.04	.29	.35	2.29	12.08***
4) Costs	.02	.31	-.19	-2.34	10.71***
5) Communication Channels	.02	.33	.11	1.70	9.16***

(3) Dependent Variable - Aggregate Measure of Success

Variable Loading By Step	adj. r^2	adj. cum. r^2	Beta	T-Value	Equation F
1) Benfts to Extended Users	.15	.15	.78	4.65	21.61***
2) Ease of Use	.05	.20	.45	2.23	13.56***
3) History of Success & Fail	.04	.24	.16	2.08	10.86***
4) Costs	.04	.28	-.16	-2.11	8.79***

Note: * = $p < .05$
** = $p < .01$
*** = $p < .001$

2.4 DISCUSSION

This study produced a number of significant findings. One finding concerns the empirical results attained from testing the effects of the various content factors on the two outcome variables: *system use* and *perceived value* of the GIS. Using the survey response sample from U. S. local government jurisdictions (See Tables 8 & 9), of the possible ten content factors that were originally identified through the factor analysis, only six were found to be predictive of some form of adoption success. The most important factor across all three dependent measures was the one labeled "Benefits to Extended Users." This factor refers to the outreach advantages of the new system, including usefulness to multiple units within the organization, the usefulness for customers and citizens as opposed to usefulness for internal

Table 9. Stepwise regressions: total sample - excluding recent (last 12 months) adopters (N = 151)

(1) Dependent Variable - System Use

Variable Loading By Step	**r^2**	**adj. cum. r^2**	**adj. Beta**	**T-Value**	**Equation F**
1)Benfts to Extended Users	.11	.11	.49	3.81	12.31***

(2) Dependent Variable - Perceived Value

Variable Loading By Step	**r^2**	**adj. cum. r^2**	**adj. Beta**	**T-Value**	**Equation F**
1) Benfts to Extended Users	.14	.14	.68	4.73	22.39***
2) Hisory of Success & Fail	.08	.22	.15	3.08	15.23***
3) Utility	.03	.25	.32	2.26	11.28***
4) Cost	.03	.28	-.17	-2.30	10.04***
5) Communication Channels	.02	.30	.11	1.78	8.67***

(3) Dependent Variable - Aggregate Measure of Success

Variable Loading By Step	**r^2**	**adj. cum. r^2**	**adj. Beta**	**T-Value**	**Equation F**
1) Benfts to Extended Users	.14	.14	.73	4.68	20.98***
2) Utility	.08	.22	.37	2.29	12.61***
3) Costs	.04	.26	-.13	-1.74	9.28***
4) Hisory of Success & Fail	.03	.27	.08	1.72	7.99***

Note: * = p < .05
** = p < .01
*** = p < .001

tasks, the ability of the GIS to be adapted for the organization's needs, and the existance of a system "champion." The fact that the "champion factor" grouped with three other outreach-type factors through the factor analysis process evidences confirmation of past studies that conclude that "system champions" are far more important to selling new innovations throughout an organization and to the community as a whole than they are to the technical implementation of a system.

The appearance of "Benefits to Extended Users" as a significant and dominant predictor of success (or system "satisfaction") for U. S. local government users contrasts with the results we obtained when foreign local government responses were also included in the sample (See Onsrud and Pinto 1993). In the larger sample, the most important factor across all three dependent measures was the one labeled "Utility." This factor is made up of a combination of internal and external advantages of the new system, including advantages provided by the GIS over current processes, data accuracy adequate for the organizations needs, consistency with organizational goals, and the ability to expand the types of uses in the future. Thus, "Benefits to Extended Users" were of greater importance and perhaps considered more important indicators of success in the U. S. sample than they were when foreign responses were included.

Perhaps the difference can be explained by cultural factors. However, we suspect the primary reason for the difference is that the foreign responses in the previous sample tended to be from institutionally stable bureaucratic operations. These agencies tended to be better able to implement GIS with their own internal resources and therefore were more focused on the utilitarian tasks of improving internal tasks through the use of GIS and less on meeting the needs of those external to the immediate organization. A similar conclusion is suggested from the results of the aggregate measure of success for the U. S. sample that had more than one year of experience with their GIS operations (i.e. See Table 9). For these more mature or experienced systems, the ability to meet utilitarian organizational needs as well as provide benefits to extended users are the two most important factors which arose in predicting success in use of the GIS.

Because of the strong showing of both "Benefits to Extended Users" and "Utility" in predicting success of use, further inquiry into the meaning of both variables is probably warranted. One approach would be to investigate what other factors might be contained within the bounds of those factors listed under each. For instance, under "Utility" one might inquire whether the importance accruing to "consistency with organizational goals" is due primarily to GIS being able to respond to government mandates and reporting requirements as opposed to being able to respond those goals which are optional but only preferred on the part of government management. Another approach would be to hypothesize and explore additional aspects of the "Utility" and "Benefits to Extended Users" variables. A yet further approach would be to investigate the potential interdependencies, counter relationships, and commonalities among these two variables. Perhaps this should be done with a focus on isolating the specific user characteristics which give rise to the importance of one over the other in achieving use success.

Under "Benefits to Extended Users" one might further investigate the importance of the factor "existance of a champion" in GIS acquisition, implementation, and use environments. A great deal of research has examined the importance of champions and their positive effect on gaining organizational acceptance of innovations (Chakrabarti, 1974; Maidique, 1980; Meredith, 1986). Likewise, from our informal discussions with GIS practitioners and the "Tracer" case studies, the importance of having an internal champion to aid in system acquisition appeared to be profound. However, when foreign responses were included in the sample the factor Benefits to Extended Users was found to be non-predictive of any of the three forms of the dependent measure (i.e., see Onsrud and Pinto 1993).

In speculating as to the cause for this seeming discrepancy between qualitative interviews and the quantitative analysis on the importance of the champion, it is interesting to consider the distinction that is being made between "acquisition" of a GIS and "adoption" by the organization of that system. When asked to indicate the importance of an in-house champion for making the decision to acquire GIS technology, the mean score of the responses was 4.26 on a five-point scale where five is labeled "very important." At the same time our results using the entire data set demonstrated no direct effect of champions on either system use or perceived value of the GIS. One conclusion that seems to arise from this analysis is that champions are very important for influencing their organizations to acquire a system but tend to have less impact on getting the organization to fully use (i.e. successfully adopt) the new technology. An alternative conclusion, is that champions have an effect on the ultimate use and value of a system but this becomes an apparent predictor of success only when a focus is placed on subgroups of users and the effect of champions on those subgroups. For instance, "benefits to extended users" which included the "champion" factor arose as a significant predictor of adoption success when we included only U.S. respondents in the sample. This result may be due to the generally smaller and less institutionalized U.S. local government installations having greater need than the Canadian, European, and Australian respondent installations for individual champions or having a greater need to reach out to users beyond their own agencies. The disparities resulting from varying the sample are worthy of future study.

On another interesting note, an additional independent variable that was significant across both aggregate measures and across both perceived value measures, as shown in Tables 8 and 9, was "History of Past Success and Failure." This variable referred to the organization's existing computer skills and to its history of other, previously successful and unsuccessful computer systems experiences. Those respondents with higher success measures considered previous successful and unsuccessful computer systems experiences in the organization to be of importance in their adoption decisions whereas those with lower success measures considered such past experiences to be of low importance in their adoption decision. This finding suggests that organizational concern with existing computer skills, past successess, and past failures helps ensure success and is actually predictive of success.

Another important finding of the research regards cost considerations. In previous work by others it has been found that "costs", presuming that slack in organizational resources is available, have often been found to be non-critical in predicting adoption success. However, that past research has often treated adoption success and acquisition success as being synonymous with each other. By defining "adoption success" to include "use success" as a prerequisite and critical element, cost considerations become a noticeably present factor and predictor of success. However, they do so in a negative sense. As illustrated by the negative values in Tables 8 and 9, considerations of cost in the acquisition of a GIS are ranked as relatively unimportant by those perceiving significant use and value in their operational systems. The reverse is true for those perceiving low success in their operational systems. These findings support the results of adoption research in other fields discussed previously which suggest that cost is typically not a critical factor in the actual decision to acquire a new technology. Our findings suggest further that lack of pressing concerns about costs in the acquisition of GIS is actually a significant predictor of the perceived success of the eventual operational GIS. This makes sense from the perspective that an organization with significant "slack" in its monetary resources is more likely to be able to provide adequate support and maintenance for its GIS over time. However, alternative hypotheses might just as reasonably explain the predictive phenomena observed and they suggest an area of research inquiry.

Another issue to be considered concerns the results uncovered during the stepwise regression analysis. While several of the independent variables were found to be significant predictors using various measures of adoption success, the total adjusted r-square (coefficient of determination) values were somewhat low, although not unusually low for the type of analysis pursued. As Table 8 shows, when *perceived value* was employed as the dependent variable, five factors were significant predictors: including, benefits to extended users, history of success and failure, utility, costs, and communication channels. However, the total adjusted r-square value for the five independent variables was .31, suggesting that approximately 69% of the causes of adoption success were still unidentified. This was not an unexpected result for a first attempt at statistically verifying the causes of adoption success. Although a total of 34 different hypothesized predictors were uncovered and tested, it may be that the additional causes of adoption success within local governments may be identified in other than the diffusion of innovations literature base. Further literature research is indicated in order to determine other potential predictors of successful geographic system adoption. More intensive and direct forms of observation, such as by case methods, may also be appropriate for exploring for these additional predictors. The importance of any such factors exposed would then necessarily need to be evidenced as verified/falsified through follow-up quantitative research methods.

It has been suggested that a more likely cause of system adoption success may have to do with internal organization and team dynamics, culture, and inter- and intra-organizational politics than with many of the better-known factors as tested in this study. The organization's specific external environment, GIS user characteristics (e.g., clerical, professional, or managerial), and GIS builder characteristics (e.g., computer science, cartographer, or MIS professional) all offer obstacles and/or contingencies that must be taken into account in tailoring an adoption process for each government organization (Azad, 1991). Kraemer et. al. (1989) argue that "management" rather than "environment" is the primary determiner of success in implementation and use of technologies in organizations. If their hypothesized nine stage model of information management is shown valid through further empirical research, evaluating adoption factors and processes by management stage groupings might shed greater insights on more-successful versus less-successful users of GIS. At a minimum, numerous additional contingency factors and several alternative research frameworks need to be tested to determine how they may moderate the direct relationship between a set of independent variables and system adoption success.

As a result, this study should be viewed, as our results indicate, as a call for additional work. That work should incorporate formulation and testing of new hypotheses, in attempts to determine a more comprehensive set of predictors in the adoption of geographic information technologies within local governments.

3. Correlations between User Characteristics and GIS User Satisfaction

In this segment of the study we compared selected characteristics of individuals using GIS, the organizations where housed, and the uses of GIS with the varying degrees of satisfaction expressed by the users of the systems. As mentioned previously, respondents were asked a series of questions on their perceived use of GIS within their organization and another series on the perceived value of the GIS to the organization. Additionally, respondents were asked a series of background questions on themselves and their organizations. Because the primary focus of our survey work was directed towards isolating adoption factors, this meant that eliciting characteristics of GIS users and organizations was a secondary goal. As a result, the scope of our inquiries on user and organization characteristics were rather narrow in order to retain a reasonable length survey instrument.

The purpose of these backgound questions was to isolate subclasses of responses to determine whether there were significant differences in satisfaction with the use and value of GIS due to personal or organizational characteristics, i.e. engineer versus planner respondents, county versus municipal jurisdiction, large versus small jurisdiction, etc. We did not attempt to question respondents on their reasons for their degree of satisfaction or dissatisfaction. Nor was it our intent to hypothesize potential reasons for their level of satisfaction or dissatisfaction. Our objective was simply to identify the existance of disparitys in satisfaction or patterns of dissatisfaction. Should such be exposed, the results might lead to later investigations into potential causes. (e.g. Hypothesis 1: Users at the operations level are dissatisfied because tasks and issues to which the GIS is being applied are incongruent with the concerns of managers having control over the system (Kraemer, et. al. 1989), Hypothesis 2: Users are generally dissatisfied because the technology is not yet mature enough for effective use by this subgroup of users attempting to use it, Etc.)

For each background question we totaled responses in each of three categories. The three categories were determined by tri-section of the response scale. Those with an aggregate success factor less than or equal to 2.33 were designated as *dissatisfied users*, those with an aggregate success factor between 2.33 and 3.66 were designated as *marginally-satisfied users*, and those with an aggregate success factor greater than or equal to 3.66 were designated as *satisfied users*.

In our evaluation we used only that subset of respondents who had acquired the technology for greater than one year. As stated earlier, the rationale for using this subset argues that new adopters may not be in the position to provide as accurate and meaningful an assessment of the level of use and perceived value of the system as would a more experienced respondent.

For the following figures we offer little or no explanation as to the results obtained, only that the results exist. However, several of the results beg potential explanations and those potential explanations may prove fruitful for further investigative work.

Figures 1 through 4 provide background on the organization or jurisdiction using the GIS and on the person who completed the quesionnaire. All of the figures are easily misread. The degree of satisfaction with the GIS expressed in each figure is that indicated by the respondent answering the questionnaire. For instance, in Figure 1, one should not read the second set of columns in the figure to mean that 20% of cities expressed dissatisfaction, 37% of cities expressed marginal satisfaction, and 40% expressed satisfaction with the use and value of their GIS. The people answering the survey questionnaire were typically individuals involved in operating or supervising the current GIS operations or were instrumental in acquiring the GIS capability. One may not reliably assume that the responses of these individuals are representative of members of the organizational unit as a whole. Many of the respondents had a vested interest in thinking about their GIS operations in as positive a light as possible. We attempted to lessen this bias in the responses by ensuring anonymity for the respondents and by asking several questions requiring itemization of the performance of their GIS before asking questions on the respondent's satisfaction with the use and value of the system. Regardless, we suspect some positive bias in the expressions of satisfaction by the respondents. Thus, for the second set of columns in Figure 1, one should read it to mean that of the respondents, all of whom are affiliated with the operation or acquisition of a city GIS, 20% expressed general dissatisfaction, 37% expressed marginal satisfaction, and 40% expressed satisfaction with the use and value of GIS to their city. As the last set of columns in Figure 1 shows, the respondents as a whole were generally satisfied or marginally satisfied with the use and value of the GIS to their jurisdictions.

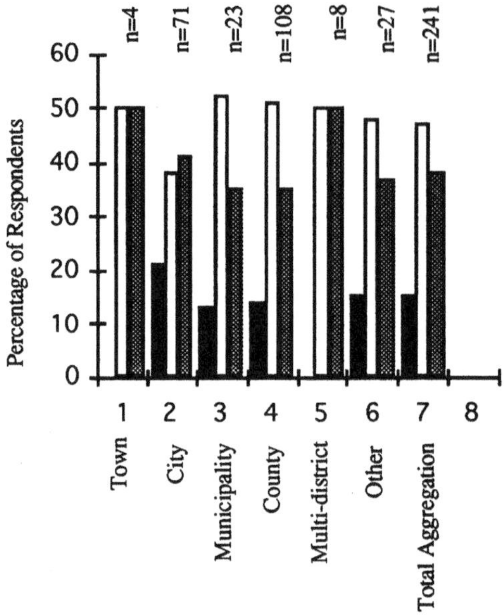

Legend
Satisfied with use and value (upper trisector of scale)
Marginally-satisfied with use and value (middle trisector of scale)
Dissatisfied with use and value (lower trisector of scale)

Figure 1. Local Government and Public Organizations: Aggregate Measures of Satisfaction with the Use and Value of GIS

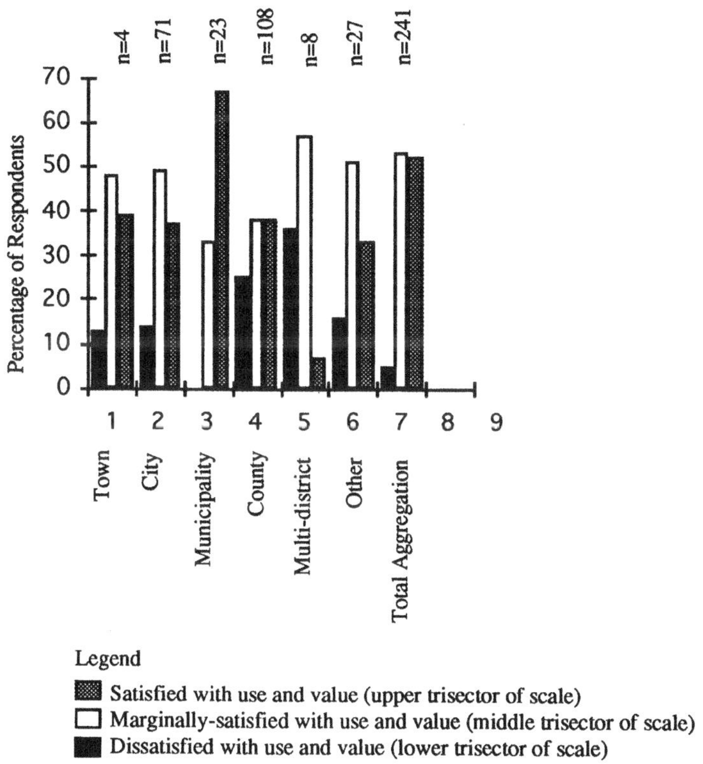

Figure 2. Primary Activities of Responding Individuals: Aggregate Measures of Satisfaction with the Use and Value of GIS

Figure 2 shows the responses in satisfaction relative to the primary job activities of the responding individuals. For the larger n values there again appears to be general satisfaction with GIS. It might be noted however that there were fewer dissatisfied administrator, planner, public works, and computer processing respondents than dissatisfied engineer and assessor respondents. This may of course be a simple reflection of the maturity of the technology and its ease of applicability to those tasks of greatest concern to the individual categories of respondents.

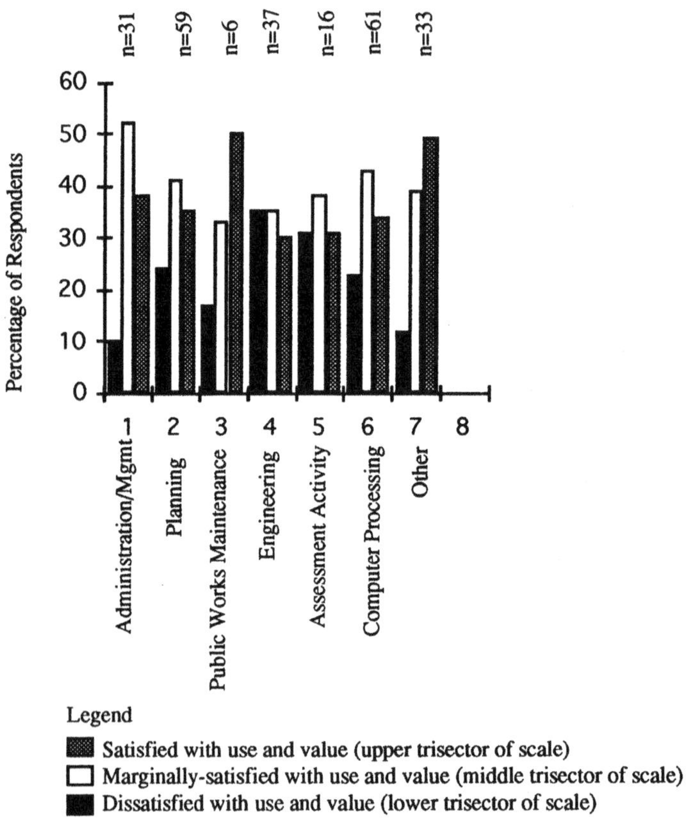

Figure 3. Functional area in which responsibility for the GIS resides aggregate measures of satisfaction with the use and value of GIS.

Figure 3 shows the degree of satisfaction relative to the functional area in which responsibility for the GIS resides. From our sample, we discern no distinct pattern of satisfaction from having GIS operations centered functionally in one administrative unit over other units.

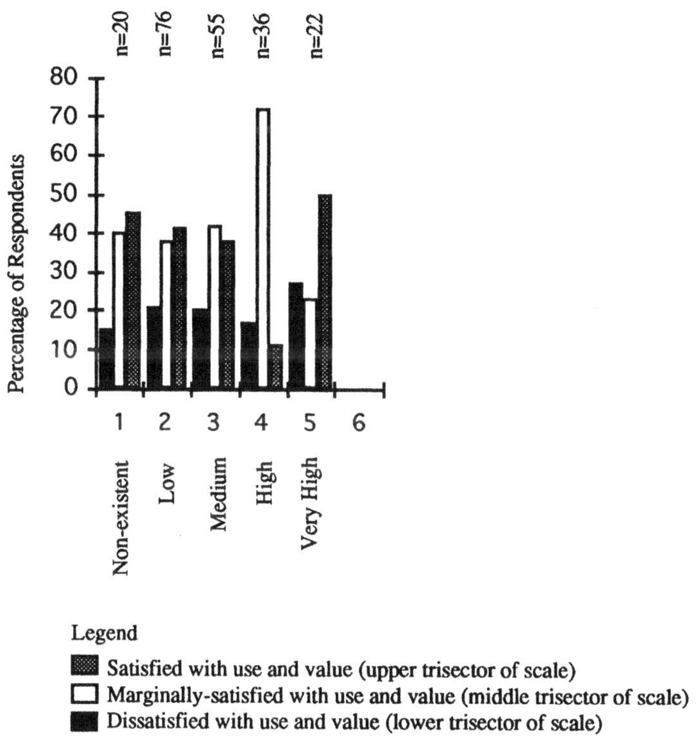

Figure 4. Computer usage skills of staff before organization considered adoption of GIS.

Figure 4 shows the degree of satisfaction relative to the computer usage skills of staff before the organization considered adoption of a GIS. It is interesting to note that the highest level of dissatisfaction with GIS (i.e. 27%) was expressed by those respondents who possessed very high computer skills prior to their organiation's acquisition of GIS (See the columns in group 5). Yet, the highest level of satisfaction with GIS (i.e. 50%) resulted from this same group of respondents.

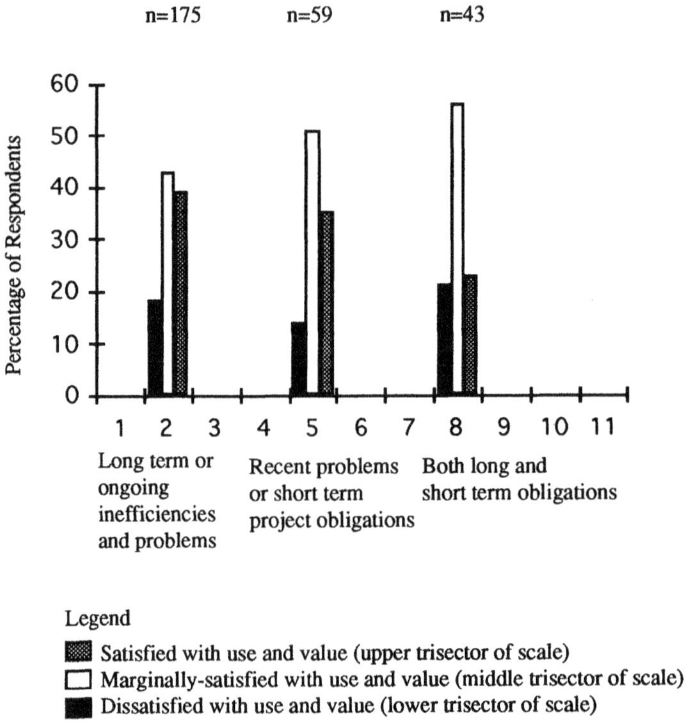

Figure 5. Issues which encouraged organization to acquire GIS.

Figure 5 is interesting in that most respondents indicated that their organization acquired GIS capabilities in order to deal with long-term or ongoing inefficiencies and problems. However, a significant number of respondents indicated that their organization acquired GIS capabilities primarily in order to respond to some short term need or project obligation. Only 15% of respondents indicated that their organization had both long term and short term needs in mind when they initially acquired the capability.

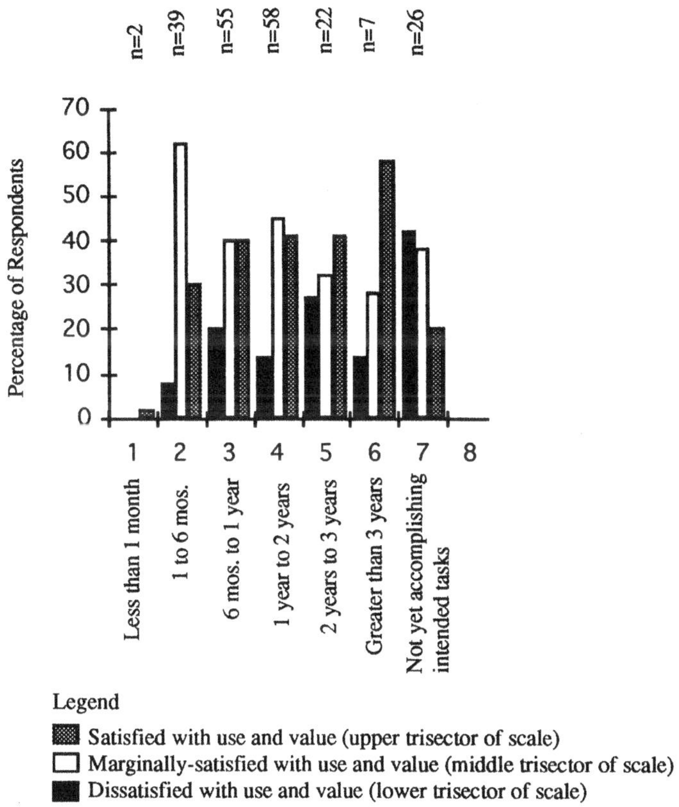

Figure 6. Length of time which transpired from the date of purchase to the time the GIS started to accomplish at least one or more intended tasks.

Figure 6 indicates the length of time it took for the local government jurisdictions of the respondents to be able to accomplish at least one of the tasks intended for the GIS. The majority met this level of performance within two years of acquisition but it is interesting to note that even those who took greater than three years to perform their first intended task still express general satisfaction with their GIS. As expected, those respondents from jurisdictions where the GIS has yet to perform any of the originally intended tasks (i.e. whether acquired one month previously or four years previously) expressed the highest degree of dissatisfaction with their GIS.

Figures 7 and 8 are of greatest interest when compared with each other. Figure 7 shows the tasks and activities the GIS were intended to address when acquired and Figure 8 shows the tasks and activities the GIS were actually addressing at the time of the survey. Figure 7 includes the responses from those who have not yet been able to use the GIS for any intended purpose and therefore a larger percentage of dissatisfied respondents is reflected in each category as compared to Figure 8. In Figure 8 the organization has been able to

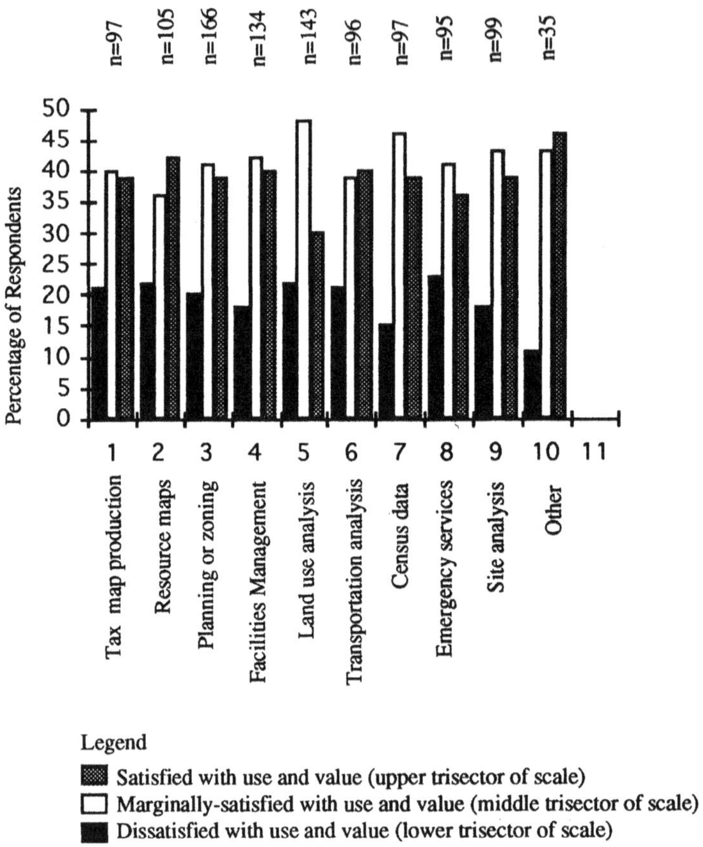

Figure 7. Tasks and activities the GIS was intended to address.

actually use the GIS for the task or activity indicated and the percentage of dissatisfied respondents is much lower. From that figure, it is interesting to note the wide range of tasks to which GIS is being applied and the general satisfaction with the GIS-method of obtaining results as compared, presumably, to the previous methods used.

Planning and zoning (n=166), land use analysis (n=143) and facilities management (n=134) are the leading tasks the GIS were originally intended to address and also the leading tasks actually accomplished; although a lesser percentage have yet been successful in utilizing their GIS for facilities management purposes. Comparisons between individual groups are also interesting. For instance, although only 43% (42/96) of those intending to use their GIS for transportation analysis were yet able to accomplish that activity, those actually accomplishing the activity are represented by a high degree of satisfied respondents. A similar result is obtained in observing the site analysis category.

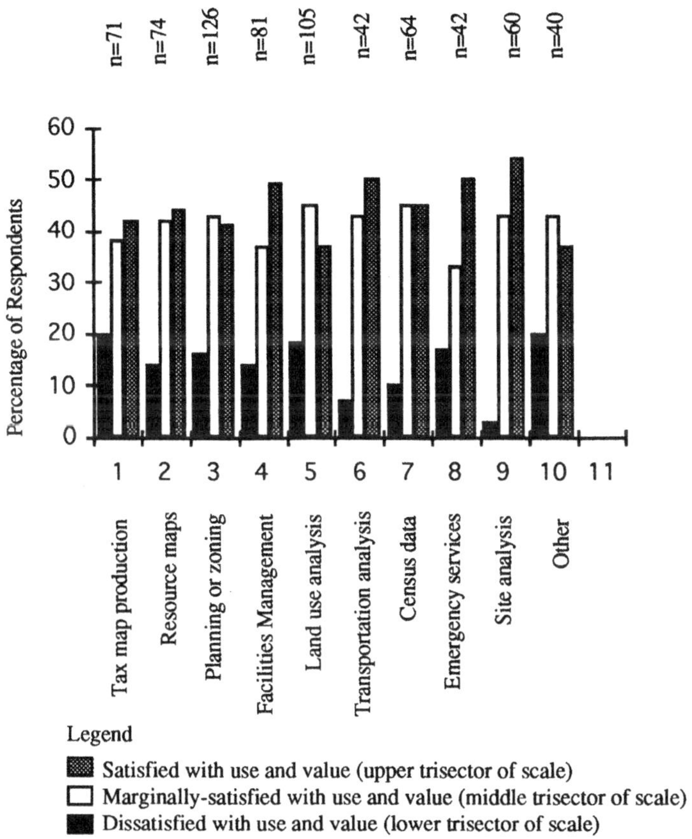

Figure 8. Tasks and activities the GIS is currently addressing.

Figures 9 and 10 give general indications of the range and extent of internal and external users of the local government GIS included in the survey. As expected, those indicating satisfaction with their systems are indicated by the figures as providing the greatest range of GIS products and services to other internal and external users. By comparing the n values in the two figures, the relative ratio of internal to external uses is evident.

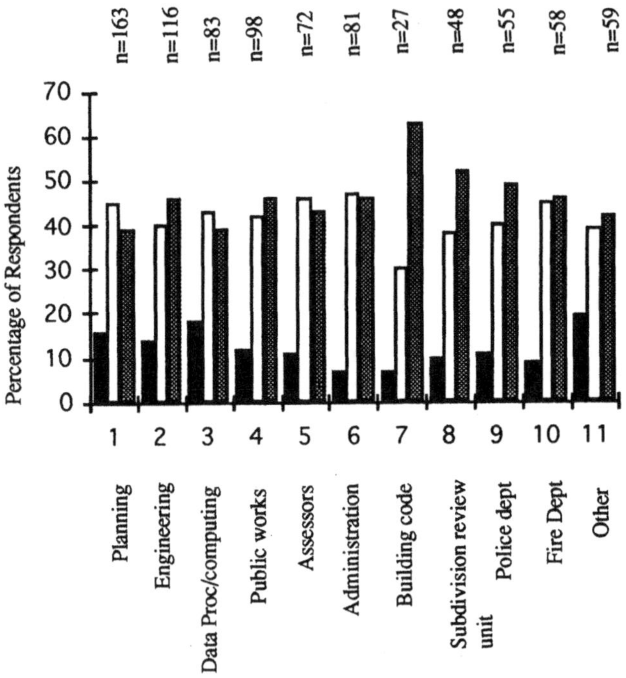

Legend

Satisfied with use and value (upper trisector of scale)

Marginally-satisfied with use and value (middle trisector of scale)

Dissatisfied with use and value (lower trisector of scale)

Figure 9. Organizational units which have utilized the GIS or its products

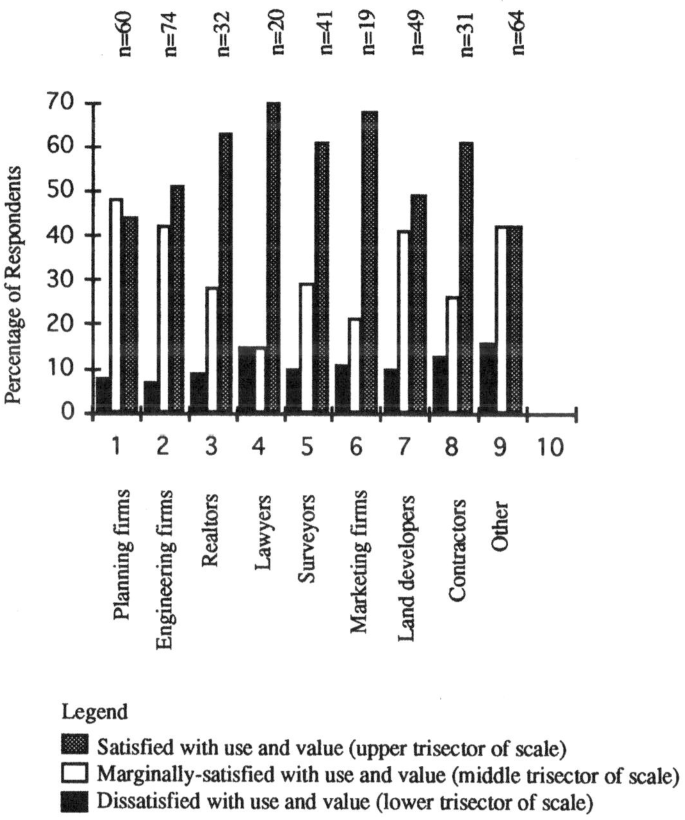

Figure 10. External organizations and individuals that have made use of the GIS or its products.

4. Conclusions

One of the most challenging tasks for researchers and developers of geographic information systems lies in gaining a better understanding of the processes by which an organization adopts and diffuses these geographic information innovations. Although technology transfer principles have been successfully applied within a broad context to a number of fields, the extension of such principles toward determining those critical activities and processes involved in successfully implementing geographic information innovations is yet underdeveloped. The purpose of this study was to take an incremental step toward gaining a better understanding of critical factors necessary to gain successful adoption of geographic information innovations within local governments.

In earlier work we gathered and evaluated initial information on specific interpersonal, organizational, and institutional variables affecting an organization's decisions to acquire, implement, and use GIS (Onsrud and Pinto 1993). In the present investigation we considered a sub-sample of the previous data (i.e. responses from only U. S. local governments) and compared the results with our earlier analysis. In addition, we extended the work through comparison of background characteristics of respondents and their local government organizations with the degrees of use and satisfaction of their GIS.

While this research may have shed some new light on the area of GIS diffusion, much additional work needs to be done. In efforts to elaborate on initial findings, we and other authors in this volume have offered directions for future research in the field. It is our hope that this specific study will spur interest on the part of other academics and practitioners to engage in additional large-scale, quantitative research programs aimed at gaining an increased understanding of the problems and approaches in successfully diffusing geographic information innovations.

5. Endnote

1. This article is based on work partially supported by the National Science Foundation under Grant No. SES-88-10917. Any opinions, findings and conclusions or recommendations expressed in the material are those of the authors and do not necessarily reflect the views of the National Science Foundation.

6. Acknowledgments

The authors wish to acknowledge all the vendors who cooperated with this research by mailing an initial questionnaire and two follow-up letters to all their local government clients. We also thank Gary Jeffress for working with the vendor organizations and aiding in the preparation of the survey questionnaire. Further, we wish to thank all those individuals working with GIS operations in local government who took their valuable time to provide us with the information for this work. Finally, we thank URISA for granting permission to include in this derivative article extensive verbatim excerpts from Onsrud, H.J. and J.K. Pinto, Evaluating Correlates of GIS Adoption Success and the Decision Process of GIS Acquisition, (1993) *Journal of the Urban and Regional Information Systems Association.*

7. References

Azad, B. (1993) Implementation of Geographic Information Systems, Ph.D. Proposal, Department of Urban Studies and Planning, MIT, forthcoming.

Brancheau, J.C. and J.C. Wetherbe (1990) The adoption of spreadsheet software: testing innovation diffusion theory in the context of end-user computing. *Information Systems Research* 1(2): 115-143.

Chakrabarti, A.K. (1974) The role of champion in product innovation. *California Management Review* 17(2): 58-62.

Davis, F.D. (1989) Perceived usefulness, perceived ease of use, and user acceptance of information technology. *MIS Quarterly* 13(3): 319-339.

Dilman, D.A. (1978) *Mail and Telephone Surveys: The Total Design Method* (New York: Wiley-Interscience).

Didier, M. (1990) *Utilite et Valeur de L'Information Geographique* (Paris: Economica).

Galletta, D.F. and A.L. Lederer (1989) Some cautions on the measurement of user information satisfaction. *Decision Sciences* 20(3): 419-436.

Gillespie, S. R. (1991) Measuring the benefits of GIS use. *Technical Papers of the ACSM-ASPRS Fall Convention*, Atalanta, Georgia (Bethesda, Md: ASPRS and ACSM), 84-94.

Goodman, P. (1993) Implementation of new information technology. In Masser, I., and H.J. Onsrud, editors, *Diffusion and Use of Geographic Information Technologies*. Kluwer Academic Publishers, Dordrecht, The Netherlands.

Hair J.F., Jr., R.E. Anderson and R.L.Tatham (1987) *Multivariate Data Analysis, Second Edition* (New York: MacMillan).

Igbaria, M. and S.A. Nachman (1990) Correlates of user satisfaction with end user computing: an exploratory study. *Information and Management* 19: 73-82.

Ives, B., M.H. Olson and J.J. Baroudi (1983) The measurement of user information satisfaction. *Communications of the ACM* 26: 785-793.

Kraemer, K.L., ed., (1991) *The Information Systems Research Challenge: Survey Research Methods* Volume 3 (Boston: Harvard Business School).

Kraemer, K.L., J.L. King, D.E. Dunkle, J.P. Lane (1989) *Managing Information Systems: Change and Control in Organizational Computing* (San Francisco: Jossey-Bass Inc.).

Maidique, M.A. (1980) Entrepreneurs, champions and technological innovation, *Sloan Management Review* 22: 59-76.

Meredith, J.A. (1986) Strategic Planning for Factory Automation by the Championing Process. *IEEE Transactions on Engineering Management* EM-33(4): 229-232.

Norusis, M.J. (1985) *SPSSx Advanced Statistics Guide* (New York: McGraw Hill).

Onsrud, H.J, and J.K. Pinto (in press) Evaluating Correlates of GIS Adoption Success and the Decision Process of GIS Acquisition. *Journal of the Urban and Regional Information Systems Association.*

Onsrud, H.J. and J.K. Pinto (1991) Diffusion of geographic information innovations. *International Journal of Geographical Information Systems* 5(4): 447-467.

Onsrud, H. J., H. W. Calkins, and N. J. Obermeyer (editors), (1989) *Use and Value of Geographic Information: Initiative Four Specialist Meeting Report and Proceedings*, National Center for Geographic Information and Analysis, Technical Paper 89-7, University of California, Santa Barbara.

Onsrud, H.J., Pinto, J.K., and Azad, B. (1992) Case study research methods for geographic information systems. *Journal of the Urban and Regional Information Systems Association* 4(1): 32-44.

Pinto, J.K. and Slevin, D.P. (1988) Project success: definitions and measurement techniques. *Project Management Journal* 19(2): 67-72.

Rivard, S. and S.L. Huff (1988) Factors of success for end-user computing. *Communications of the ACM* 31:552-561.

Rogers, E.M. (1983) *Diffusion of Innovations, Third Edition* (New York: The Free Press).

Rogers, E.M. (1993) The Diffusion of Innovations Model. In Masser, I., and H.J. Onsrud, editors, *Diffusion and Use of Geographic Information Technologies*. Kluwer Academic Publishers, Dordrecht, The Netherlands.

Tan, B.W. and T.W. Lo (1990) Validation of a user satisfaction instrument for office automation success. *Information and Mangement* 18: 203-208.

Wellar, B. (1988) Institutional maxims and conditions for needs-sensitive information systems and services in local governments. *Proceedings of the 26th Annual Conference of the Urban and Regional Information Systems Association* 4: 371-378.

Zaltman, G., R. Duncan, and J. Holbeck (1973) *Innovations and Organizations* (New York: John Wiley & Sons).

EMBODIED GIS - A CONCEPT FOR GIS DIFFUSION

PETER R. ZWART
Centre for Spatial Information Studies
University of Tasmania
GPO Box 252C
Hobart, Tasmania 7001
Australia

ABSTRACT. Our most widely used processes are those that provide background services for the completion of some desired task. Such enabling technologies tend to go largely unrecognised and therefore all but disappear from view; they become embodied in the task. This chapter proposes some conditions for GIS to attain a similar status. These are compared with prevailing GIS products. Suggestions on how GIS may change into a ready-to-hand embodied technology are made.

1. Introduction

To be accepted, an innovation needs to adapt to the requirements and situations of its potential adopters. That is, successful diffusion implies as a precondition that the users of an innovation have a major say in its form, its functions and the environment in which it resides. If the innovation is responsive to its users, then, as some of our more profound technologies (like the wheel, or writing) have demonstrated, the innovation may be so highly successful as to become integral to its user community. Under such conditions it will tend to disappear - it will no longer be recognised as a discrete technology, but rather become part of the everyday life of its users. These diffused technologies lose their separate identities and requirements for specific recognition. They become background, enabling technologies, unnoticed, and embodied in other tasks and processes.

It is suggested that this too should be the objective of the GIS community - to embed GIS into everyday life until it is indistinguishable from it - until people fail to recognise it as a separate technology. This chapter is an initial exploration of this notion; it outlines some of the conditions necessary to achieve embodiment, and compares this with where we are today. The supporting empirical data are derived from two projects at the Centre for Spatial Information Studies at the University of Tasmania. The first is an annual survey of the sales characteristics by major GIS and desktop mapping vendors in Australia (Figure 1). The second dataset is derived from GIS workspace logs, at the individual command level from three organisations collected as part of a GIS network study reported in Zwart and Coleman [1992]. The data in Figures 2 and 3 is from one of these organisations, the Tasmanian Forestry Commission, a near mature ARC/INFO GIS for forest planning, operations and management.

I. Masser and H. J. Onsrud (eds.), Diffusion and Use of Geographic Information Technologies, 195–204.

2. Embodied GIS

Whenever a technology "disappears" it is fundamentally as a consequence of human psychology and learning rather than as some particular trait of the technology itself. Essentially, if we learn something sufficiently well, we cease to be aware of it, we no longer consciously invoke the technology; it becomes a background, enabling technology. The German philosopher, Heidegger, calls this "ready to hand technology", in a sense transparent "... in my ordinary dealings ... I 'see through it' to the work that has to be completed" [Guignon 1983]. A ready to hand entity becomes a means of doing something within a hierarchy of predetermined goals and purpose. Thus, if our goal is to communicate the ideas contained in this book to people other than the authors, the ready to hand technology of books has to be available. Importantly, few of us understand or have even thought about how this book has been produced and manufactured. We merely wish to use the technology to communicate our ideas. For our higher purpose of communicating, the technology of the book has disappeared from our thoughts, it has become embodied in the process of communicating by the written word.

Embodied GIS is an analogous concept. Our aims should be to diffuse GIS to such an extent that it is incorporated into, and subsumed by, the particular task at hand. While a GIS has to be present to complete the task, the fact that a GIS is being used should be a totally transparent, unrecognised and unconscious event. What is acknowledged at all times, however, are the produce of the GIS; what disappears are the means, the GIS technology of its production.

There appear to be a number of conditions which need to be satisfied to make GIS such an enabling technology. Some of these are evident from the work of Weiser [1991] and his colleagues at Xerox in their attempt to bring about what they term ubiquitous computing, "... the idea of integrating computers seamlessly into the world at large". To achieve this, they believe a number of conditions need to be satisfied.

1. Computers will need to be invisible, in fact as well as in metaphor. They will therefore need to be small, so that they can be embedded in other devices. As well they will need to be available in a variety of sizes and forms to suit a range of individual tasks from the complex to the simple. Contemporary examples are microprocessors in watches, microwave ovens, and light switches.

2. Each component part will be dedicated to execute a specific or limited, well defined range of operations common to everyday life, as well as more esoteric tasks.

3. Unlike today's computing devices, components of a ubiquitous computing system will not be "owned" by any one individual or remain fixed in any one location. The system components "are intended to be 'scrap computers' (analogous to scrap paper) that can be grabbed and used anywhere; they have no individual identity or importance". Computers will therefore need to know where they are, where in a wired or wireless network they are located at any given time, in order for them to recognise the user and the user them, as both must be allowed to move around freely.

4. The real power of a concept does not come from any one type of device, but from the interaction between all of them.

The vision of ubiquitous computing is diametrically opposed to the concept of virtual reality. The latter attempts to simulate the world in a computer - the computer becomes the focus, whereas ubiquitous computing attempts to invisibly enhance the real world. Some of Weiser's colleagues therefore use the term "embodied virtuality" to distinguish it from virtual reality indicating that ubiquitous computing is designed to "bring the 'virtuality' of computer readable data - all the different ways in which they can be altered, processed and analysed" out into the physical world, "out of their electronic shells".

Weiser believes that ubiquitous computing will gradually emerge as the dominant mode of computer access over the next 20 years. Given Xerox's track record of pioneering major computer developments (like the Xerox Star - the precursor to graphic user interfaces and object oriented constructs) this prediction should be taken seriously. There therefore seems to be a reasonable probability that at least significant elements of ubiquitous computing concept will be available as a means of fulfilling the parallel notion of embodied GIS.

The work on ubiquitous computing provides some empirical ideas on how the concept of embodied GIS could be implemented. It suggests that, *inter alia*, the conditions for embodied GIS are:

1. In the majority of cases the operation, technology and products of the GIS will provide a background service only. To do this it will need to be totally subjugated to, and subsumed by, the task or process to which it is coupled.

2. There will be a number of kinds of GIS differing in size, form and function, with most dedicated to performing only a limited but well defined set of operations on limited and defined types of data.

3. They will be ubiquitous, having multifarious users, none of whom has a proprietary or usage right to, or necessarily understands the operations of a particular embodied GIS device.

4. They will be affordable and not noticeably different in cost to other support role technologies.

Undoubtedly these initial conditions for embodied GIS will prove to be neither sufficient nor complete, or appear to be particularly profound. But like the embodied virtuality concept, no new revolutions in either hardware or software are required, nor will its realisation produce anything fundamentally new. Rather it will help to make GIS fit into the human environment, make it an unconscious part of every day life instead of attempting to mould humans to it. Some possibilities on how this may eventuate will be the subject of the following discussion.

3. Present Day GIS

From an embodied GIS point of view, the present state of GIS technology may be characterised as follows. Firstly, as the 1991 GIS World software survey [GIS World 1991] indicates, the number of systems on offer continues to expand, reflecting an increasing and more diverse market for GIS products. A close examination of the types of software on offer in Australia supports the trend evident in other countries (Figure 1) towards an increasing market share, at least numerically, by systems with less than full spatial analysis capabilities - the so-called desktop mapping systems. In general these systems employ simple data structures, limited and more intuitive functionality leading to simpler command structures and user interfaces plus lower system costs.

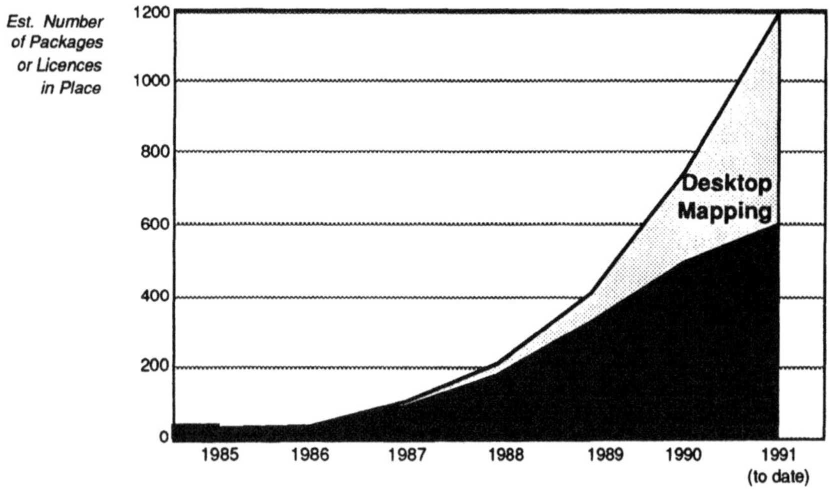

Figure 1. GIS and Desktop Mapping In Australia 1985-1991
(Estimates Based on Figures Supplied by Selected Software Vendors in Australia)

Secondly, there is an emerging trend towards systems being directed to particular market/user segments based on functionality and use, e.g. TRANSCAD for transport planning, TACTICIAN for retail marketing. Whereas conventional GIS software does, or at least strives towards providing functions to satisfy most conceivable query, mapping and analytical needs - to be all things to all men - many of the desktop mapping systems are now designed to target specific niche markets. These products, and desktop mapping systems in general, are one of the main contributors to GIS gaining acceptance as a tool with application well beyond the "traditional" land based professions in the wider community of general economic, social and community planning, management and operations. As Dangermond [1991] notes, and the sale of desktop mapping systems supports, the growth of these commercial GIS applications "in the decade ahead will dwarf the developments of the past".

Closely related to the market success of the desktop systems is the growing stock of digital data on a wide range of topics available on conditions that makes their use comparatively attractive and simple. This not only applies in the socio-economic arena, but also in the land based agencies where many systems are now reaching maturity in terms of data and application development.

The availability of near complete data sets is in turn changing the way most GIS systems are used. Not only will end users (as opposed to systems developers or system managers) begin to dominate (Figure 2), the functions the system needs to perform will also be inclined to change, and are in fact becoming more selective and restrictive. To illustrate this point, Figure 3 represents three months averages of ARC/INFO commands invoked by a system manager and a systems user. They show that not only are the commands that each group uses different, but so are their variety and range. End users tend to use fewer commands, mainly for textual queries and display with little or no use of spatial operators. Full details of these user characteristics may be found in Coleman and Zwart [1992]. These user characteristics, together with the functionality built into most desktop systems and products like ARCVIEW or GENABROWSE, support the notion of offering GIS with limited options, designed to fulfil specific tasks for a given purpose.

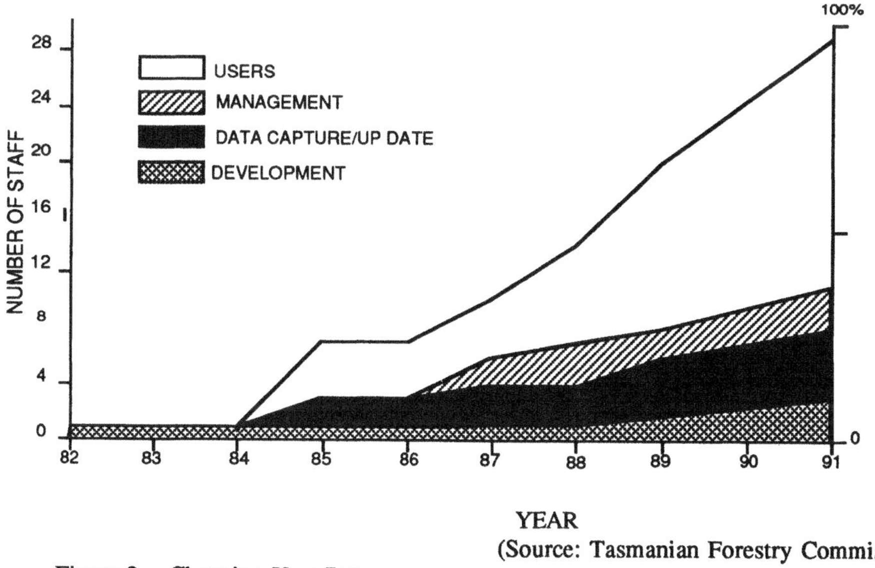

(Source: Tasmanian Forestry Commission)

Figure 2. Changing User Patterns

In summary, GIS is moving away from its initial base in the landed sciences, where location is the major determinant, to a less cohesive, heterogeneous range of applications where location may be just another factor for consideration. With an increasing consolidation of data bases and their contents becoming more readily available, the use to which data are put and the number of people using them may be expected to grow. There is evidence, however, to suggest that these users may be satisfied with relatively simple tools to undertake comparatively well-defined tasks on straightforward and simple systems.

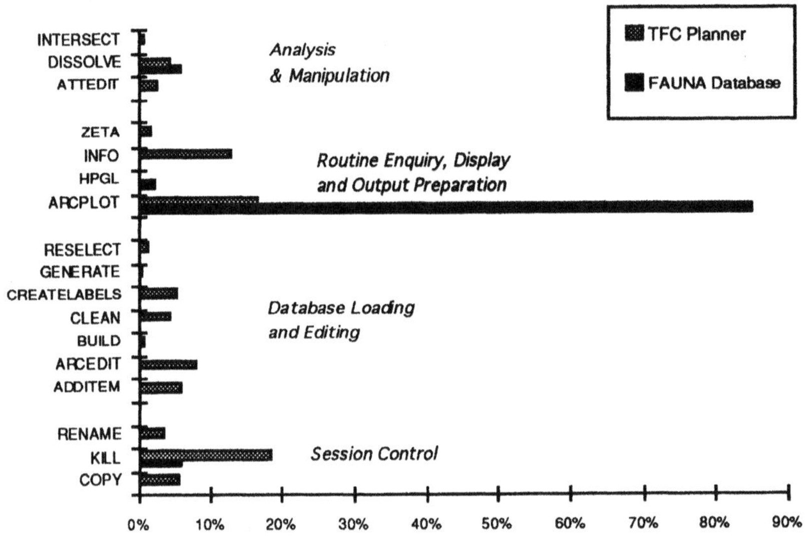

Figure 3 Comparative Summary of ARC/INFO GIS Usage
TFC Forest Planner and Routine Database Enquiries

4. Towards Embodied GIS

If we compare current conditions with those proposed to make GIS an everyday, ready-to-hand technology, it appears that some underlying conceptual, if not physical changes in the way we perceive, implement and use GIS may be required. The key change is the transformation of GIS, within the bounds of a particular application, into a subservient background technology as opposed to the foreground stand-apart technological profile so commonly adopted today. Some ways of facilitating this change are proposed below.

4.1 BACKGROUND GIS

To date the GIS community has taken great pains to differentiate GIS from the rest of information technology by emphasising its unique ability to store and manipulate data in the spatial domain. It is this difference that is stressed by GIS cognoscente when promoting the advantages of their technology and its potential applications. Their deliberate emphasis on GIS spatial prowess creates the perception that GIS is a distinctly different technology; sets it apart from other information systems, heightens its exotic nature and thereby reduces its everyday acceptance.

It is true that GIS is a different information technology - but not so different. What is required to make GIS an everyday tool is to emphasise its similarity with other data base and information technologies, to highlight its commonality; to indicate how it may fit seamlessly and transparently with existing information systems and applications. The extra functionality that GIS offers may then be described as an addition to, not as something separate from, present day computing.

The required shift in emphasis called for is more cultural than physical; a change that would also bring the GIS community to more fully integrate with the overall information science community, thereby helping to break down the artificial, intellectual and disciplinary barriers that are unnecessarily growing up and restricting the development of GIS in many areas. Systems like SIRO-DBMS [Abel 1989], and experimental systems which provide spatial extension to standard data base models and operations, as well as the increasing interest being shown in spatial data bases by computer scientists [e.g. Guenther and Buchman 1990] are positive signs in this direction.

A more significant change, however, will need to take place at the application level. In future, as the analysis of user profiles in Coleman and Zwart [1992] indicates, it is likely that - proportionally - less and less *ad hoc* use of GIS systems will take place - their use will tend to become a mainstream, rather than an exotic activity, close-linked, and an integral part of the routine business fabric. In this environment GIS will then begin to operate within the confines of systems and processes external to itself, be they vehicle despatch systems, forest planning or urban infrastructure maintenance programmes. It will be these activities which determine what is produced and how it is produced; the GIS will merely become the component part that handles and processes the spatial phenomena. Under these conditions GIS functionality and operations would all but disappear, tend to lose their separate identity and requirement for specific recognition. This in turn implies a change of creed, a change in the belief by hoist organisations and the GIS community in general that GIS is a stand-alone and different technology to one that explicitly acknowledges its role as a complementary but subjugate tool, embodied with in some larger system. There is evidence to suggest that this is already occurring in organisations with near mature GIS [Zwart 1992].

At the operational level, bringing about a reduction in GIS as a distinct technology may require, amongst others, geography to be conceived and implemented as an attribute of thematic data rather than the theme being an attribute of its spatial entity. Most spatially intelligent GIS packages today automatically reflect changes in the spatial domain with corresponding changes in the textual domain. Emphasis on text as opposed to space implies that changes in the textual domain should be automatically reflected in the spatial domain - quite a different approach and one for which few theories are available. Given, however, that in an embodied GIS environment such operations would be tightly constrained and defined by their subsuming technology and its goals, operators of this kind appear to be - at least *prima facie* - possible. An acceptance of this notion would imply considerable change or extensions to most current software.

There are a number of reasons for proposing such a 'text before space' concept, the principal one, as discussed earlier and illustrated by Figures 1 and 3, being that geography in general and spatial analysis based primarily on adjacency and connectivity in particular are not, paradoxically, items of major concern to many GIS users. Future generations may require more extensive spatial operators, but until they have achieved an everyday standing, their incorporation as an embodied tool is unlikely to reap many benefits. In the meantime, geographical operations, of even the simplest type, tend not to be familiar or everyday skills. Current GIS software requires users to identify spatial entities such as areas and arcs and select the spatial operators they require to complete the task at hand. If GIS is to become a ready-to-hand technology, then such spatial operations need to become transparent and background tasks triggered largely unknowingly by the user through an operation with which he is familiar. In short, what is proposed is that users be unaware of the geographic operations occurring in the background, but the results of which they may display on request.

4.2 DEDICATED GIS

One of the suggested conditions for converting GIS to a ready-to-hand technology, is its availability in a variety of guises, dedicated to performing specific tasks on specified types of data. A number of present day GIS' are already moving in this direction as noted above.

With GIS data sets maturing, and experience in their use accumulating, it will be possible to define more and more operational and management tasks. What was an *ad hoc* operation may well become routine as the organisation and users adjust to daily dealing with GIS. Consequently, it will become practical to tightly define, and separately programme, the GIS operations associated with an increasing number of everyday tasks. Embodying specific GIS operations within a range of routine tasks therefore seems desirable and distinctly possible.

To effect this, however, simplified and reduced GIS instruction sets, and in particular straightforward system interfaces, are needed. Reduced GIS functionality like those employed in desktop mapping systems with extensive use of window technology are only part of the way to achieving readiness-to-hand status. We need to reach a point similar to that illustrated by Winograd and Flores [1986, p164], where, like while driving a car you do not think about how far to turn the steering wheel to round a corner.

> "In fact, you are not even aware (unless something intrudes) of using a steering wheel. Phenomenologically, you are driving down the road, not operating controls. The long evolution of the design of automobiles has lead to this readiness-to-hand."

Similarly a successful embedded GIS device should also allow a user to operate on spatial data displayed on the screen, or from within a coupled task, without explicitly being aware of formulating or issuing a GIS command. The users should remain in the flow of their work and not be disrupted by the spatial or mapping process unless it is the work on which they are engaged.

Creating systems of this kind is a general problem, not one unique to GIS, and already is being addressed by groups like Weiser's [1991]. Reducing geographical operations to the level of unconsciousness is, however, very much a task for the GIS community, one which perhaps may necessitate a different approach. For example, since tasks and functions in embodied GIS will need to be definable, it may be fruitful to investigate how systems functionality could be improved through alternatives like incorporating spatial intelligence to the way the data is organised and structured [Driessen and Zwart 1989].

Secondly, even though there are signs that GIS vendors are starting to develop more specialised products for specific applications, there are few signs to indicate a move away from utilising general purpose CPU's processing software encoded GIS functionals. GIS specific hardware, or firmware to perform embodied GIS functions and processes are still a long way from realisation. But, as the diffusion of GIS to date has resulted as much, if not more, from the supply of appropriate software and data rather than one of demand [Brown 1981], the shortfall of hardware and software for embodied GIS technology is perhaps not surprising. With more digital data becoming available, the pressure to exploit this resource may lead through demand side pressures to more specialised systems. Given the vitality and number of players in the GIS software market, it seems unlikely that opportunities of this kind will long remain unrecognised.

4.3 UNRESTRICTED GIS

Assuming GIS does become an integral, unobserved background means of achieving some pre-determined objective, one further condition to achieve every day familiarity seems necessary; within the context of the task, and as far as the user is concerned, it must become a free commodity or service. When we buy a computer we buy it complete. The purchaser is not concerned about where the components came from, how they are licensed, or the conditions applying to resale. Unless we are interested in computer manufacturing, all the technical details of the system and their workings are of little concern.

If GIS is to become part of some larger system, a similar approach should be adopted. Hence, in embodied GIS, GIS vendors will supply components to be coupled with others to make up some new system. The GIS component parts (in either soft, firm or hardware form) may consist in some instances of just spatial functions and operators, or in other cases bundled products comprising both function and data. Where data is widely used, such as census boundaries and census attributes, the bundled approach, together with simple I/O and system interfaces, would seem an appropriate method to embed census spatial operators into business and marketing decision systems, for example. In other instances, complete systems as we know them today may be incorporated as a component of some larger system, such as environmental monitoring decision support systems.

Superficially perhaps these conditions do not appear to differ greatly from moves towards value adding and software integration being offered by a number of today's suppliers. They differ significantly, however, in their starting point.

The component (embodied) philosophy takes it for granted that GIS is not an end in itself. It is assumed that it operates within the confines of systems and processes external to itself, and that it will be these outside processes that specify the GIS task functions.

This also means that, just like other enabling technologies (the book, steering wheel and computers), the user will be disinclined to want to understand, consider or consciously recognise the component parts making up the system. For all practical purposes they will have disappeared from view.

5. Summary

This chapter has advanced the idea that for GIS to obtain an everyday acceptance it will need to diffuse to such an extent that it disappears, recedes from conscious view as a separate technology. To achieve this level of diffusion, it is further proposed that GIS needs to be transformed into a ready-to-hand technology, embodied and unnoticed inside some higher level task or process. A preliminary formulation of the conditions required to attain these ends where also given.

As the overview of present day systems trends showed, it appears that in some ways GIS is already moving tentatively in this direction, and thus critics may argue that the concept of embodied GIS contributes little except to formalise, with hindsight, what is already taking place. Such a view, however, misses the central idea of embodied GIS, namely, the diffusion of the technology will only be complete when those in the community gaining advantage from its use, know it so well that they stop being aware of it; when they unknowingly use it in their daily activities merely as a means to an end. Bringing this about implies a change of creed, a change in the belief that GIS is a stand alone and different technology to one that explicitly acknowledges its role as a complementary but subjugate tool.

This has, and with few exceptions will always be the role of GIS; an enriching enabling technology by which to gain insight into geographical phenomena and processes to assist in arriving at some desired outcome. The concept of embodied GIS is an attempt to explicate and highlight this role and thereby possibly providing an alternative framework and perspective on how GIS could, or even should, diffuse.

6. References

Abel, D.J. (1989) SIRO-DBMS: A Database Toolkit for Geographic Information Systems, *International Journal of Geographical Information Systems* 3(2): 103-115.

Brown, L.A. (1981) *Innovation Diffusion*, Methuen, London.

Coleman, D.J. and Zwart, P.R. (1992) Modelling Usage and Telecommunication Performance in Real Time Spatial Information Networks. *Proceedings 5th International Symposium on Spatial Data Handling* 144-153.

Dangermond, J. (1991) Commercial GIS will be Driven by Diverse Applications, *International GIS Sourcebook*, GIS World, Fort Collins.

Driessen R.J. and Zwart, P.R. (1989) A Model for Representing Cultural Parcels in a GIS. *Proceedings of URPIS 17*, Australasian Urban and Regional Information Systems Association, 443-451.

GIS World (1991) *International GIS Sourcebook*, Fort Collins.

Guenther, O. and Buchman, A. (1990) Research Issues in Spatial Databases, *Data Engineering* 13(4): 35-42.

Guignon, C.B. (1983) *Heidegger and the Problem of Knowledge*, Hackett Publishing, Indianapolis.

Weiser, M. (1991) The Computer for the 21st Century, *Scientific American* 265(3): 66-75.

Winograd, T. and Flores, F. (1986) *Understanding Computers and Cognition*, Ablex Publishing, New Jersey.

Zwart, P.R. and Coleman, D.J. (1992) *Trade-Offs towards Real Time Spatial Information Networks*. In Newton, P.W., Zwart, P.R. and Cavill, M. *Networking Spatial Information Systems*, Bellhaven, London.

Zwart, P.R. (1992) Mature GIS - or Taking the G out of GIS. *Proceedings of SIRC '92 Colloquium*, University of Otago, New Zealand.

COMMENTS ON 'EMBODIED GIS - A CONCEPT FOR GIS DIFFUSION' BY PETER R. ZWART

MICHAEL WEGENER
Institute of Spatial Planning
University of Dortmund
Postfach 50 05 00
D-4600 Dortmund 50
Germany

1. Intuitive Computing

The paper by Peter Zwart evokes one of the most popular metaphors of science fiction: the 'intelligent' computer environment which invisibly but sensitively responds to the needs, emotions and desires of the human being. There are other metaphors for the relationship between man and machine such as HAL, gentle companion, suffering creature and deadly enemy, or the Orwellian surveillance machine inflicting obedience or terror. But none is so alluring as the 'disappearing' computer indistinguishably blended into everyday life.

The metaphor has already been taken up by the computer and software industries and created its own market. In Japan and elsewhere 'intelligent spaces', 'intelligent buildings', 'intelligent cars' and 'intelligent roads' culminate in the 'information city' pervaded by ubiquitous computers and computer programmes. The slogan that computing can be intuitive is behind the success of the Apple Macintosh and the even greater success of *Windows*: "The first word processor that works like second nature: Nothing could be more natural than using the all-new Microsoft Word for Windows" (Microsoft advertisement).

2. Intuitive GIS?

So why not adopt the metaphor of the disappearing technology also to GIS? The paper by Peter Zwart does an excellent job in explaining the principles and describing the benefits of future 'embodied' GIS which are so intuitive to use that users become unaware that there is a GIS between them and the task at hand.

In brief, according to the paper the principles of future 'embodied' GIS vs. conventional present GIS are as follows:

- The dichotomy between the spatial and attribute domains characteristic for present GIS will be abandoned in favour of a more general data base organization where topological features become attributes like any other.
- Future GIS will be dedicated to specific tasks rather than multi-purpose systems and will

I. Masser and H. J. Onsrud (eds.), Diffusion and Use of Geographic Information Technologies, 205–208.

be accessed via intuitively navigated window-based user interfaces.
- GIS functionality will be ubiquitously embedded as subsystems in other hardware or software systems or bundled with spatial data bases and become transparent to the user of these systems.

The author claims that the implementation of the above principles will immensely enhance the acceptance and utilisation of GIS in a much wider range of fields than today and that it will unleash the potential of GIS as "an enriching enabling technology by which to gain insight into geographical phenomena and processes to assist in arriving at some desired outcome".

The first reaction of the reader to this proposition is approval. Yes, there are tendencies away from large multipurpose command-driven GIS towards 'leaner', less complex desktop GIS with reduced functionality but more intuitive interfaces. Customised GIS dedicated to a single standardised task such as mapping or route planning address a broad commercial market and set new standards of ease of use unknown to the earlier research-oriented GIS. Such dedicated, user-friendly decision support systems are a prerequisite for the effective utilisation in non-research environments where the typical user is not an expert in computing and data handling.

Perhaps the most advanced examples of 'embodied' GIS (not mentioned by the author) are the GIS functions embedded in air traffic control systems or new car navigation systems under development in many countries. In the former the results of complex spatial operations (in 3D-space) are displayed in real time in highly condensed form to controllers in order to prevent information overload of the operator. In the latter in-vehicle computers perform sophisticated spatial operations (e.g. route search) on signals received from external transmitters and use them to display route choice recommendations in real time to the driver. In either case the user does not need to *understand* how the result is arrived at by the programme, it is sufficient if the user has sufficient confidence that the results are reliable.

3. Routine vs. Innovation

The common feature of the above examples for 'embodied' GIS is that they deal with routine operations. Air traffic control, car navigation, route planning or forest maintenance are functions performed everyday. These systems solve a narrowly defined and never changing task. The input data are updated following a strictly regulated protocol; the classes of data to be processed are fixed, surprises do not occur. Even the desktop mapping packages referred to in the paper are likely to fall into this category. Although some of them may be capable of doing more sophisticated jobs, more likely than not these packages are bought and used for routine and repetitive mapping tasks such as, for instance, showing quarterly sales by state on a polygon coverage of state boundaries that came with the package. This may be immensely useful for many people, and since there seems to be sufficient demand for such routine applications, the software industry will take advantage of it and provide the range and variety of programmes it has provided for other tasks such as word processing or business graphics.

However, routine operations are only one segment of possible GIS applications and not the most interesting. If indeed GIS will always be "an enriching enabling technology by which to gain insight into geographical phenomena and processes", single-purpose dedicated GIS suited for routine operations will not do. This is especially true if GIS are used in research or for decision support. Scientific inquiry, by definition, is concerned with situations in which the range of answers, and sometimes even the range of questions, is not known in advance. Decision-making in spatial planning has to deal with a universe of possible solutions in a turbulent environment of rapidly changing problems. GIS for research or planning cannot be constrained by a limited set of pre-programmed operations. They are only useful if they are open for *innovation*.

This requires that a comprehensive set of basic spatial and logical operations must be available. GIS for innovative spatial analysis have to become more general rather than more specific. Or they have to become more modular and more open to be interfaced and linked with other existing or user-written methods of analysis. In this respect current GIS leave much to be desired. Non-standard proprietary, undocumented file formats, calling conventions and memory management techniques make it difficult if not impossible to integrate existing GIS into problem-specific modelling frameworks or, vice versa, to add customised user-written commands to GIS command languages.

This is not a defense of cryptic, cumbersome and error-prone command-language user interfaces. Well thought-out, efficient and flexible menu systems, self-explanatory point, click and drag conventions and, above all, powerful macro languages are highly desirable and likely to become a standard in future releases of commercial GIS packages. The point is that these features are not so important for computer-literate users as the paper suggests.

4. Understanding

Fundamental objections need to be raised where the paper claims that the users of 'embodied' GIS do not need to *understand* the operations within the GIS. The analogy with driving a car without understanding how it works is misleading. First of all, it certainly helps to know about the risks of defective brakes, worn tires or slippery roads. Second, what are driving lessons for? Obviously driving a car is far from intuitive and needs to be exercised extensively before driving becomes a second nature. Similarly, reading a book without understanding how it was printed? Yes, but in order to understand the text, we need to understand how to read.

So while it is true that the users of GIS do not need to know all the technicalities of data storage and retrieval within a GIS, they do need to thoroughly understand the operations they want the system to perform. The operations in a GIS, and in particular spatial operations, are not trivial and, if poorly specified, can lead to grossly misleading results. Just as even sophisticated statistics packages do not relieve the user from understanding the possibilities and pitfalls of the methods they offer, also GIS should be designed in a way which elucidates rather than hides or obscures the nature of the data transformations performed.

5. Spatial Data

A final comment on data for GIS. The paper suggests that with growing availability of commercial GIS packages, digital spatial data on a wide range of topics also will become widely available. There are many indications that this will be only partly true.

- One reason is the growing 'commodification' of information. As the value of stratified socio-economic data for marketing is becoming recognised, the generation of such data is becoming an industry. The result is that such data are becoming affordable only for corporate users with a strong economic interest, whereas non-commercial researchers and public agencies are unable to pay the market price. On the other hand, public agencies generating their own spatial data, who used to distribute their data for a nominal fee, are now required to cover an increasing share of their costs from sales revenues.
- Another reason is the growing awareness that GIS offer unprecedented possibilities of violating privacy rights. The effect is that in many countries the restrictions on the production, distribution and use of digital spatial data are more rigorously enforced.
- A third reason is that in general that the costs of updating and extending GIS tend to be underestimated. The consequence is that many ambitious digital spatial data bases have not been updated or have even been abandoned.
- Finally, the availability of digital spatial data is further reduced by the lack of generally accepted standards for storing and exchanging spatial data. The consequence is the coexistence of a large number of incompatible or proprietary file formats.

The most likely prospect emerging from these observations is that digital spatial information will be highly concentrated in the hands of large corporations and hardly accessible for non-commercial users. Because of this there will a trend to re-digitise the same data over and over again by a variety of users with different interests and skills and in all sorts of incompatible formats.

SECTION 3:
ASSESSING CULTURAL AND INSTITUTIONAL ISSUES

PETER A. BURROUGH
Faculty of Geographical Science
University of Utrecht, POB 80.115
Utrecht, 3508 TC
Netherlands

and

KEN JONES
Mariesgate 8
0368 Oslo
Norway

Software developers and vendors of geographic information technology products, including specific GIS packages, have tailored their products to meet the needs of a particular, but general international market. Ten years ago the variation of software functionality in commercial software was considerable, and comparisons of the range of functions supplied by any given system were important items in benchmarking and comparative evaluation. Today, most established GIS cover a broad range of functionality (from raster and vector display to map overlay and relational databases) that users everywhere are perceived to need. This range of functionality includes all important methods for data capture, data storage, data retrieval and display, but today embraces a wider range of options for carrying out spatial analysis. These options include buffering, polygon overlay, logical modeling, networking, interpolation and digital terrain model generation. Specific user requirements, however, are still usually addressed by interfacing standard products with specially developed software.

Because functionality requirements have stabilized, the accent of developers has been on improving ease of use of the command system, with features such as menus, icons, multiwindowing and multitasking. Many systems now permit users (or specialist experts) to customize the user interface so that a given GIS can be configured or tailored to a wide variety of work situations. These adaptations include the ability to define data structures, design menus, and build links to other software. The vendors rarely permit users access to the original source code and often provide their own "meta-language" which can be used to make modifications or write local "macro" programs.

In spite of all these efforts by vendors to make their products as general as possible, there are still several important cultural and institutional barriers to the adoption of standard geographic information technology products in many countries, particularly because many such products have been developed to meet the specific requirements of users in North America. The papers in this section address some of these problems, which are the following:

I. Masser and H. J. Onsrud (eds.), Diffusion and Use of Geographic Information Technologies, 209–212.

Cultural Problems

Do cultural issues influence the diffusion of geographic information technologies? Is culture a factor for the design, implementation and use of geographic technologies? Or does the diffusion of the same GIS packages across countries and cultures mean that technology is a homogenizing factor? (Craglia 1991).

Of special interest to this workshop is whether there are cultural differences between southern Europe and the rest of Europe which affect the diffusion and use of geographic information technologies (GIT). (A taste for wine and a love of opera should surely not create any issues, despite the opinions of some of our brothers. Some other cultural differences may have more effect however.)

The papers by Antonio Arnaud and Massimo Rumor suggest, directly or indirectly, several cultural characteristics which may be significant. For example, attitudes of citizens toward local government and the distribution of power and functions within government determine the environment (applications) in which GIT are introduced. Similarly, lax attitudes toward comprehensive planning, environmental monitoring and the enforcement of development regulations influence the governmental context so that some GIT applications may be less interesting, have lower benefits, and result in a reduced likelihood of diffusion. Personnel policies and attitudes in local government are also cultural factors of significance. A tradition of low salaries and negative attitudes to staff training may lead to poorly educated and poorly motivated personnel which hinder both diffusion of innovative ideas and implementation of new technologies. Other personnel factors are recruitment procedures, hiring restrictions and opportunities (or lack of them) for staff advancement. The availability of personnel with certain professional skills may also reflect cultural preferences.

The feasibility or choice of GIT applications can be affected by the political culture, such as the "national political culture of using public resources and office for political and personal patronage". "At the municipal level, projects concerned with infrastructure, social facilities and leisure provide greater mileage and are more "visible" than information systems. Land information systems can be a thorny political issue in a country where tax avoidance is high or land ownership is concentrated in a small percentage of the population.

Cultural differences in higher education can affect the diffusion and acceptance of new technologies. In some countries, such as the Netherlands, courses in geographic information systems are taught in some universities as part of the undergraduate curriculum; in others (e.g. UK) the shorter length of the course (3 years) may preclude the introduction of GIS before postgraduate levels. In other countries the inflexibility of traditional subject definitions (geography is a humanity in Italy) or conservatism in associated professional guilds may slow the introduction of the subject into the curriculum. Differences in funding and provision of resources can also greatly affect the ability of any institution to teach GIS as a practical subject. Consequently there can be many reasons why one country may have a large pool of persons who "know about GIS", while others have only a few specialists.

Attitudes toward access to information and privacy can also limit applications and affect diffusion and use of GIS. There are large differences between countries concerning the sharing of information , and whether databases should be linked and made available for uses other than that for which they were initially collected.

In their paper, Michael Wegener and Hartwig Junius point out some of the cultural impediments experienced in Germany concerning the adoption of standard GIS, which in most cases is a North American product. Among the cultural barriers they discuss is first and foremost the strong surveying tradition in Germany with a requirement for high geometrical accuracy. This, linked with a unique "object-oriented" land parcel system that demands a unique data structure meant that most foreign GIS were considered unsuitable. Because the Germans had developed their own systems for land information, which were distributed at low cost , the opportunities for vendors of foreign products in certain parts of the German GIS market were strongly limited. In other areas, such as utilities, this is not the case and both foreign and home grown systems can be found next to each other. The German surveying tradition of defining and measuring objects with a high degree of accuracy may in fact be counterproductive in areas such as planning and environmental assessment, where complex, but poorly definable spatial phenomena are paramount. (e.g., see Burrough 1992).

Institutional Issues

Many institutional aspects can enhance or retard the adoption of geographic information technologies. In countries with a strong tradition of mapping and surveying one might expect that the reservoir of expertise and data would facilitate the introduction of GIT. In fact, institutional inertia, commitment to long-term investment in other methods, departmental and institutional rivalry, and national and institutional attitudes to data security and copyrighting can be of overwhelming importance. In developing countries, the long-term political and institutional stability is extremely important for promulgating the necessary data collection, building institutes and accomplishing training necessary for the successful use of GIT. The role of international agencies, such as the United Nations Environmental Programme - Global Environment Monitoring System (UNEP-GEMS), in providing assistance, training, stimulation, low-level software and international data sets to developing countries is often more important than any single software package. The presence of key persons in positions of power, both in developed or developing countries, can be crucial for the structural commitments necessary for geographic information technology adoption.

The organization of local government in a country is probably one of the most important institutional factors affecting the diffusion and use of GIT in local and regional administration and planning. Where local governments are small, capital resources are generally fewer and GIT may be too expensive. This may be the case in southern Europe; for example 75% of the municipalities in Portugal and 80% in Italy have a population of less than 10,000 inhabitants.

The presence and functions of national institutions are also important, affecting for example the existence of up-to-date analog or digital map data (mapping agencies), statistical data (statistical offices), land parcel data and building data (Cadastral offices). In both Portugal and Italy there are no central agencies which map at large scales and the medium and small scale maps are out of date.

The first three papers which follow provide a sampling of the institutional and cultural problems and attempts at overcoming them in implementing geographic information technologies in varying cultural settings. The fourth paper by Timothy Cartwright discusses factors affecting the use and diffusion of geographic information technologies in third world countries generally.

References

Burrough, P.A. (1992) Are GIS data structures too simple minded? *Computers and Geosciences* 18: 395-400.

Craglia M. (1991) GIS in Italian Urban Planning. *Department of Town & Regional Planning Faculty of Architectural Studies*, The University of Sheffield, Sheffield, UK.

'UNIVERSAL' GIS VERSUS NATIONAL LAND INFORMATION TRADITIONS: SOFTWARE IMPERIALISM OR ENDOGENOUS DEVELOPMENTS?

MICHAEL WEGENER and HARTWIG JUNIUS
Institute of Spatial Planning
University of Dortmund
Postfach 50 05 00
D-4600 Dortmund 50
Germany

ABSTRACT. There are large differences in the diffusion of geographical information systems in European countries. Whereas in some countries, notably in the United Kingdom and the Netherlands, GIS technology has already become standard practice in cartography, local land management and planning, other countries have been extremely slow in adopting the new technology. This paper suggests the hypothesis that national land information traditions determine the institutional environment and procedures into which the new technology needs to be integrated, and that this restricts the transfer of 'universal' GIS software developed in different national settings. The paper discusses approaches to deal with this problem and draws conclusions for a European GIS research agenda.

1. Introduction

Over the last two decades the rapid development of digital computers has brought increases in computing speed and memory size and reductions in hardware costs beyond all expectations. One of the many impacts of this explosive development is the ability to efficiently and economically handle large amounts of spatial data. The organization of these data in *geographical information systems* (GIS) has the potential to revolutionize all fields in which geographic data are handled such as mapping, land management and planning.

And indeed there is a veritable boom in GIS. What used to be an exclusive field of specialization of a small number of university experts has developed into a multimillion dollar market. International GIS conferences draw hundreds if not thousands of participants. There is a growing flood of GIS books and journals. And whereas a few years ago GIS software was a matter of two or three high-priced mainframe programmes, today there is a host of competing GIS packages of all levels of complexity and for a variety of hardware platforms including low-cost workstations or personal computers, and new entries appear on the market in rapid succession.

So it appears that the GIS revolution is part of the general trend towards the information society, and as that trend is produced by the revolution in computer hardware, one might expect that it would take place at about the same speed in all countries where computers are

I. Masser and H. J. Onsrud (eds.), Diffusion and Use of Geographic Information Technologies, 213–228.

freely available. However, whereas office automation, computer-assisted manufacturing and advanced logistics are rapidly spreading across all countries with market-oriented economies, there are large and surprising differences in the diffusion of geographical information systems in European countries that cannot be explained by differences in affluence or technological advancement.

In some countries, notably in the United Kingdom and the Netherlands, GIS technology has already become standard practice in cartography, local land management and planning. These two countries have launched extensive national research programmes to promote the expertise in handling digital geographic data. In Britain, the Economic and Social Science Research Council allocated £2 million for funding eight Regional Research Laboratories over a five-year period. In the Netherlands, the Dutch Science Council spends 1.8 million guilders over a four-year period for a National Centre for Geographic Data Processing (Masser, 1988).

Not surprisingly, in both countries, the official base map systems have already to a large part been digitized across all map scales. In Britain, for example, the Ordnance Survey has used digital mapping techniques since 1972 and now digitizes some 5,000 of its 1:1,250 and 1:2,500 maps annually. It expects to have completed the digitising of all major urban areas by 1995 and all other warranted digital conversions within a further decade (Ordnance Survey, 1987). Ordnance Survey digital map data can be ordered by anybody for an individual charge or by subscription. On the basis of Ordnance Survey map data and the Post Office's Postcode Address File (PAF), which contains the postal addresses of 24 million locations in the UK, an immense commercial GIS industry has developed, of which *Pinpoint* is only the most spectacular example. *Pinpoint Analysis Ltd.* offers its customers a wide range of information services centred around the *Pinpoint Address Code* (PAC), a geographic reference attaching to every single address in Britain a twelve-digit *National Grid* reference and its precise geographical location.

In contrast to this, other European countries have been extremely slow in adopting the new GIS technology. In some of them this was obviously a consequence of insufficient public funds for research and development or of lack of skilled staff or hardware. In other countries this simple explanation does not hold because they belong to the more prosperous countries in Europe and yet have surprisingly failed to take full advantage of the immense technical potential of computerization for the handling of geographical data. Because the authors are most familiar with the situation in their own country, the following discussion focuses on the slow diffusion of GIS technology in Germany, which has made only regrettably slow progress in the adoption of GIS technology.

This paper discusses why this is so. The hypothesis is that national land information traditions determine the institutional environment and procedures into which GIS technology, if it is to be really successful, needs to be integrated, and that this restricts the transfer of 'universal' GIS software developed in different national settings. The paper discusses approaches to deal with this problem and draws conclusions for a European GIS research agenda.

2. Early GIS in Germany

The history of GIS in Germany is one of early enthusiasm, misconceptions, exaggerated hopes and great disillusions, many failures and few successes.

The first advances into the field were connected to the perceived needs of a particular kind of end users, the planners. Already in the 1960s, some information about the use of computerized 'planning information systems' in large cities of the US swept across the Atlantic. Stimulated by the USAC research programme of the US government to establish pilot information systems for urban planning, the West German government funded a pilot project for the development of an urban planning information system in the city of Cologne. Similar projects were launched by the cities of Munich and Berlin. However, inadequate computing power, batch processing, primitive mapping facilities and the lack of efficient software made these systems unattractive, so the planners returned to their traditional ways of information collection and processing. The Cologne project was completely abandoned during the 1970s, while the Munich system, though it still exists today, has never played an important part in the planning process.

The basic misconception underlying these systems was that, understandably, they started from the users' end but neglected the basic processes of data collection and updating. The Munich system, for instance, did its own large-scale digitization of the whole city area but had to find out that the task of keeping up with the constant flow of alterations made necessary by explosive city growth in the early 1970s was far beyond the capacity of even the most enthusiastic planning department. The same applied to the nongraphical data associated with (in the Munich case) individual addresses. As it turned out, only the most rudimentary population data could be routinely updated using periodical excerpts from the city's population register. However, as public sensitivity for the privacy issues brought up by computerization grew, even that became infeasible for political reasons. More interesting information, e.g. about household composition or the distribution of workplaces, either proved too costly to be periodically collected or increasingly became off-limits for the same privacy reasons that eventually toppled the 1980 West German census.

The importance of privacy issues for the slow progress in GIS applications in planning in West Germany cannot be overestimated. In effect they turned out to be a serious restriction for any advancements in two ways:

- For once, through excesses of data collection and surveillance revealed during the anti-terrorism campaigns of the 1970s, the West German public became more sensitive in this respect than probably in any other country in Europe. Even well-intended proposals for the utilization of personal data for public service provision or planning became stigmatized as potentially dangerous and open for misuse.
- Second, exactly for that reason, new planning information systems were not established, and existing ones were not updated and hence quickly became outdated and void of any meaning, and this erosion of the significance of the information provided by the systems made it difficult if not impossible even for their most ardent defenders to demonstrate their usefulness.

Today the situation comes close to the absurd. When the census originally scheduled for 1980, after severe legal disputes and only with strict privacy safeguards, was finally conducted in 1987 (seventeen years after the previous census of 1970), the results were not published except in highly aggregate form and not made available to researchers. Even worse, the census revealed that the local population registers and the employment statistics maintained by the Federal Labor Agency, which the opponents of the census had claimed made a census unnecessary, were widely unreliable - if the census is considered reliable, which itself is questionable given the large number of responses that had to be enforced by court decisions or to be substituted by assumptions. The city of Dortmund, for instance, learned by way of the census that it had 11,000 more inhabitants and 30,000 more jobs than the population register and employment statistics contained, whereas other municipalities discovered they had less. To complete the folly, the census law expressly prohibited local governments to update their population registers using census data thus perpetuating the inconsistency between the two types of information.

However, even if the early planning GIS had succeeded in overcoming the privacy problems associated with disaggregate planning data by appropriate safeguards, the failure to be firmly linked to the digital base map system of the city or country would have been sufficient to kill them. In this respect, only a close cooperation with the institutions responsible for surveying and mapping could have helped (cf. Junius, 1988). This cooperation, however, did not work because the surveying people played their own cards.

3. Land Information in Germany

To understand this, it is necessary to briefly review the history and main characteristics of the land information system in Germany and compare them with those of other European countries. Some significant differences appear.

The first written documents on land in Germany are treaties between territorial rulers about border lines dating back to the 16th century (Kloos, 1990). As land taxes were the principle source of income of these princes, some of the larger territories developed land registers in which individual lots were listed by area, soil quality and owner. Later the land registers also contained information on titles such as mortgages. However, the first systematic inventory of all properties in the form of a cadastral register and maps based on surveying property boundaries was only introduced in the 19th century after the Napoleonic wars and the consolidation of the great number of small territories to larger states such as Prussia. Now the installation and maintenance of the cadastre was assigned as a permanent task to specialist surveying departments.

With the decreasing role of land taxes as government income and the growing importance of land as an economic asset, the principal function of land information became the registration of property rights. For this the cadastre (which essentially was a map) was complemented by a written land register containing information on land ownership that could not conveniently be placed on a map and was liable to frequent changes.

This led to the dual mode of land information existing in Germany today. There are two registers belonging to this system: the cadastre (*Liegenschaftskataster*) and the land register (*Grundbuch*). The cadastre contains information on the physical characteristics of properties such as size, inclination and location by geometric coordinates. The land register, on the other hand, contains information on the ownership of properties and all property rights, especially mortgages. The two registers are linked by a common referencing system based on a hierarchical numerical code for each property, and each change in one of them may affect the other. This makes updating procedures cumbersome and slow but guarantees a high level of reliability of the cadastre as it participates in the official status of the land register. Liegenschaftskataster and Grundbuch might, directly or indirectly, be linked to the files of the local population register, planning and budgeting departments and the commission for land value assessment. Figure 1 visualizes this complex set of relations.

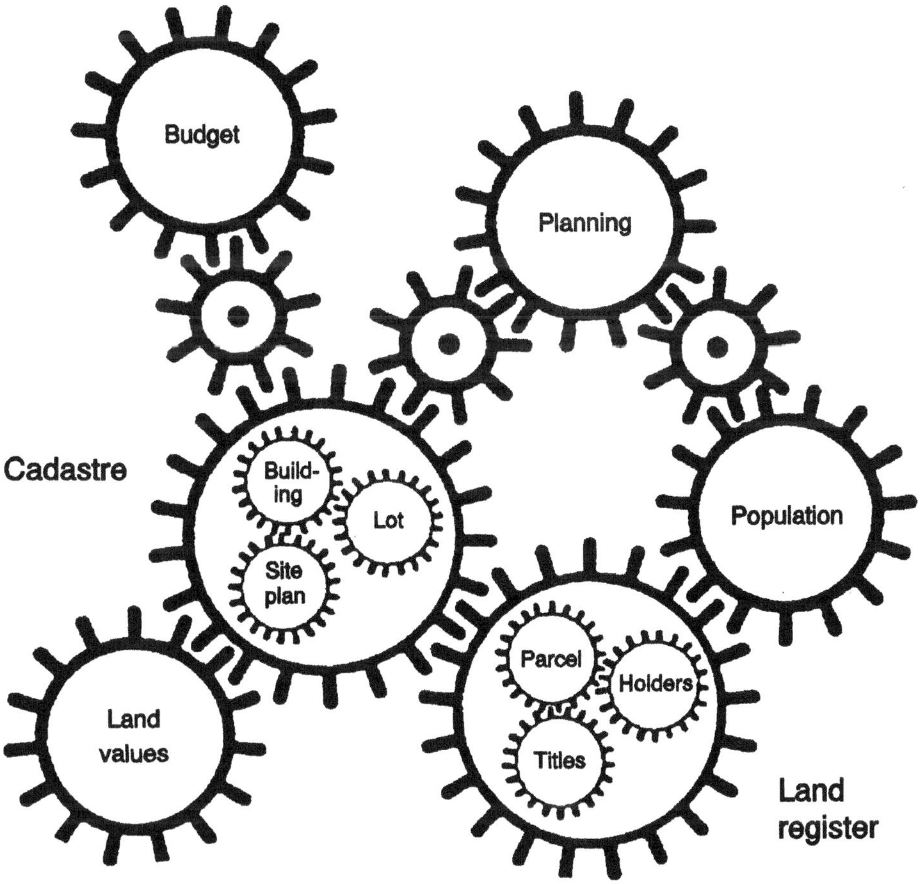

Figure 1. Cadastre and land register and the network of local government files (adapted from Stöppler, 1983). The dual mode of land information in Germany consists of two registers: the cadastre (Liegenschaftskataster0 and the land register (Grundbuch). They may be linked to the files of the local population register, planning and budgeting departments and the commission for land value assessment.

The German land information system (which is also representative for the Austrian and Swiss systems) differs from those in most English- and French-speaking countries in that it registers *titles*, i.e property rights belonging to a specific parcel, whereas in France, most of the US and parts of Canada and Australia traditional land information systems register *deeds*, i.e. documents on land transactions. In other words, the German land information system is organized by parcel, whereas the traditional British land information system is organized by *owner* (cf. Kloos, 1990). It is easy to see that the German system is much closer to the structure of GIS which by definition are organized by location.

While this affinity between the logical structure of land information in Germany and a relational GIS database should have facilitated the introduction of GIS technology, the decentralized institutional setup and distributed responsibility for land information made it more difficult.

Unlike Britain, West Germany has no central surveying and mapping institution. Rather, for historical reasons, surveying and mapping in Germany is highly decentralized or, more precisely, fragmented, reflecting the federal organization of government in the country. Local or county government surveying departments are in charge of 1:1,000 *city maps*; medium-scale maps, such as the 1:5,000 German Base Map (*Deutsche Grundkarte*) and 1:25,000, 1:50,000 and 1:100,000 topographical maps are prepared by the surveying departments (*Landesvermessungsämter*) of the states (*Länder)*, and the 1:200,000 and 1:1,000,000 maps by the Institute of Applied Geodesy (*Institut für Angewandte Geodäsie*), a Federal agency in Frankfurt and Berlin.

In addition, several Federal and state agencies such as the Federal Railways or the state highway departments are involved in specific tasks of surveying and mapping. To make things worse, in some states medium-level regional planning authorities are in charge of maintaining a special set of registers and maps, the Regional Planning Inventory (*Raumordnungskataster*). In line with this setup, the cadastre is maintained by the local government, i.e. municipal or county, surveying departments (in some states by the state surveying agencies). The land register, however, is administered by special authorities at courts of law (*Grundbuchämter*), but in some cases by local governments.

To make this multi-level system work, a complicated set of procedures is necessary which guarantee that queries and updates performed by a multitude of users in a wide range of agencies and institutions are processed according to defined rules with access control and data integrity being maintained. In short, if such a system is to be computerized, it requires a multi-user, multi-agency database management system.

4. Endogenous GIS: the German LIS

In order to develop such a system, already in 1971 a working party with representatives of state surveying departments submitted a proposal for automating both registers. The first project started was the Automated Cadastral Register (*Automatisiertes Liegenschaftsbuch*), in short ALB, designed as a standard database management system (DBMS). The second project, started in 1976, is the Automated Cadastral Map (*Automatisiertes Liegenschaftskarte*), or ALK. That the two projects were conducted separately was due to the fact that data structures to efficiently link attribute and geometry data as in today's GIS were simply not available at that time. As DBMS technology became available, work on the register was started earlier, whereas map automation was a byproduct of surveying. Today, however, both systems together form the integrated land information system (LIS), which includes a complex set of procedures and administrative processes for generating and updating land information in a multi-user, multi-agency environment. This system is characterized by a strict separation of the users from the database management system through a standardized interface or data transfer format, the EDBS (*'Einheitliche Datenbankschnittstelle'*).

The implementation of the EDBS is very similar to a graphics metafile format consisting of records identified by a record identifier, a header, and any number of data fields. Figure 2 illustrates how a typical ALK object is coded into the EDBS format and decoded at the user's application. In the ALK terminology, an object is a collection of attribute and geometry data. Attribute data of a building object, for instance, are its label, shading, or all sorts of annotations, while geometry data are its label coordinate and the coordinates of, say, its perimeter: in other words, object primitives such as line objects can be nested within more complex objects.

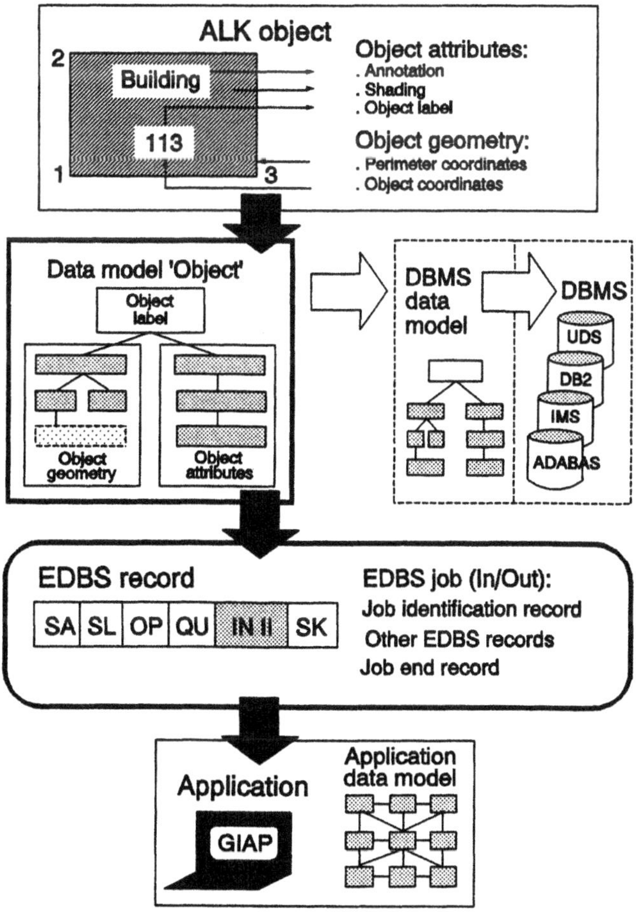

Figure 2. ALK objects and data models (adapted from Stöppler, 1989). ALK objects are collections of attribute and geometry data. Object primitives such as line objects can be nested within more complex objects. The figure illustrates how the object 'building' is mapped to the logical data model of the DBMS but also to the data transfer format of the EDBS and decoded at the user's application.

This data model 'object' is mapped to the logical data model of the respective DBMS but also to the data transfer format of the EDBS. At the application end, the data are decoded and translated into the representation required by the specific application. It is assumed that the application consists of an interactive session at the GIAP (*Graphischer Interaktiver Arbeitsplatz*), the graphical workstation functionality defined by the ALK system (which may be implemented on any kind of real workstation). Each query or update request by the user is translated into the EDBS by the EDBS interpreter and put into the job queue of the ALK-DBMS. After processing, the response or acknowledgment is sent back to the EDBS interpreter which reconverts it into the format required by the application.

The ALK system as it has been sketched out above, is supposed to be the basic GIS standard not only for the local cadastre and associated base map system, but, in a bottom-up fashion, also for the middle-scale topographical maps of the state surveying agencies under the acronym ATKIS (*Amtliches Topographisch-Kartographisches Informationssystem*).

Today, two decades after the initiation of the project, only first steps have been made towards implementation of the ALK system. Only three of the twelve Federal states of the former Federal Republic (Hessen, Niedersachsen and Nordrhein-Westfalen) have actually started to implement the data collection component of ALK (digitising of existing analog maps or converting existing non-compatible digital map formats). A few other Federal states are preparing themselves to join the ALK system. Some Federal states have already decided not to join. Needless to say that the five new Federal states (the former GDR), where a different, much more centralized surveying system had developed since the war, present new unforeseen problems.

Understandably, the system components implemented so far concentrate on the primary functions necessary for data collection, updating, map editing and data communication between the subsystems. More complex GIS functions such as union or intersection of coverages, buffering, digital terrain models or network analysis are still under development. More specialized applications for statistical analysis, thematic mapping or land use planning are still far away in the future.

Also on the state level only first experiences with the ATKIS mapping standard have been made. Only two states, Niedersachsen and Nordrhein-Westfalen, started actual map production before 1991 (Grünreich, 1990). The original plan to generate ATKIS maps by generalization from ALK base maps has been abandoned in most states yet digitization from existing analog maps takes longer than expected. The idea that one could order any scale digital map from the *Landesvermessungsamt*, as it is common practice in Britain, would appear quite inconceivable here. Also the data exchange between the different agencies and institutions of the ALK system with the EDBS data transfer format, although in principle it is designed to be interactive, to date is used only off-line, i.e. via batch processing; more far-reaching interface problems such as, for instance, the exchange of geographical data between mapping systems in different European countries or between European institutions, have hardly been taken into view. Most importantly, the whole system is totally under control of the surveying departments without broad consultation let alone participation of other potential users.

5. 'Universal' GIS?

Considering the small progress made in the introduction of the ALK so far, one might ask why the responsible actors decided to go through all the pains of developing their own 'endogenous' GIS instead of buying a commercial GIS software product off-the-shelf. This strategy has, for instance, been adopted by the Hong Kong government, which after an international competition between major vendors of GIS selected ARC/INFO of ESRI Environmental Systems Research Institute, Redlands, California, USA for its surveying/mapping and planning information system. A comparison between Hong Kong and Germany reveals the main reasons why in Germany a different decision was made.

-*Language*. Because the most commercially available GIS (ARC/INFO) is of US origin, all its command language, user interfaces, manuals etc. are in English. This presents no problems as long as either the country is English-speaking (as Hong Kong), or the users are multilingual academics. For the average officer in a municipal surveying department in Germany the foreign language presents a serious barrier.

-*Data models*. The cadastre system in Germany suggests complex object types such as the 'building' object used as an example in Figure 2, which themselves contain geometrical primitives as objects, such as the perimeter line object. Such a data object can only be emulated with, say, ARC/INFO, which only knows point, line and polygon objects.

-*Algorithms*. There are significant differences in the measurement techniques and algorithms used in surveying in Germany and in the US which make American surveying programmes "practically useless as components of land information systems" (Kloos, 1990, 131.)

-*Map projection*. The German base map system uses the Gauß-Krüger transversal Mercator projection similar but not identical to the UTM projection used for NATO and civil aviation maps (see Figure 3). Neither ARC/INFO nor any other US-based GIS support this projection.

-*Networking capability*. The multi-level and multi-user organization of surveying and mapping and the associated cadastre and land register require more elaborate support for access control, data communication and data integrity than a single client or a centrally organized city state such as Hong Kong. Presently only few commercial GIS take account of this need.

-*Data interface standards*. The decentralized system of map production and maintenance requires standards for the exchange of cartographic and non-cartographic data. Most commercial GIS use their own proprietary data formats and allow data exchange with other software only by way of the lowest common denominator, i.e. ASCII files. Needless to say that the German EDBS is not directly supported but requires special conversion programmes.

-*Cost*. Commercial GIS software is expensive to buy and to maintain. In contrast, the true costs of developing an endogenous GIS such as the ALK system are hidden in general administrative budgets. Once existing, endogenous solutions have to be used. For instance, in Nordrhein-Westfalen the municipalities get the ALK software free which means that they simply no longer have the option to choose an expensive external system.

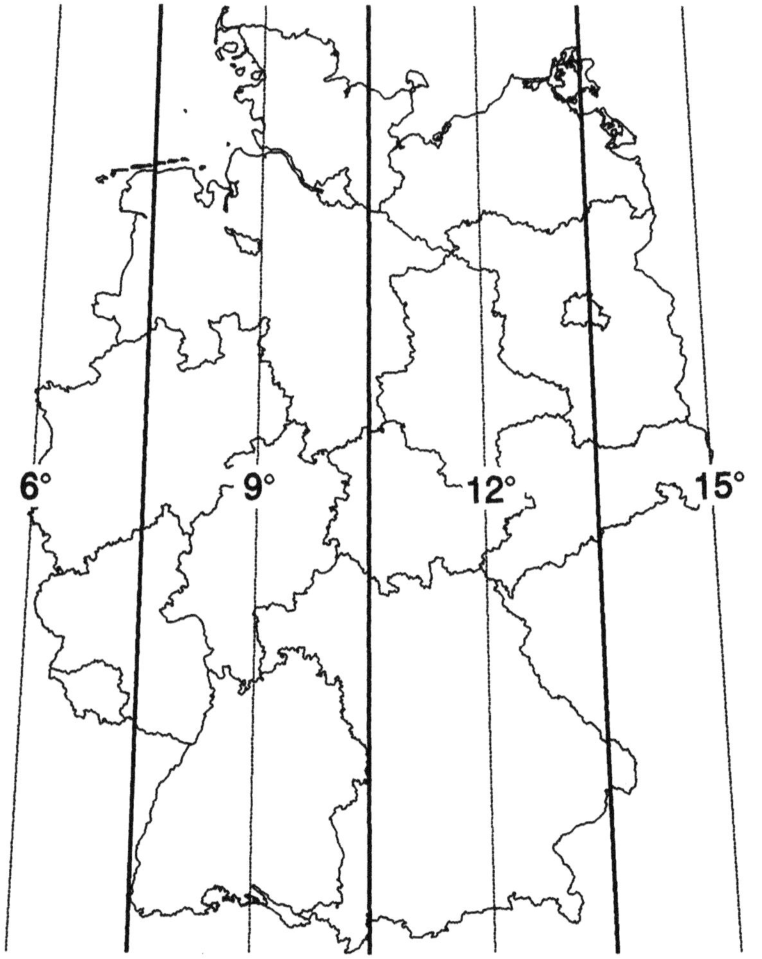

Figure 3. The Gauß-Krüger map system. The transversal Mercator projection of German basemaps uses 3°-wide bands centred on the 6°, 9°, 12° and 15° meridians. Neither ARC/INFO nor other US-based GIS support this projection.

As a result of these considerations, the GIS scene in Germany is divided into two worlds. Public surveying and mapping, and hence most native GIS developments, are committed to the endogenous development of ALK and ATKIS, whereas universities and other public and private research institutions are linked to the international world of 'universal' GIS. This has serious consequences for the diffusion and acceptance of GIS technology and for GIS research and education.

6. Alternative Solutions

To be sure, Germany because of its size is a large GIS market, and there is no reason to assume that private companies in Germany should be less interested in using GIS technology for truck routing, real estate management or market research than firms in other countries. However, the diffusion of GIS technology in Germany might have been much more successful if the ALK/ATKIS system had been implemented on schedule and as universally as originally intended and if it had been made publicly available as a national standard to software producers and GIS research and education. However, as it was delayed beyond all expectations, even states and Federal agencies that originally were sympathetic to its objectives turned away to look for alternative solutions.

One example is the *Institut für angewandte Geodäsie* which recently chose ARC/INFO as GIS for the preparation of its 1:250,000 and 1:1,000,000 maps, which obviously creates problems of data exchange with the larger-scale maps produced at the state surveying departments. A second example is the 'Statistical Information System on Land Use' (STABIS) launched by the Statistisches Bundesamt (Federal Statistical Office). Although ATKIS, when it eventually is implemented, will provide more land use information than required by STABIS, the Bundesamt has started its own digitization of land coverage from aerial photographs and intends to update it every five years, although for a later stage utilization of ATKIS data is intended (Rademacher, 1990). Remarkably, the Office has never developed a publicly available georeferencing system for census information like the US TIGER files, be it because of privacy concerns or because of lack of demand or both. Other Federal and state agencies such as the Federal Railways or the state highway departments face the problem of integrating their digital databases with the ALK/ATKIS system.

Other GIS projects have simply ignored the problems of integration with the ALK/ATKIS system and prepared their own digital database. There is the large category of environmental projects which typically use a combination of satellite data, aerial photographs, digitized existing analog maps and special measurement or data collection programmes (cf. Ashwood and Schaller, 1990). Obviously, as in these projects the users tend to be the researchers themselves, problems of language and user-friendliness are less relevant, so in many of them commercial GIS such as ARC/INFO or ERDAS are used. However, there is at least one local government environmental department (Dortmund) which has purchased ARC/INFO for implementing an environmental information system. Although some of these projects aim at long-term environmental monitoring, none of them seems to have addressed the problem of updating their base maps.

In local government, GIS technology has proved indispensable for planning and maintaining utility networks. Pioneered by private utility companies, in most cities the cadastre of utility networks is more advanced than the digitization of the cadastre, which will produce serious integration problems in the future, although there are some recent examples of cooperationa between utility companies and the cadastre. Many of utility GIS, if they do not use custom-tailored software, are based on SICAD, a CAD package by Siemens Nixdorf, for which modules for cartography, network documentation, etc. have been developed. Because of its popularity, SICAD file formats have become a de-facto standard in this field. In addition, SICAD supports the EDBS and so is a serious contender for the ALK-GIAP.

GIS applications for local planning, after the negative experience of the early predecessors, are practically nonexistent. One notable exception is the Greater Frankfurt Area Association of Local Governments (UVF) which has since 1977 developed a sophisticated planning information system for preparing a joint land use plan for 43 municipalities (Rautenstrauch, 1989). The digital base maps for this system were extracted from aerial photographs and have since been updated three times. There is no link to the ALK/ATKIS system. The software used is ARC/INFO.

The example of the UVF seems to suggest that 'universal' GIS can be effectively applied in local planning. However, as the UVP is unique by its large staff of academics and its showcase position in Germany, its example cannot be generalized to small and medium-sized cities with less sophisticated staff.

One alternative is to custom-tailor a German-language user interface shell around the original GIS software. In a project conducted at the University of Dortmund two simple evaluation tasks were selected for a prototype of a planning information system for small cities (Altenhoff et al., 1991). The first evaluates local street design alternatives with respect to criteria such as cycling lanes, pedestrian crossings, traffic noise, landscaping, activities spaces, car parking and access for delivery. The second evaluates land use alternatives with respect to accessibility, provision of services and environmental constraints. Both applications make use of topological information such as adjacency, distance or domain. The prototype system was implemented using pcARC/INFO combined with a menu-driven user interface in German.

The user interface was written in FORTRAN with additional assembler routines for operating system and display management functions. It consists of a hierarchy of character-based windows, dialogue and alert boxes. Figure 4 shows the hierarchy of windows. The interface shields the user from all operating-system and English-language ARC/INFO commands. It has been programmed in a way that makes user errors as unlikely as possible without restricting the freedom of choosing any reasonable sequence of operations.

7. GIS Research and Education

The division of the GIS diffusion in Germany into two separate streams, the endogenous GIS centred around the ALK/ATKIS system and the 'universal' GIS linked to the international GIS world, puts GIS research and education into a dilemma.

In research, to develop GIS applications exclusively for the domestic standard would probably make it easier to get research contracts from local clients but also entail the risk to be separated from the international state of the art. The slow progress made by the ALK/ATKIS project and the lack of interest on the side of the participating agencies to publicize their standards and interface conventions have contributed to make this alternative less than attractive. Conversely, to entirely ignore the standards set by the domestic market would mean to have no access to it and hence no access to data and no clients for one's products.

In education, the choice is not easier. On the one hand, state-of-the-art GIS such as ARC/INFO are the natural choice for introductory GIS courses. On the other hand, students would be poorly prepared for the GIS environment they are likely to encounter in their future practice in public agencies if they would not get exposed to the ALK/ATKIS system at all. This, however, would make it necessary for universities to install an ALK implementation on the campus, something to date only few surveying departments and no planning schools at German universities have yet achieved. A survey among German-speaking universities revealed that ARC/INFO is by far the most frequent GIS software used for teaching (Bill, forthcoming).

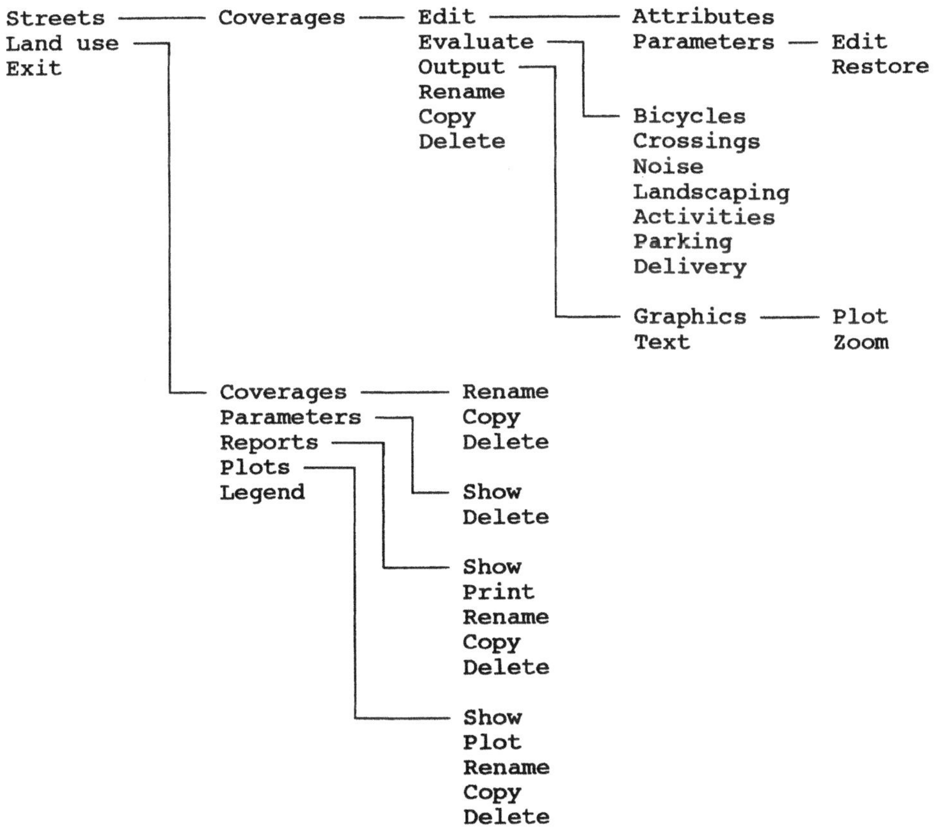

Figure 4. Hierarchy of windows of user interface for pcARC/INFO. The interface shields the user from all operating-system and English-language ARC/INFO commands. It has been programmed in a way that makes user errors as unlikely as possible without restricting the freedom of choosing any reasonable sequence of operations.

8. Conclusions

The answer to the initial question: 'Universal' GIS or endogenous developments? can therefore only be: both. In non-English-speaking countries with a decentralized system of surveying and mapping like Germany custom-tailored endogenous solutions combining software *and* procedures seem indispensable. At the same time in the research field 'universal' GIS reflecting the international state of the art allowing for a variety of possible approaches and giving room for innovation and experimentation are likely to be the more appropriate choice.

From a research point of view, three areas seem to have been neglected in the past and to deserve particular attention:

The first addresses the question of better user interfaces reflecting the needs of local users including native-language shells. If the GIS market continues to expand as it has in the recent past, part of this can be expected to come from the software suppliers themselves, just as word processors appear in different languages. If not, at least the vendors should provide adequate documentation of their file formats and programming conventions to aid developers of such interfaces. The past policy of ESRI has been regrettably deficient in this respect, although the multi-language user shell of ARCVIEW is a first step in the right direction.

The second field regards implementation issues. The problems discussed in this paper suggest more emphasis on the organizational and institutional environment of GIS in the land information field taking account of the different traditions and restrictions of each particular country. Many of the issues discussed after the failure of the first attempts to establish planning information systems in the 1960s and 1970s, such as the information needs of planners and the way planners use information, are equally relevant today yet tend to be forgotten over the enthusiasm of the new GIS boom (see, for instance, Wegener, 1983; 1987; 1988).

The third neglected research area concerns data exchange between different GIS systems within and between countries. With the Single European Market coming closer and further advances in European integration in sight, the problems arising from the integration of incompatible geographical data bases between European countries or between European institutions deserve increased attention. Although significant progress in the harmonization of data exchange standards have been made (cf. Brüggemann, 1990), the multitude of incompatible data formats used by different countries, European agencies, industry groups or NATO is still staggering and makes isolated national efforts towards standardization, such as the EDBS, questionable.

GIS education in non-English-speaking countries will become more difficult. Except for the most basic concepts, the unmodified transfer of foreign teaching material and software will not be sufficient if the claim of practice-orientation of the course is taken seriously. Planning schools in Germany, for that matter, will have to install at least one implementation of the ALK-GIAP in their GIS teaching laboratory, even though the maintenance of more than one complex GIS systems may put a heavy strain on their financial and staff resources.

9. Endnote

This chapter is in part based on a contribution by the authors to the European Science Foundation Workshop on European Research in Geographic Information Systems held in Davos, Switzerland, 24-26 January 1991 (Junius and Wegener, 1991).

10. References

Altenhoff, D., Varlemann, D. and Wegener, M. (1991) Evaluation of local street design and land use alternatives using pcARC/INFO, Paper presented at the Second International Conference on Computers in Urban Planning and Urban Management, Oxford, 6-8 July 1991.

Ashdown, M. and Schaller, J. (1990) *Geographic Information Systems and Their Application in MAB Projects, Ecosystem Research and Environmental Monitoring*, German National Committee for the UNESCO Programme 'Man and the Biosphere' (MAB), Bonn.

Bill, R. (forthcoming) Zur GIS-Ausbildungssituation an deutschsprachigen Universitäten (GIS education at German-speaking universities), *Geo-Informations-Systeme*.

Brüggemann, H. (1990) Exchange formats for topographic-cartographic data, *Geo-Informations-Systeme* 3(4): 27-32.

Grünreich, D. (1990) ATKIS - Amtliches Topographisch-Kartographisches Informationssystem der Landesvermessung (ATKIS - the official topographic-cartographic surveying information system of the surveying and mapping administration), *Geo-Informations-Systeme* 4(1): 4-9.

Junius, H. (1988) Landinformationssystem - eine geeignete Grundlage für die Planung? (Land information system - an appropriate basis for planning?) *Vermessungswesen und Raumordnung* 50(1): 1-7.

Junius, H. and Wegener, M. (1991) Universal GIS versus national land information traditions: internationalization of software or endogenous developments? *Computers, Environment and Urban Systems* 15: 219-227.

Kloos, H.W. (1990) *Landinformationssysteme in der öffentlichen Verwaltung* (Land Information Systems in Public Administration), Decker & Müller, Heidelberg.

Masser, I. (1988) The Regional Research Laboratories in perspective, *Mapping Awareness* 2: 5-12.

Ordnance Survey (1987) *Ordnance Survey Information*, Leaflet 48. Ordnance Survey, Southampton.

Rademacher, W. (1990) Das Statistische Informationssystem zur Bodennutzung STABIS der amtlichen Statistik (The statistical information system on land use STABIS of official statistics) *Geo-Informations-Systeme* 3(4): 9-15.

Rautenstrauch, L. (1989) Electronic data processing in planning at the Umlandverband Frankfurt, *Ekistics* 338(9): 254-258.

Stöppler, H.W. (1983) Automation des Liegenschaftskatasters (Automation of the cadastre), *Nachrichten aus dem öffentlichen Vermessungswesen Nordrhein-Westfalen* 16: 129-137.

Stöppler, H.W. (1989) Die Einheitliche Datenbankschnittstelle (EDBS) im fachlichen Überblick (The common data base interface (EDBS): Technical Overview), *Zeitschrift fuur Vermessungswesen* 12: 576-597.

Wegener, M. (1983): The impact of systems analysis on urban planning: the West German experience, in G.B. Hutchinson and M. Batty (eds.), *Systems Analysis in Urban Policy Making and Planning*, Plenum, New York, 125-152.

Wegener, M. (1987) Spatial planning in the information age, in J.F. Brotchie, P. Hall, P.W. Newton (eds.), *The Spatial Impact of Technological Change*, Croom Helm, London, 375-392.

Wegener, M. (1988) Information technology, society and the future of planning, in M. Giaoutzi and P. Nijkamp (eds.), *Informatics and Regional Development*, Avebury, Aldershot, 42-55.

THE USE OF GEOGRAPHIC INFORMATION TECHNOLOGY IN THE CITY OF PADOVA

MASSIMO RUMOR
Comune di Padova
Via Municipio, 1
35100 Padova
Italy

ABSTRACT. This article describes GIS diffusion in Italy with particular regard to local authorities. The paper then illustrates the main objectives of the Urban Information System of Padova, the approach for that system's implementation, and the systems general architecture. The conception of the system, the most important stages in its implementation, and the use of the system are outlined and the system's present situation is described in detail. The future development programme is presented together with some conclusions based on the experience in Padova.

1. Introduction

The opinion that GIS technology is going to be of great benefit to public administration is widely held in most of the developed world. In particular, great interest has been shown and initiatives have been taken in some countries. A few example initiatives in Europe include those by the British Government, a framework for which was established by the Chorley Report and whose findings have come to be considered as essential reading in the field; the Joint Nordic Project whose activities have assisted much in the diffusion of GIS; and the Nexpri Centre in the Netherlands which is described later in this volume by Henk Ottens.

Yet the predicted benefits of GIS technology are yet to arrive in most jurisdictions. In fact, the use of GIS technology is at the early stages even in countries which have taken the lead with it. In the European context, the experience of GIS in the UK, a leading country in the field, confirms this. The British situation has been illustrated by Professor Ian Masser in his paper "The Diffusion of GIS in British Local Government" and by Dr. Heather Campbell in "GIS Implementation in British Local Government" (Section 2 of this volume). Research work and some operational initiatives continue to assist in the diffusion of geographic information in many countries. Interest extends further than to just the technical aspects of the technology and this broad interest can only but further aid its diffusion.

I. Masser and H. J. Onsrud (eds.), Diffusion and Use of Geographic Information Technologies, 229–244.

2. The Italian Situation

2.1. THE USE OF INFORMATION TECHNOLOGY BY PUBLIC AUTHORITIES

The use of information technology by the various sectors of the public administration in Italy has increased greatly in recent times. Generally available statistics on information technology (IT) expenditures confirm this trend: central government and local authority expenditures on IT in 1990 accounted for 17% of total IT expenditures in Italy. This was a 12.3% increase over 1989 and was 2% higher than the growth of total IT expenditures in Italy (10%). However, these growth rates are still rather low when compared with figures of 29% in the USA, 19.5% in the UK and 20% in France.

More detailed figures for Italian Municipalities and public utility companies are available in two recent surveys: a 1989 survey by ANCI (Association of Italian Municipalities) and a 1990 survey by CISPEL (Association of Italian Public Utility Companies). According to the ANCI survey only 54% of the approximately 8000 Italian Municipalities had access to computing facilities and, of those Municipalities that had access to such facilities, 61% were in Northern Italy and only 39% were in the South. The CISPEL survey showed that 81% of utility companies in Italy have access to computing facilities and some 32% also use computers for the automation and control of plant. The CISPEL survey does not compare figures for the North-South divide.

In my opinion these figures have improved significantly, at least in Northern Italy, since the ANCI and CISPEL surveys were carried out in 1989 and 1990. At the present moment, I am not aware of any Municipality or public utility company in North-East Italy which does not have access to computing facilities. Yet, looking beyond this mere quantitative data, it should be noted that the full benefits from the expenditure on IT have not been realized to date. Two important factors have diminished the effective return on these investments:

-the underestimation of organisational constraints affecting the introduction and use of IT;
-the lack of skilled staff in public administrative agencies.

These two factors have yet to be given their correct weights in the application of IT.

2.2. THE DIFFUSION OF GIS

As already pointed out, the use of GIS is in its early stages even in countries which have taken a lead in its use. In Italy there is a very high level of interest in the application of GIS technology causing the Italian GIS market to be considered as the third largest in Europe; after Germany and the UK. Thus, Italian expenditures on GIS technology are much higher than would be expected from observing the position of IT in general within the country. For example, the 1990 Cispel survey states that 23% of utility companies in Italy have implemented or are in the process of implementing a GIS application for network management. This fairly healthy state of affairs can be explained by considering the very high benefits that are expected from the use of GIS technology, and I contend it is also partially due to the fact that GIS has been seen as 'fashionable'.

But the situation appears better than it is in reality. We observe currently that many GIS implementation studies have been carried out, several projects seem ready to be started and many experimental applications have been implemented. However, few genuinely operational systems can be found. Most of these operational systems are installed in departments of Central Government or in other national agencies. In part, this is due to the fact that in Ministries and generally in Central Government Offices GIS is applied only to support the decision-making process and such applications are easier to implement than applications for operational management use. Among national agencies or government owned companies, such as the Telephone Company, the Energy Agency, and the National Gas Company, several operational systems are found. It must be emphasized that these organizations, even if considered as part of the public sector, act as private companies and therefore have fewer problems with recruiting trained staff and introducing required organizational changes.

As we approach the more local level of organization, the operational management use of GIS increases and consequently the complexity of the required systems increases substantially. It is this required complexity which is mainly responsible for the small number of genuinely operational systems in regional, provincial and municipal governments. The situation is better in public utility companies because the required applications are better known and more clearly identified. Utility companies also have a higher degree of organizational flexibility than other public bodies.

2.3. FACTORS INFLUENCING GIS DIFFUSION

A series of factors can be identified which restrict the diffusion of GIS technology in Italy and which are seriously delaying the realisation of the full benefits from the investments already made.

2.3.1 *Data availability*. Italy lacks a single, national agency which is responsible for the creation and maintenance of large scale maps for technical use by local authorities and utility companies. The Military Geographic Institute (IGM), produces and tries to maintain small scale maps (the great majority of the Italian 1:25000 and 1:50000 scale maps are some 20 years out of date). The IGM makes use of GIS but currently no IGM map is available in digital form. The Italian Cadaster produces large scale maps of the country and about 30% of the national territory is already available in digital form but parts of this map record need major updating. The problem with these maps is that they are not suitable for use in local authorities and utility companies: essential features are not represented, the precision and accuracy of the maps are very poor, and the maps cannot be arranged in a continuous mapping space.

Regional government has started to produce regional maps at scales of 1:5000 and 1:10000 for planning purposes in order to have a common cartographic base for planning at the regional level. Most of Italy is covered but only very small portions are available in digital form (in Lombardia, Veneto, Piemonte and Umbria). Moreover the precision is not suitable for other uses except for the representation of master plans in urban areas. As a result, many municipalities and utility companies have prepared their own traditional maps using their own resources and the same will apply to the acquisition of digital cartography. It is easy to understand that this imposes a very serious limit to the further diffusion of GIS in Italy.

2.3.2 *The need for organizational change.* Organizational changes are required to realise the full benefits from the general application of IT and this is particularly true for GIS applications. An integrated, interdisciplinary approach is required from the beginning of the implementation process and only with coordination between the relevant departments or bodies involved can a GIS application hope to function properly.

2.3.3 *The necessity for long term investment plans.* It is now widely recognized that the GIS technology itself is not the major cost element of a project. Data and education are the major cost elements involved in the implementation of GIS and both these elements need to be carefully planned over a period of many years. Unfortunately, long term planning is not normally favoured by the average politician (at least in Italy) who prefers to see results before the next election. Additionally, long term planning is hindered by difficulties in financing long term investments.

2.3.4 *Lack of skilled personnel.* Staff education has never been considered important in public administration and very little money has been used to finance skills development projects. Moreover the necessity to cut public spending on salaries has resulted in a prohibition on replacing staff who leave and on creating new positions. Thus the general quality of the public administration staff has declined to a very low level. Additionally the few high level positions which have been created in recent times have not been attractive to suitably skilled applicants in comparison to equivalent positions in the private sector. The levels of salaries offered by the public sector are still lower than those offered by the private sector for the equivalent higher positions and the private sector also offers a merit approach in human resource management. Thus the problem of finding suitably skilled staff in the face of competition with the private sector is substantial. As a result, most IT projects and installations of IT, both in general applications and for GIS in particular, are handled by external private contractors with no equivalent department or agency internal to the client's organization able to contribute to the development of the applications and to keep the system working after implementation.

2.3.5 *Lack of standardization and coordination.* Standardization and coordination have never been Italian fortes. In Italy, GIS applications have developed since the late 1970's without any serious attempt to coordinate the experience gained or to set standards. Only lately has it been realized that lack of coordination and standardization results in wasted investments and delayed achievement of the full benefits of GIS. Consequently some coordination efforts have now been started by central government. Also some of the regional administrations have undertaken the task of coordination, sometimes adding uncertainty and confusion to the process. Some kind of agreement, possibly only a partial solution, is expected to be reached soon regarding the problem of standardization at the European level.

3. Geographic Information in Padova

3.1 GENERAL INFORMATION

Padova is located in northern Italy, 30 km west of Venice. On record since the 10th century B.C., it was an important city within the Roman Empire. Additionally, Padova has been a Bishopric since the 3rd century and a university town since 1221. The City is a commercial and industrial centre and is also an important centre for culture and the arts. The Municipality of Padova has a residential population of about 220,000 inhabitants with a weekday population of some 350,000 and covers an area of 93 square km. Taking into account the residential towns surrounding the city the residential population for Padova and its hinterland rises to more than 400,000.

3.2 THE BEGINNING OF THE PROJECT

The idea to use information technology to support urban planning related activities was initially suggested in the mid-70's by a city councilor who was the politician responsible on the City's executive council for both the City Information System Department and the City Planning Department. Based on this city councilor's initial suggestion, a project to set up a register of the buildings was launched. A group of surveyors was recruited and financed through a special law intended to favour the employment of young professionals in civil services. This team of eighteen worked with rather poor guidance and was hampered by political changes in the ruling majority grouping resulting from municipal elections. The councilor who had initially proposed the project did not retain his position on the executive council and the project was left without political support. It was close to being cancelled when the responsibility for the project was taken over by the City Information System Department which embraced the original idea and decided to continue work on the project. This initiative was permitted by the new executive council but there was a lack of real political support. So the project continued but with severe budget restrictions. Moreover, the lack of support put the City Information Systems Department in a very weak position when dealing with other departments within the city administration.

3.3 THE PROJECT'S MAIN AIMS AND GENERAL APPROACH

In 1982 a preliminary study was carried out and it was realized that the project's objectives were very wide in their scope and rather poorly defined. We soon understood this was one of the main characteristics of the project and the design of the system had to be developed accordingly.

The system which we had envisaged was a non-specialized general purpose geographical information system, which would be utilized by a heterogeneous group of users either within the municipal government or outside of it. We were aware that this objective was very broad but at the same time we were deeply convinced that this was the only possible serious approach. This approach was considered ambitious at the time and it forced us to search and adopt solutions that were able to provide a reasonable probability of success. We decided to embrace the following features for the system's design and implementation:

- an open architecture which would allow the implementation of new functions avoiding any significant interference with the parts of the system already in existence;
- a step by step implementation which would allow both the release of ready-to-use results at each step and, even more importantly, a graded investment;
- the maximum use of existing data to minimize data collection needs and to guarantee data maintenance.

3.4 THE URBAN INFORMATION SYSTEM'S ARCHITECTURE

The backbone of the system is the Territory Description Data Base (TDDB): (see Figure 1). The TDDB functions as a base map yet has some significant improvements. It has been designed to:

- store all geometric information together with the basic descriptive data of each defined entity and produce automatic maps;
- supply spatial references to data contained in the currently used and updated administrative data bases which can consequently be cross-referenced and referred to the proper geographic location.

The TDDB is based on the following entities, called Basic Territorial Entities :

- the block: a parcel of land limited by roads or physical barriers (rivers, channels etc.) or administrative boundaries;
- the cell unit: part of a block which generally contains a building and its surrounding area;
- the elementary unit: each building or open area, i.e. the smallest part of the territory we need to deal with.

It is important to emphasize that with the use of street codes and house numbers it is possible to link data from each administrative data base to the corresponding Basic Territorial Entity and then refer them to a geographic location.

3.5 THE FIRST STAGES OF IMPLEMENTATION

It was necessary to demonstrate early on in the project some practical results and thus we decided to start with thematic map production. We needed to represent information retrieved from already existing administrative files and we concluded that a simplified cartography was sufficient for this purpose. Such simplified cartography allowed us to achieve the first results without large investments for which money was not available.

Consequently we decided to use photogrammetric maps, which were already available, and to digitize only the indispensable elements which described the city at the block level. There were technical problems and we had to push our GIS software supplier to develop a package with a continuous mapping space. Moreover much work had to be done to make the edges of the sheets match. Yet eventually we created a map covering the whole of

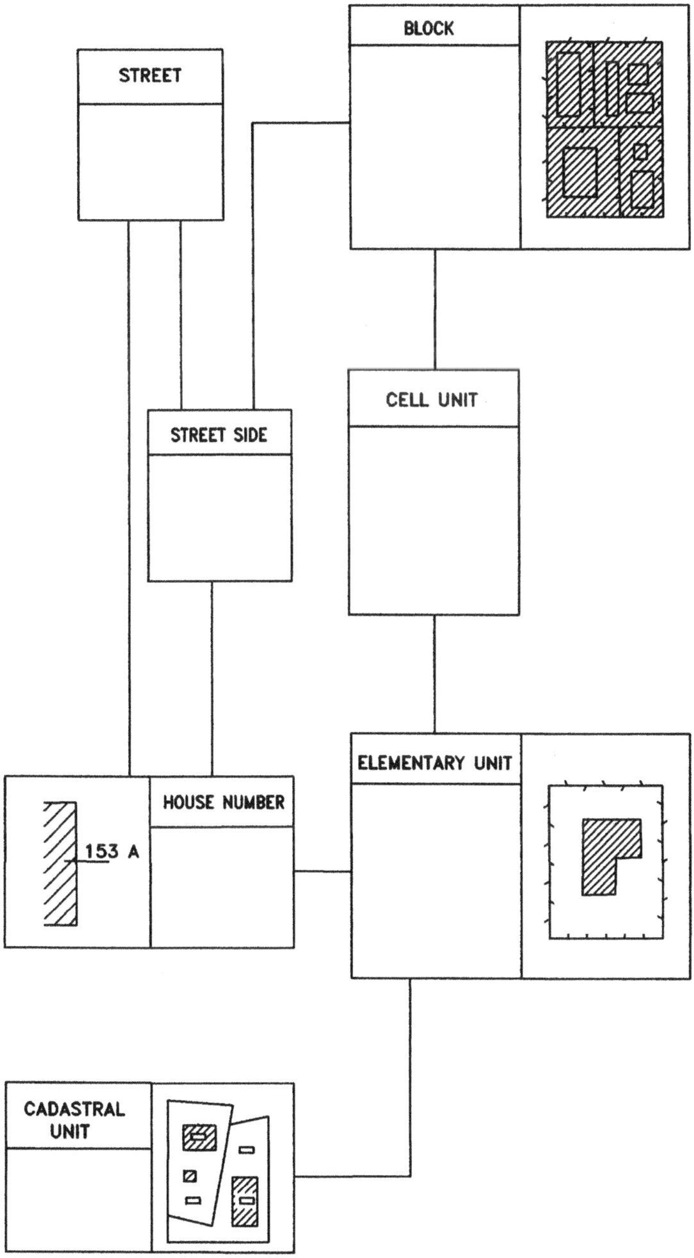

Figure 1. The Territory Description Data Base (TDDB).

Padova at the block level and we were able to produce the thematic maps we needed. Users could produce maps directly from the workstations using interactive functions driven by personalized menues. They could define the part of the city to be analysed and obtain the map on the screen; users could then select the theme of interest and the data with the desired graphic representation to be represented on the screen.

A more detailed analysis of the city centre of Padova was also carried out to evaluate some alternative redevelopment plans as part of the initial implementation of the GIS project.

3.6 THE CONTINUATION OF THE PROJECT

After the initial stages of the project, we came to the conclusion that our Urban Information System (UIS) should be able to support all activities at all levels, from operational management to decision-making support. Additionally, we wanted it used by most if not all the authorities and agencies operating in the municipal territory. This would make it possible to:

- share a single base map;
- share information;
- coordinate operations;
- coordinate plans.

Integration of data and coordination of actions were considered key factors to city management. Indeed we realized very soon that one of our major concerns would be to convince the authorities or agencies operating within the city boundaries to join us in the project. A joint project would also make it possible to share the high investment needed among all the agencies and departments concerned.

The right opportunity to integrate all the municipal cartographic records came in 1986 when the Municipality decided to start the production of a new set of 1:1000 scale city maps. These were to replace the obsolete 1:1000 city centre plans and the old 1:2000 plans which covered the rest of the municipal area. We decided to evaluate the possibility of managing the new cartography automatically within the UIS. We realized very soon that the required investment would be great and that the moment had come to reach an agreement with other relevant authorities within the framework of our proposed integrated strategy. We contacted the City Water and Gas Company which was, at this time, considering the reorganization of its network documentation service. After a series of meetings they decided to join with us and we started to discuss the technical problems concerning the new city map.

A great deal of the discussion related to content, precision, representation, and similar matters. Eventually, we agreed on an aerialphotogrammetric base with a precision of 1:1000. Our task was then to store in the TDDB a base map able to:

- produce automatic maps which could replace the traditional maps irrespective of the map's end-use;
- continue to supply spatial references to data within the system;
- satisfy the special requirements of the City Water and Gas Company.

The existing set of graphic elements was used in order to enable all users to continue to utilize the new maps as if no changes had been introduced. Additional data needed to be added on the photogrammetric maps in order to identify the basic territorial entities, and each entity needed to be coded according to the defined structure. Fig.2 is an example of a standard basic map produced by the system which shows the topography of part of the city centre.

We are still working on the base map of Padova and we aim to complete the base map for the whole of the municipal area by the end 1992. The new automatically drawn maps have replaced the traditional maps without problems. The accuracy of these new maps are far greater than the maps they replaced and, moreover, we can produce maps which better suit users' needs (i.e. simplified, coloured, any scale and any sheet-cut of map). But more importantly, we are now able to link alphanumeric data with spatial data. This has been a significant new benefit. We can interactively access cartographic data via coordinates, traditional sheet number, parcel number, street name and house number, and similar attributes and then retrieve descriptive data about each entity. For instance, the age, the number of floors, measurement information, and the state of preservation about a particular building can be accessed through this system. It is also possible to use more complex access criteria combining stored data by means of logical operators.

Maintaining up-to-date cartography is going to be a very important task. We have set up a procedure on the basis that all planning permissions for building works or for a change of use for both buildings and areas must be authorized by the Municipality. A very important source of information for updating the maps is the maintenance service of the City Gas and Water Company. This service reports all variations between the cartographic map record and the actual layout on the ground (not just variations in the gas and water infrastructure) noted during the course of routine maintenance works.

4. The Present Situation

4.1 EXISTING APPLICATIONS

Before illustrating in more detail some of the applications available within the Urban Information System (UIS), following is a list of all applications which are either in operation in Padova or are being developed or which are simply at the planning stage:

-automatic mapping (operating)
-thematic mapping (operating)
-town planning (operating)
-water network (operating)
-gas network (operating)
-sewer network (under development)
-road and traffic (under development)
-environment (under development)
- archaeological map (operating)
- public health (feasibility study)
- parks and gardens (operating)

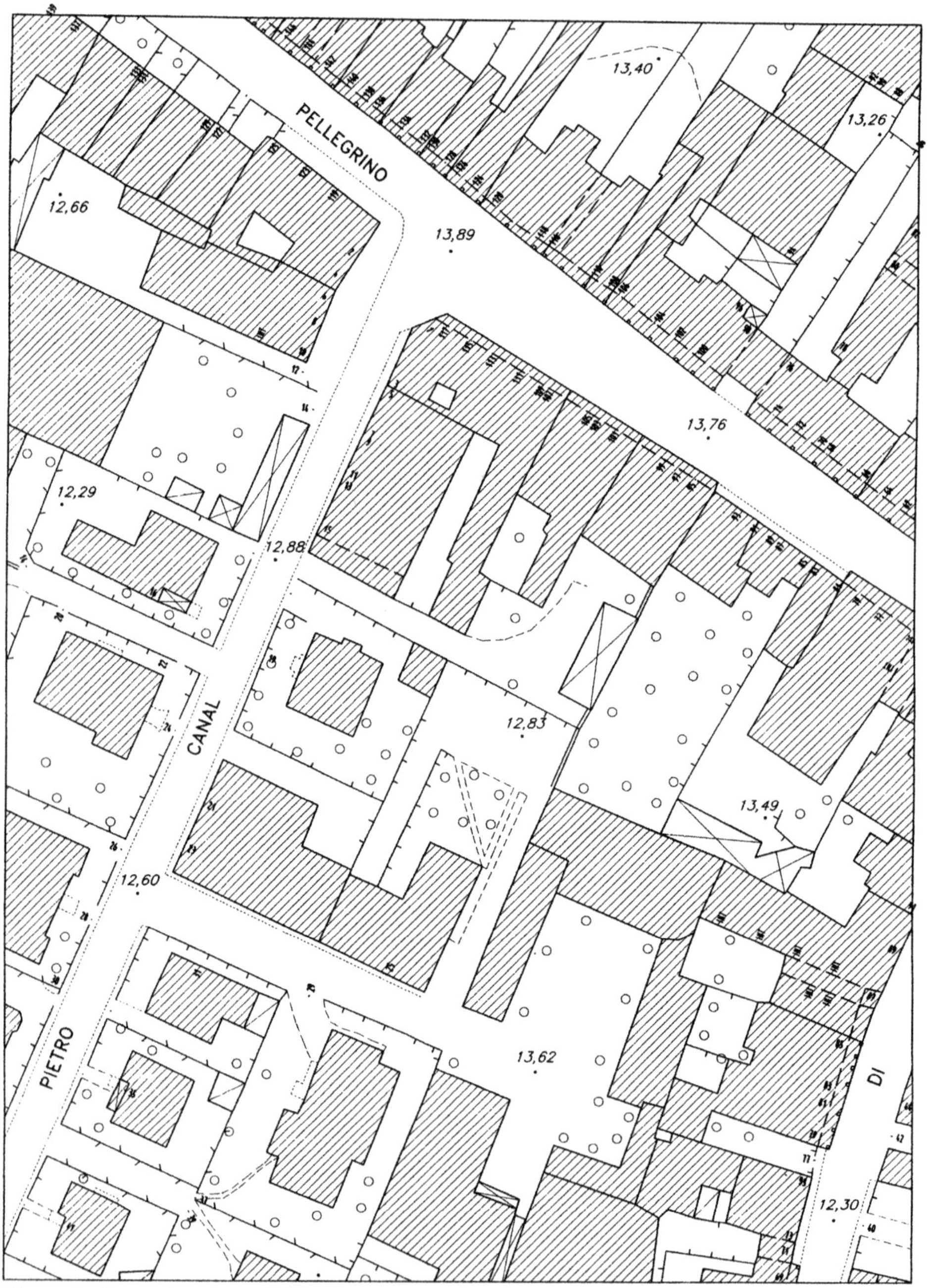

Figure 2. Example of a 1:1000 Digital Map of Padova.

4.2 THEMATIC MAPPING

Data, which had already been on computer files, is available within the UIS, provided that a connection to one of the elements of the TDDB can be established. This data can then be processed by appropriate programs and the results can be represented on thematic maps. The connection to the TDDB is normally made via the street and house number. Parcel number, entity code, coordinates are among the many other possible means which enable the UIS to locate any required data. Users can produce thematic maps directly from the workstations using completely interactive functions. The user can define the part of the city to be analysed and obtain the map of this desired area on the screen. The data to be processed may be selected, the arithmetic or logical operations to be applied may be written, and the results may be represented on the map with the desired graphic illustration.

Our experience has shown that interactive functions are very useful when selecting a combination of data that is intended to best describe the situation under investigation and when deciding on the analyses and representation methods to be applied. Hard copy maps are largely preferred for everyday work, in discussions and for decision making. The following lists the most frequently used thematic maps:

- Population distribution
- Use of buildings and areas
- Building age and characteristics
- Building permit distribution
- Economic activities
- Building regulations
- City owned real estate
- Distribution of parks

4.3 TOWN PLANNING

The Historic Centre Preservation Plan has been recently completed using the UIS. The City Executive realized that the General Urban Development Plan in force could not guarantee the preservation of the historic centre. Under the old general plan, the works permitted in the historic district risked the possibility that new works might:

- change the function of the historic City Centre with the traditional residential functions being replaced by more speculative tertiary activities and with the consequent alteration of the traditional urban balance and the increase of motor traffic circulation;

- cause the disintegration of the historical building typology in order to adapt existing buildings to new needs and new technological standards.

To confront this situation the Municipality decided to revise the General Urban Development Plan in force, drawing up a specific plan for the historic centre aimed at its "conservative renewal". The new technical regulations are tailored to the specific characteristics of each individual building or area.

In order to draw up these regulations a great amount of data had to be processed, which made it almost impossible to consider carrying out the task without the help of the computer. More than 12,000 elementary units, buildings or open areas, in an area of 7 square kilometres were taken into account. Relative to its overall size, Padova has a very large historic centre.

It was first necessary to identify the information that had to be considered in order to draw up the City Centre Plan to meet the objectives of this plan. Consequently we had to see whether data was available within the UIS itself or somewhere else within the Municipal Information System. Obviously not all the required data was available and this had to be collected. Once gathered from its source, all data was stored within the UIS and properly referenced to the corresponding elementary unit of the TDDB.

At this point it became possible to start analyzing the urban entity which is the historic City Centre. The Planning Department has used all the functions provided by the thematic map application. During this analytical phase various simulations were made in order to find the best solution for the planning problems of the City Centre. Eventually analyses and representational methods were selected for the problem and a set of maps was batch produced to be used in the following planning stages and also to be included with the final documentation.

It was then necessary to create a synthesis of the analytical results and to assign each elementary unit within an appropriate classification system. For this purpose a combination of basic information, defined as the "cultural value" of the unit, was taken as the fundamental variable in the determination of the regulations for each elementary unit. In order to determine the criteria for the cultural value of the elementary unit, the basic data was processed statistically in a classification exercise. Finally eight classes were identified and each "planning unit", generally corresponding to a "cell unit" in the TDDB, was assigned to one of these classes. For each class the regulations determine:

- the type of building operations permitted and the type of works to be carried out in order to improve the urban environment.
- the type of use allowed with the aim of controlling land use change and of ensuring a well-balanced distribution of residential, commercial, productive uses.

In addition to the regulations established for building planning units, the plan has defined a few basic criteria to be followed in the layout of the public open spaces (streets, squares, gardens, rivers, etc.).

4.4 WATER AND GAS NETWORK APPLICATIONS

In Padova, AMAG is a public company, owned by the Municipality, which distributes both water and gas to the whole city and to many of the neighbouring towns. Together with AMAG we have developed two specific databases, linked to the TDDB, for the management of the water and gas networks which contain administrative and geometric information describing each network. Figure 3 illustrates the structure of the database for the management of the water network although it should be noted that the databases for both the water and the gas networks are similar. Geometric data is divided into layers and symbols are used to represent network components. The symbols have been selected such

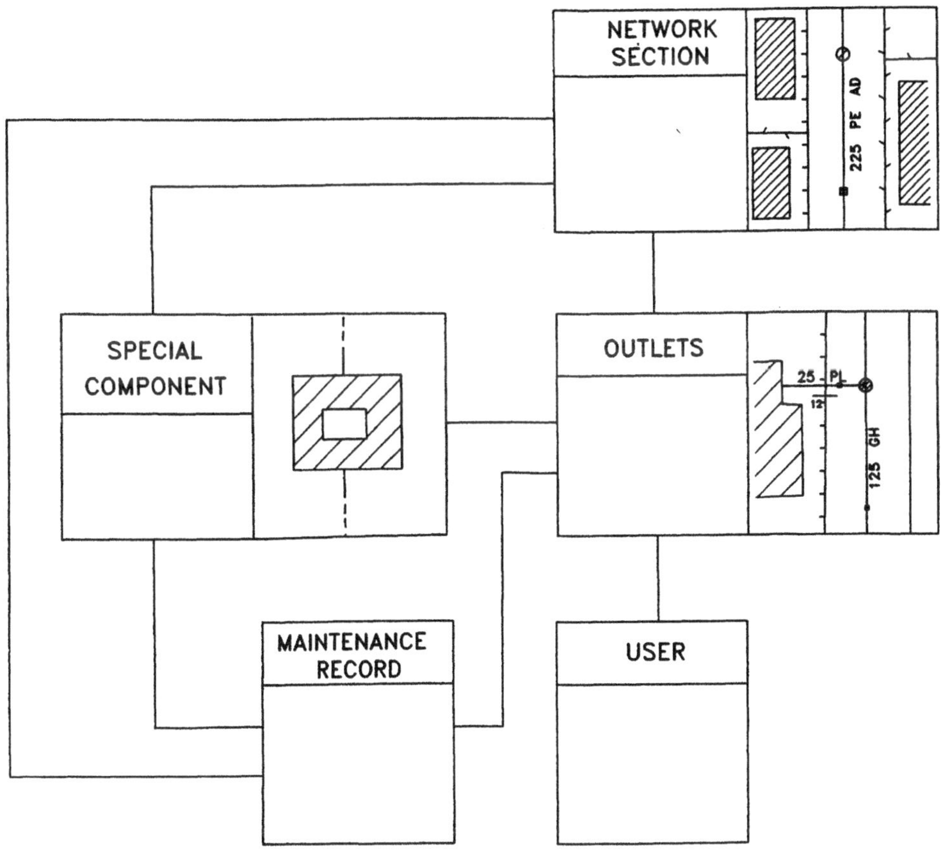

Figure 3. Database Structure for the Operational Management of the Water Network.

that both networks can be shown on either the same map or the same screen without confusion.

The main problem was data. The original documentation for both the gas and water networks was not suitable for easy input into the system (for example, a variety of scales had been used on plans over the years and the documentation was physically in a bad condition, etc.) and much work was needed to prepare the documentation before it was acceptable for the database. AMAG is still in the process of entering data but the work is scheduled to be completed before the end of 1992.

5. Future Developments

We are, of course, working to complete the cartographic base map of Padova and we continue to implement and to put into operation the applications indicated in our programme of works. At the strategic level we are pursuing two important aims :

- geometric correspondence between our maps and
- extension of the use of our maps by other relevant agencies.

We have already started working with the Italian Cadaster and a joint working group is currently at the stage of testing the technical steps necessary to adjust cadastral maps to our block contours. If these tests confirm our assumptions then the two cartographic bases will be automatically superimposed which would finally allow the easy cross referencing of cadastral, physical and planning information.

As far as the extension of the use of digital maps is concerned, some success has been achieved selling our digital maps to the National Telephone Company. This has meant that the water, gas and now the telephone infrastructural networks are represented on the same maps. The sewer network application is almost complete, thus the sewer network data can soon start to be entered into the system. However the service infrastructure map still lacks information about the electricity network. Responsibility for this network rests with the National Electricity Company with whom we have been in discussion since the outset of the project. At a recent meeting a decision was taken to consider the possibility of sharing the base map and a working group has been set up which has already started work. On the basis of this start we have asked central government to finance a project which would aim to design and implement new organizational models capable of supporting a more effective and a more efficient integration between the many agencies that operate in the City.

6. Conclusion

On the basis of our experience in Padova, even though we encountered numerous technical problems, we have realized that the most critical problems were not technical in nature. Rather we found that organizational and political issues were the most critical problems to be overcome.

From the outset we were aware of the organizational problems and we tried at all times to take any measures we could to achieve favorable responses from the potential users of the system. Involvement of potential users in application design and implementation was always sought. At the beginning, widespread involvement met with poor results yet as the project proceeded those potential users who had initially resisted sometimes became enthusiastic participants. Involvement of potential users proved to be essential in gaining success of the applications. But organization is more than just people. Other organizational components, such as organization structure and procedure, also had to be confronted. The introduction of organizational change has proved so difficult that, to date, no major changes in the organizational arrangement have been carried out.

On the political front there has been a lack of real interest. Beyond the few occasional words of appreciation at official meetings and the support of one or two enthusiast supporters, nothing serious has been done to support the project. There have been other major problems which have kept our politicians busy during the last ten years. Since the beginning, in our dealings with politicians it has been convenient not to emphasize our overall grand design but we have always sought to underline the short term objectives and to keep a relatively low profile. Yet now things have started to change and some politicians have become aware of the enormous benefits GIS technology can bring to local government and we hope to complete the project with far greater political support.

In drawing some general lessons from our experience in Padova and from observations of other GIS applications, I would like to draw to the readers' attention the following considerations. In my opinion, these considerations may be useful for other Italian authorities and may also be useful to authorities outside of Italy, in Europe and even overseas.

1. GIS technology has often been considered mainly as a tool to support management decision-making processes but it is also possible to use it for operational activities. Applications can be developed which aid at departmental levels in many technical activities. The use of GIS for everyday functions can be a key factor in the success of the introduction of GIS technology into organizations.

2. The "grand overall design" approach is rarely the appropriate way to introduce GIS technology into public organizations (and it probably is not a good approach for introducing any kind of innovation). An incremental approach has demonstrated itself to be far more effective. The retention of the global design as a framework within which each single application has its place and the development of well defined, specific, rather simple yet integrated applications can prove to be the best strategy.

The lesson, which I probably consider as most relevant, is well expressed by the saying: "horses for courses". Every case is unique because of its national context, the local traditional and cultural attitude, and (last but not least) the people who develop, install and use the system. Thus, consideration of the Padova experience or any other successful implementation experience can never be followed mechanically. The local situation must always be considered in the application of GIS technology.

7. References

Barrett, S. and M. Rumor (1989) Computers in urban planning. Tutorial series of the 13th UDMS Symposium, Lisbon, Portugal, 29 May-2 June.

Chorley Committee (1987) Handling Geographic Information, Her Majesty's Stationary Office, London.

Craglia, M. (1991) GIS in Italian Urban Planning. Department of Town & Regional Planning Faculty of Architectural Studies, The University of Sheffield, Sheffield, UK.

Kvistad, Ann (1991) Range of GIS application in Norwegian local authorities. Experiences, Common Problems and Benefits. *Proceedings of the 14th UDMS Symposium*, Odense, Denmark, May 29-31.

Rumor, M. and R. Gonzato (1989) Computer-aided urban planning: the experience of the city of Padova, *Proceedings of the 13th UDMS Symposium*, Lisbon, Portugal, 29 May-2 June.

Scholten, H. and M. van der Vlugt (1989) Application of Geographic Information System in Europe, *Ekistics* 56: 304-311.

IMPACT OF GIS ON A SOUTHERN EUROPEAN COUNTRY: THE CASE OF PORTUGAL

ANTÓNIO M. ARNAUD
Faculdade de Ciências e Tecnologia
Universidade Nova de Lisboa
2825 Monte de Caparica
Portugal

ABSTRACT: This paper reviews the Portuguese administrative structure and organizational environment. The governmental environment is a major factor in the use and development of geographic information technology (GIT) in Portugal. This paper includes a description of activities in primary governmental organizations, universities, research institutes, and in the private sector, including activities of vendors and consultants. Some adoption and rejection actors and factors are enumerated and proposed for consideration in a strategy for diffusing knowledge and use of GIS in Portugal.

The coordinated action individuals, municipalities, the National Center for Geographic Information (CNIG) and the National Statistical Institute (INE) is proposed. In this conceptual framework municipalities are seen as potential statistical data suppliers and are increasingly GIT users. CNIG is seen as a main coordinator of geographic data production and promoter of cartographic data sharing initiatives, rather than as an information concentrator. INE is seen as the coordinator of socioeconomic and demographic data and a promoter of uses of its data with geographic information technologies,, namely with Desktop Mapping packages and micro based starter GIS for Municipalities.

Using these institutional champions as a base, proposal of key actions to foster the use of GIT in primary GIS application areas is proposed. The approach makes possible use of a system dynamics model.

1. Introduction

In the fifteenth century, the nautical astrolabe introduced the practice of navigation by latitudes using graduated charts drawn up by Portuguese cartographers. The celebrated discoveries that followed, not only led to a better understanding and representation of the globe, but also a dramatically changed world economic order and new geopolitical powers. The development of maritime empires, the emergence of new countries, and communication between civilizations, owed much to this important nautical instrument.

Nowadays, powerful computing and geographic information systems (GIS), Global Positioning Systems (GPS), and remote sensing capabilities are creating new methods to further improve our knowledge and representation of the Earth. They promise to help us better understand the environment and manage scarce resources. The countries which now lead in using geographic information technologies (GIT), and other communication technologies, are at a considerable economic advantage. A redistribution of power is occurring, without any need for armament.

I. Masser and H. J. Onsrud (eds.), Diffusion and Use of Geographic Information Technologies, 245–260.

Recent historic events have shown how quickly social phenomena can evolve, even in apparently stable democratic parts of the world. Social and environmental changes and pressures demand improved knowledge and timely decisions regarding resources and infrastructures. The need for geographic information systems (GIS) and (GIS)-based decision support tools is readily apparent.

A number of GIS for microcomputers with new user-friendly interfaces are allowing for greater accessibility to information about our earth. Geographic data viewers and desktop mapping tools are just mature enough to become as widespread for decision-makers as spreadsheets (Melgoza 1991). The GIS boom is in progress: "by the year 2000 there probably will be from 1 to 5 million GIS users in the world" (Dangermond 1991) (These figures would be considerably higher if desktop mapping users were included). How can a country best deal with such growth? How can it manage the information used and produced by such tools? How can it optimize resulting benefits?

The diffusion of geographic information technologies outside of the leading countries, although strongly dependent on application areas, is also dependent on digital data availability and on timely access to remote sensing data. Where there is current availability of digital data for the most important frameworks, such as cadaster, base maps, land use and road networks, the speedy diffusion of GIS-related applications is improved.

It is expected that utilities and cadaster application areas (i.e. those area where benefits are more evident) will play a leading role as first users. Initial progress in these application realms should spur digital base data production and increase data availability.

The administrative structure and political idiosyncrasies of each country are major factors with respect to the use and diffusion of the technology. Therefore, this chapter begins with an overview of the Portuguese administrative structure. A description of the distribution of information empowerment in Portugal is presented followed by a description of the main organizations most likely to be involved with GIT. These groups are more or less aware of benefits to be gained from their involvement. Finally I describe the organizational environment within which GIT is likely to be diffused in Portugal.

2. Administrative Structure of Portugal

Portugal covers 92,000 sq Km and has a population of about 10.4 million. The country is one of twelve European Economic Community (EEC) members. It has five continental administrative regions and two autonomous regions, the islands of Madeira and the Azores. Portugal is administratively divided into 305 Municipalities, ranging from 7 to 1,720 sq Km (mean area: 323 sq Km). These are divided into a total of 4208 districts, the smallest administrative units, with boundaries based on parishes.

Comissões de Coordenação Regional (CCR) are decentralized bodies of the central government in each region. In the future, each region will have an elected regional board and a regional assembly, in accordance with the Constitution of the Republic. The two Metropolitan areas, Lisbon with 2.5 million inhabitants and Oporto with 1.5 million inhabitants, are now organized in two supra-municipal governmental institutions. It is expected that transport management and utility data sharing agreements will be developed among the various agencies within these new institutions. The Ministry of Finance, Tax Directorate, has a delegation in every Municipality. The National Bureau of Statistics (INE) now has delegations in each region which are working increasingly with the Municipalities.

The Ministries of Health, Education, Employment, and Agriculture are decentralizing their services and their regional directorates are requesting information tools. The Post Office has divided the whole country into 459 distribution areas. However, this level of division is inadequate and there is still no nation-wide street directory or address coding system.

3. Organizations involved with Geographic Information Technologies

The organizations involved with GIT are the primary producers of cartographic and alphanumeric data, and the users and potential users in central, regional and local government. Some international GIT projects also involve research institutes and Universities. The private sector, beginning mainly with data acquisition and conversion in circumstances where there is a lack of governmental data, is also expected to be a major player in the near future.

The institutional framework of potential users and clustering in that framework by common data sets, interest and complementarity, are the grounds whereby diffusion takes place. Therefore, an understanding of this framework and an examination of possible links among users within the framework seems to be crucial.

3.1 CARTOGRAPHIC DATA PRODUCERS

The main cartographic data producers in Portugal are the National Mapping Agency, *Instituto Geográfico e Cadastral* (IGC), and the Army Cartographic Services, *Serviços Cartográficos do Exército* (SCE). IGC is responsible for the geodetic, leveling and gravimetric networks; the land cadaster (1: 2 000 and 1: 5000) which covers 50% of the country and 10% of the properties, but unfortunately is now somewhat outdated. IGC is making orthophotomaps in conjunction with digital terrain models at 1: 10000, mainly for Municipal plans. Approximately 65% of the country is now covered. IGC is also responsible for topographic mapping at 1: 50000 and for derived cartography.

SCE is also involved in map production. The accessibility to their digitized maps has so far been limited to military users but efforts are now being made to increase their availability. Their 1: 25000 coverage is the most widely used source for planning.

A number of institutions, such as the National Geological Services, the Ministry of Agriculture (land use) and the Directorate for Natural Resources (National Environment Atlas), produce thematic maps. The National Physical Planning Directorate (DGOT), produces large scale maps for urban areas to support traditional needs. In addition to these institutions, eight private companies are providing digital maps, mainly using AutoCad, Intergraph microstations and raster/vector conversion,

3.2 ALPHANUMERIC DATA PRODUCERS

The main alphanumeric data source is the National Bureau of Statistics (INE), now a public company, which is responsible for the coordination of all statistical data production and distribution of the data. New data collecting and diffusion methodologies (using decentralized administrative data processing and statistical data standards) are being introduced to replace the traditional paper-based data flows.

Activities and possible roles of some of the government departments that are potential sources of geo-coded alphanumeric information are described as follows:

- The Ministry of Health has 21 Regional Centers in which decentralized data files are being organized with, typically small investments in hardware. Progressive alignment with the Health Information Systems in more developed countries in the European Community is expected.
- The Ministry of Education is being decentralized and is improving its own Statistical Information System. The Planning Department within the Ministry is starting to use GIS.
- The Ministry of Employment has its own Statistical Department and the Institute for Employment and Professional Training has 5 regional delegations with a national network of 80 Employment Centers.
- The Ministry of Agriculture has 7 Regional Directorates and some (e.g. the Azores) are quite advanced in the organization of farm databases.
- At the Tax Directorate, central tax files relative to income tax (IRS) and VAT, are important potential sources of geographical information should there be an improvement in address quality and a national street file.
- The Electoral Roll, as might be expected, is strictly compatible with the lowest administrative units (districts). Each district has its own traditional card file. These decentralized files include information as to the name, sex, age, address, place of birth and ability to read of electors. New file managers include the production of local statistics.

Thus, data from these Ministries, Directorates and local administrative files present various opportunities for the production of small area statistics, suitable for local planning and also for cross studies at regional and central levels. However, the usefulness of such data for statistical purposes is dependent on address quality, geocoding procedures and updating routines.

3.3 EXPERIENCES AND POTENTIAL USERS

The potential users of geographic information technologies are the General Directorates for Planning, Environmental Quality, Natural Resources, Forestry, Agriculture (Central Agencies), the Regional Planning Commissions (CCR), the 305 Municipalities, and the utility companies. For instance, the Environmental Quality Agency, participates in the EEC's CORINE project LAND-COVER, in cooperation with CNIG, and is developing an Environmental Quality Information System based on Arc/Info. Several important projects are underway on waste, water, ground water vulnerability and simulation models. Also, several governmental departments are cooperating on international projects like the DRIVE program.

The *Comissões de Coordenação Regional* (CCR) are preparing their Regional Physical Plans (PROT) with the intention that all the information gathered will become part of the Regional nodes of the National Geographic Information System which is based on Intergraph. The 305 municipalities must now finish their Master Plans (PDM) by the fall of 1992, as only five had their plans approved in the fall of 1991 (the previous deadline). In the larger municipalities, some water and sewage services are starting to use AM/FM systems and some of these facilities management focused systems are trying to coordinate data collection and conversion with their planning departments.

Benefits from adopting GIT are most evident in the utilities sector. The telephone system for Lisbon and Oporto (TLP) now utilizes a SIEMENS/SICAD system and there is an effort to digitize base maps and develop data sharing agreements with a few Municipalities. These Municipalities in turn are digitizing their large scale base maps, sometimes with the Electricity Distribution Company (EDP). The Water Distribution Company (EPAL) of Lisbon Metropolitan Area has a first generation, Autocad-based GIS.

Gas distribution, formerly confined to Lisbon, will be extended via a natural gas distribution network covering the two Metropolitan areas and Coimbra. The regional gas distribution companies involved in this project are launching large scale (1: 1000) digitizing, again in cooperation with Municipalities and other operators. They are leading GIT users and, as one of the most important basic data collectors, they can introduce many Municipalities to this technology.

Pulp and paper companies have also been a major user of GIT. The Pulp Producers Association (ACEL) did a full survey using infrared false color air photos at a scale of 1: 15 000 in 1990. ACEL is leading the production of Forest image maps, with integration of remote sensing images and GIS technology in cooperation with the National Forest Directorate and CNIG. These maps will be used for the analysis, control and monitoring of fires, clear-cut and new forestations over the entire country.

In the marketing sector, Readers Digest Portugal (SRD) and the Association of Direct Mailing have been the most dynamic potential users of small area statistics. They own one of the most comprehensive street files and a client file with more than 2 million registers. Recently, market research companies have started to use small area statistics for area sampling and determining market prospects. GIS is expected to be a natural tool for these studies.

3.4 UNIVERSITIES AND RESEARCH CENTERS

A number of universities and research centers are now planning to teach GIS introductory courses and do training with GIS software packages. They are also cooperating in the development of applications mainly in Forestry, Agriculture and Environment Impact analysis. Activities of some of the universities and research centers include the following:

- The National Geographic Information Center (CNIG), a central administration agency, has been established as the coordinating body for the National Geographic Information System (SNIG). CNIG, supported by the Secretariat of State for Science and Technology, includes a central node with 5 regional nodes in the mainland and two in the Autonomous regions of Madeira and Azores. SNIG is an Intergraph-based top-down approach. CNIG is also active in several research projects both in Portugal and in the European Community. Major research areas are remote sensing, cartographic databases, mathematical modeling of spatially distributed phenomena using GIS and multimedia technologies.
- The Statistics and Information Management Institute (ISEGI) at the New University of Lisbon runs graduate and master degrees and provides data processing services. A training program for officials is underway with INE.

- A consortium consisting of a research institute, LNETI, a private company and NATO/PO - Satcart is using ISS and Arc/Info to produce vector cartography of burned forest areas from remote sensing images.
- Participation in NATO/PO - Rivers, lead by the Faculty of Engineering at the University of Oporto, will include one of the first training programs in Portugal with the Arc-Info Lab kit.
- The Environmental Systems Analysis Group at UNINOVA, a research institute at The New University of Lisbon (FCT/UNL) was responsible for the development of the National Environmental Information System (SINAIA) and the National Water Resources Information System (SNIRH). These systems store data on air and water quality, air pollution, point sources in water resources, environmental uses and users, and hazardous waste and are manipulated with a GIS based interface for query and analysis.
- At the New University of Lisbon (FCT/UNL) and UNINOVA a pilot project for a local Statistical Information System, SIEM, covering four municipalities, aims at producing local data for integration at the national level. Some of the experiments and research in supplying data to GIS include: 1) MUNICENSO, small area statistics and thematic map production from 1981 Census data of population and housing; 2) Electoral Roll; RELOC, a micro- based user friendly application that facilitates the management of the electoral roll and allows for the production of local statistics; 3) local directories of streets and place names supported in a user friendly application, RELUR; 4) socioeconomic small area statistics from the Tax-payers central file, INCLOC, and 5) small area statistics from local administrative files, ADLOC.
- A GIS introductory course, focusing on data structures will be given this year in the M.Sc. programme in Computer Science at FCT/UNL.
- At the University of Aveiro, the Geo-sciences Department is working with remote sensing images and the Department of the Environment is planning to cooperate with Municipalities, using Arc-Info.
- The Agronomic Engineering Institute, using SPANS, and the University of Trás-os-Montes, using Intergraph, are working on forest applications.
- The Agronomic departments at the Universities of the Azores and Évora are working with Intergraph and ERDAS-PC, to produce land use maps.
- The Instituto Superior Técnico, the main Engineering University Institute, is planning to introduce GIS in the Master of Science in Urban and Regional Planning and some other Universities have applied for funding for GIS laboratories.

3.5 VENDORS AND CONSULTANTS

An increasing number of Service Companies and consultants, together with the main vendors AutoDesk, ESRI, ERDAS, Intergraph, McDonald Douglas, SIEMENS, UNISYS, and sometimes in association with computer makers, are promoting solutions in several areas. These include mainly data conversion, utilities, environment and agriculture applications. Vendors are taking advantage of the low awareness of GIT use and the appropriate steps for implementing GIT, often selling the system as a first implementation task.

There are a growing number of consultants in planning and impact analysis using GIT in Portugal, however, there is a lack of experience in the community generally in implementation, feasibility, analysis, data sharing and system selection.

4. Use and Diffusion of Geographic Information Technologies

The diffusion of GIT appears to be greatly enhanced when supported by a national institutional framework where digital data is available and flows into and between application areas driven by skilled experts. In countries like the UK (Masser 1988), the use of GIT is rapidly increasing. It is supported by universities and regional research laboratories throughout the nation and by government organizations. These have contributed to improving awareness, providing guidelines for coordination, training personal, and demonstrating new applications to an informed user community. Data availability is increasing rapidly as a result of their national geographic information policy.

In thc USA, NCGIA and NSF (Goodchild 1992) are contributing to similar goals (Batty 1992). Moreover, basic data is available from the US Census Bureau (e.g. TIGER database), US Geological Survey and the National Ocean and Atmospheric Administration (NOAA). These digital maps provide solid frameworks for GIT diffusion. These initiatives, together with the efforts of the Bureau of Land Management (US Land Information System) and the US Forestry Service (Parker 1991) are providing users with a considerable amount of basic data. A large number of vendors provide software and services taking advantage of the large number of users. A growing number of consultants play an important role by helping to save costs and reduce implementation time.

In some other countries, like the Netherlands (Scholten 1991) and Scandinavia (Østensen 1991), GIT is spreading over a large user community, mainly under the pressure of environmental control and municipal management with strong governmental support. The diffusion of GIT in these countries is in a *steady progressive state*, where greater experience and maturity of the GIS community exist.

For countries just starting, the "taking-off" effects within the overall diffusion process are of greatest interest. A particular problem is the low availability of public and commercial data products that bridge nations and cultures. Barriers found in one nation and the means of addressing the barriers, may be very different from one nation to the next. (Wegener 1991). So, diffusion models appropriate in the GIT mature societies may not be generalizable for many other countries. However, the experience of mature GIT societies is valuable and must be studied. Some of the knowledge from their diffusion experiences, such as those experiences showing data sharing benefits and implementation pitfalls are likely to be valuable for newcomers.

A methodology of technology transfer analysis suitable for some of the southern European countries, might be based on a model of the institutional framework. Such a framework should include consideration of current and future data flows using state-of-the-art data conversion techniques and should accommodate hardware trends. For those countries where availability of data and skilled personnel is dramatically low and where the level of awareness among politicians and decision makers is also low. Two models which might be followed in encouraging the more rapid diffusion of GIT are as follows:

1) Allow or promote a liberal unrestricted and possibly anarchic spread of GIS. Under this scenario we are likely to see applicability focused on short term benefits in compartmentalized areas with each department responding to those vendors which best meet their needs. Lack of coordination among all parties is likely to result in multiple personal and institutional conflicting interests.
2) Define a National Geographical Information policy and develop and implement a GIS diffusion strategy which promotes as much coordination between agencies as possible.

With the first choice it would make sense to model growth and develop diffusion strategies in a way similar to that proposed for more developed countries. This process seems to risk high information waste. The second approach seems more useful and rational for monitoring progress and improving "healthy" growth of the main application areas of geographic information. If reasonable guidelines are found, both a higher speed of diffusion and a higher level of efficiency could be attained. This should lead to large overall savings, and a higher acceptance by the user community.

Let us take, as a primary reference, Onsrud's view of diffusion: *Diffusion refers to the process of communicating an innovation to and among the population of potential users who might choose to adopt or reject it* (Onsrud 1991). Adoption and rejection actors and factors must be enumerated and considered within a population of potential users clustered in application areas, together with an analysis of the complex relationships within the organizational framework.

4.1 CHAMPIONS AND BARRIERS

Among the adoption actors are the "project champions", those persons so often referred to in talks about implementation in each different application area. There are also "institutional champions" who are typically the effective coordinators of both cartographic and statistical information among numerous division of an organization or among organizations.

Some factors affecting adoption are: awareness of the usefulness and benefits of GIS applications, existence of trusted champions, availability of finances, digital data and expertise, involvement of politicians (e.g. GIT enables shortages to be managed and growth controlled), the status of user interfaces, learning curves, and data sharing proposals; and the need to replace old fashioned filing systems.

Modeling the diffusion of GIT must also include rejection analysis. Some rejection factors include the fear of losing power (or even of being fired) and the existence of personal interests (GIS is differently related to power in the less favored regions). Institutional, technical and financial barriers include legal restrictions, cumbersome standardization or lack of it, data cost, and difficulties in data sharing or in multiparticipant projects. The lack of skilled personnel and the "invisible" benefits, are also important rejection factors, among the many others already almost exhaustively enumerated (Chorley 1987).

As in the Government sector, there are many recalcitrant organizations, which expend a great amount of energy in trying to avoid change (Ackoff 1974). This type of institutional behavioural barrier must be taken into consideration. Research institutions act as activists and their actions are important for diffusion. However, active and involved management could certainly promote data sharing and GIT diffusion.

Digital data is less available in Portugal than other European countries. This is one of the most critical bottlenecks to GIS implementation and diffusion in Portugal. Digital base maps, street center-line files, topometric data bases, timely statistical data and data standards are the basic scarce information resources that need to be managed at the national level.

Initiatives towards improving data availability could be taken by institutional champions (data actors). In Portugal the most important data actors are some of the central and regional organizations described above and the 305 Municipalities. CNIG and INE could be the natural coordinators of cartographic and statistical information resources - institutional champions of GIT diffusion - with the participation of the mapping agencies and private sector in cooperation with universities and research institutes. It is assumed that global needs would be matched among the participants and major pitfalls avoided.

In the following sections, the geographic information aspects of these organizations are analyzed, followed by a personal view of the most important actors in GIT diffusion in Portugal in six application areas.

4.2 MUNICIPALITIES

In most municipalities, as in more developed countries, automation started with primary operational needs and administrative routines. These include water billing, wages, accounting and financial applications. However, most local authorities are still facing problems with this early automation stage. They are not yet aware of the benefits to be attained by using GIT and the benefits of producing digital geographical information from local administrative files.

A survey held in 1990 on the use of computers in local authorities revealed that about 50% of the municipalities had computer resources even though only 5% of their manpower has a high school degree. Most difficulties relate to manpower, training, software functionality and, for those groups which are not yet equipped, financial resources.

Because the human resources and technical infrastructures needed for information processing, for local updating, and for use of more recent applications are not generally available, these activities are largely dependent on training through funding from the central government or the European community. This major problem is a barrier to diffusion.

Most of the Municipal Master Plans (PDM) are produced by consultants or consortia and are delivered in traditional paper form. In some cases, data cemeteries are being created. This is occurring where there is a lack of coordinated effort and regulations on digitizing standards. However, a few plans are being made in such a way as to allow information to be included in a local GIS. Some maps have been digitized under AutoCad DXF and suitable data structures could be built. A few planning consultants and service companies are using Arc/Info to prepare the plans. Unfortunately, the legal responsibility to create the PDM has not been used as an opportunity to build an updated cartographic mosaic, or even a distributed geographic database. However, in some cases, the information gathered for PDM may be integrated in local GIS, allowing for dynamic planning, if human and financial resources are available.

As much data is generated locally there is a need to implement locally updated referencing systems to allow the automatic geocoding of administrative data files. Local segment-based referencing systems within a European standard are of the utmost importance since these will allow data integration for average and large cities. In these localities, utility network management systems would be cost effective. If the conditions are met to solve these problems, local authorities could play an important role in GIT diffusion, as data producers and also as users. The referencing system is seen as a major "seed" for diffusion.

As data suppliers the data flows of authorities must be completely changed. They should keep micro data, keep it timely, maintain the required precision (referencing). Standard aggregated data seems appropriate. By adapting this approach the fear of losing power is removed and the currency of information is improved. So, with demonstration projects including local data and the demand for reliable and timely information for decision support could be created.

Past experiences in more advanced countries with respect to trends in hardware costs and software developments, point to a growing involvement of municipalities (even the smallest) in GIS use, presuming minimal data sets are available. So, it is believed that GIT will diffuse (be adopted sooner or later), according to needs and depending on size, geographic location, on the possibility to attract skilled personnel and also on politicians and internal established powers (adoption factors). However, these factors, together with the growing need for updated information in order to monitor planning and social phenomena, call for the coordination of the growth of local GIS. These elements should be consolidated at regional and central levels. In order to get municipalities generally involved in GIT, a set of demonstration projects in various application areas (see section 4.5), which include analyses of costs should be established to aid the diffusion process.

4.3 ROLES OF A NATIONAL CENTER FOR GEOGRAPHIC INFORMATION

The role of a National Center for Geographic Information is well known in countries where information resources are not so scarce, such as the USA, the UK and France. Such centers are key in the promotion of the economic use and diffusion of technology, as pointed out in the conclusions of the Chorley report (Chorley 1987). However the Portuguese situation is quite different because of its scarce information resources, namely with respect to data availability, expertise and institutional environment.

The National Geographic Information Center (CNIG) in Portugal, because of its coordinating role, could be important in fostering data sharing and exchange agreements, promoting standards (national and international), assessing the value of data and fostering new data flows. However, institutional conflicts resulting from the top-down conception of the National Geographic Information System (SNIG), must be avoided. Particularly, the integration of a number of local nodes with SNIG, which is foreseen by CNIG/SNIG. This approach may cause technical and institutional conflicts with the National Statistical System, as the local nodes are primarily actual and potential providers of statistical data (Arnaud 1987).

In order to achieve these goals a Government data survey, including cartographic and attribute data, must be made. The survey should be made in cooperation with producers of GI and with representatives from users and interest groups, namely local authorities. This could constitute a starting point for a new geographic information order among governmental organizations and private institutions.

Teaching implementation strategies and discussing data structures are also key actions, to be made also in cooperation with other data actors. The central role must be accepted by data users and producers after discussions and demonstrations of strategic advantages.

Among other concerns, a National Geographic Information Center must create an acceptable image. It should treat all vendors equitably and the center should avoid taking a position in the debate on recovering the value-added components of information. Moreover, fear and mistrust of partners is a major cause of resistance and rejection and is to be avoided.

Briefly, a National Geographic Information Center, to be effective, must be an interactivist organization which respects pre GIT administrative structures and avoids monopolizing Geographic Information. Coordination of information resources should prevail over concentration of information resources.

4.4 INE AND THE NATIONAL STATISTICAL SYSTEM IN THE GEOGRAPHIC ERA

The National Bureau of Statistics (INE), is responsible for the coordination of statistical data production and its dissemination. It is working out new data collecting and dissemination methodologies, promoting statistical data standards, decentralizing administrative data processing, and distributing data in magnetic form (complementing traditional paper publications).

The Geographic Base for Spatial Referencing (BGRE), in process since 1984 and used during Census'91, is the geographic framework for statistical information in Portugal. The Base includes 118,000 polygons (districts, wards and blocks) and maps are available at scales of 1: 2,000, 1: 5,000, 1: 10,000 and 1: 25,000. The process of digitizing BGRE, is currently driven by the pressure of local needs. Digitizing BGRE maps by local municipalities is an expensive and time consuming task. Those that are digitized often have an expected low mean rate of use, as most of the municipalities are not yet technically skilled in how to update and use those boundary files.

However, new uses of statistical data are appearing. Among the applications of GIT are census data display, demographic analysis, marketing analysis and facility siting make direct use of census data. Many other applications use census (Francis 1985) and other socioeconomic and demographic data as important layers. Many new users of statistical data often using microcomputers, carry out their own data processing using statistical data analysis and graphics packages. This emerging market needs data in suitable formats and support.

Affordable and easy to use GIS and desktop mapping programs for microcomputers require new data sets. Particularly needed are socio economic and demographic statistics. Answering to these new uses and users, INE is providing a package of small area statistics, to be used mainly by the municipalities, within MuniCenso 91 (Arnaud 1991). A starter set of 25 variables suitable for general needs will be available along with a set of around 250 variables which are geographic information suited for to local planning needs. The variables may be used with spreadsheets, desktop mapping (the simplest geographic information tool), or GIS. A set of 2000 variables will be also produced on demand for municipalities with more advanced GIS and applications. These provided variable sets should help ease the understanding of GIS.

The automation of local and regional administrative processes, sometimes with easy to use data collecting programs at the local level, should allow for national and regional consolidation of local statistics at low marginal cost. A new data approach which includes standard statistical data collecting applications, is commencing with building license records and hopefully the electoral roll.

INE, by fostering coordination and data sharing agreements and by developing an address coding system is responding to the needs of users and is improving accessibility to several administrative data sets. This is allowing local production of timely and aggregated socioeconomic and demographic information.

That production of small area statistics is therefore a diffusion factor towards the building up of coordinated cellular statistical GIS. This decentralized data collection could change national data flows in the National Statistical System (SEN) to which GIT diffusion is basically linked. GIS tools are commonly used for statistical and spatial analysis and it is now generally accepted that most statistical data is geographic. INE may play a role as a GIT diffusion champion. It can do this by supporting a strategy for delivering geographic information including such components as:

- produce digitized boundaries of districts and localities, on the largest scale possible, updated with Census operations.
- produce thematic maps with desktop mapping facilities using high quality output devices.
- promote BGRE, in cooperation with municipalities, by digitizing local geographic bases.
- promote Munistat, small area statistics, including those from Municenso 91.
- maintain and disseminate tables and norms in digital form, e.g. table of localities, European codes of professions, economic activities, and other meta-information.
- participate in a national referencing system with bottom-up updating procedures which would incorporate a toponymic data base and a segment based referencing system (similar to TIGER) .
- cooperate with administrative data owners to produce timely and quality statistical data and work towards a network of local statistical information systems.
- diffuse sets of census and other selected statistical data sets in magnetic form (Municenso) and promote CD-Rom as well as CD-I, DVI multimedia editions.
- provide demonstrations of the use of Census data for municipalities, particularly with Desktop Mapping and GIS software for microcomputers and workstations.

It is believed that this will greatly improve awareness among local authorities of data usefulness and will result in an increased use of data by the government and private companies. INE is now promoting GIT through the delivery of sets of statistical data, namely data sets from Census 91 and 1989 Agriculture Census which are ready for thematic map production, including maps which contain digitized administrative boundaries in small scales. Eurostat is now developing the European Geographic Information System (GISCO) (Cubitt 1992) which is aimed at providing the Commission with information needed for decision support (Heath 1991). This GIS is being fed with data both from the National Statistical Institutes and other National and European sources. INE is finding its place in this new information order. The production of geographic information for decision makers and the promotion of its use at national, regional and sub-regional levels are important actions in diffusing GIT. The strategy of INE might be to foster the meeting, as an institution champion of top-down coordination initiatives from European, national and regional levels with bottom-up local data production (Municipal - regional - national).

4.5 APPLICATION AREAS

GIS growth will be driven by applications (Dangermond 1991). Demonstration projects that provide solutions to key problems or demonstrate use of the technology in application areas could provide *seeds* for GIT growth. Technology must be diffused not as a multipurpose tool but as a set of tools, each one with a different scope and implementation methodology.

A GIS taxonomy proposed by Goodchild (1991), describes application areas in terms of functionality, data volume and access speed. This taxonomy seems highly suitable as a starting point for addressing GIS use:

Natural Resource Management - This area which includes Forestry Applications, and Environmental Control and Monitoring is ideally suited for a centralized diffusion model (Onsrud 1991). The key issues are timely access to remote sensing data, data sharing and cooperation with research institutes and universities. Private Pulp and Paper Companies in Portugal are natural partners that could be leaders as their benefits are known.

Infrastructure Management - Municipalities, mainly in the metropolitan areas, are starting now to use AM/FM. Some have been obliged to order digitized large scale base maps. In this area, benefits and costs are more easily calculated compared to many other application areas. It is evident also that data sharing initiatives may be successful due to the heavy costs of data collecting in this application area. The Telephone (TLP), Electricity (EDP) and Gas (GDP) companies together with Municipal Civil Engineering and Planning Services; water and sewage services and transportation departments are natural partners within the Municipalities. A few agreements have already been made and it is hoped that, as a consequence, the Municipalities in the Metropolitan areas will benefit from the operators interest. A decentralized diffusion analysis would be suitable, where EDP, GDP and TLP could act as institutional champions.

Land Information Systems - Large scale maps are expensive but local property taxes could allow decentralized, local LIS. Data sharing is possible with utilities. Standards and legal and administrative barriers are key issues to be addressed. The Tax Directorate and the Municipalities could be the institutional champions in this area.

Vehicle Routing and Scheduling - The basic needs are street and road networks and an address coding system. According to Donald F. Cooke (Antennuci 1991) 5 to 15% out of a total $ 0.5-trillion annual expenditure on logistics could be saved with route optimizing and scheduling systems. This is a strong argument for investing in the data infrastructure needed to apply such models. Demonstration projects for garbage collection, emergency dispatch, and route planning should be accomplished. Institutional champions may be those who benefit most from cooperation with basic data providers. The private sector can play an important role.

Marketing and Retailing - The amount of available data and its timeliness together with the opportunity to geocode addresses are key issues for the development of this area of applications. To be usable, census blocks must be compatible with address referenced data. Here again, the address coding system is a basic need. INE must play the key role in this area by promoting the referencing system and small area statistics together with demonstration projects. Use of census data for Municipal planning would be an important function to demonstrate.

Districting - Districting is not of much interest in Portugal since political boundaries are stable. However, the methodology might be used for school catchment and sales areas. The most interested institutions would be INE and the Department of Planning of the Ministry of Education.

It is believed that diffusing GIT in Portugal would be best accomplished by considering as many application areas as suitable and the data flows between applications should also be considered. The role of consultants and vendors in the diffusion process, as promoters of application packs, could be important but they should be subjected to a national GI policy.

5. Conclusion

It is expected that, as has happened in the USA and parts of Europe, the GIS market boom will occur in other countries and regions with a second GIS wave. In order to avoid the losses and waste which can result from vendor-driven growth, geographic information resource policies at national and international levels are needed. Such policies are particularly needed in those countries such as Portugal with a low levels of GIS use. These countries, more than others, must regard data as expensive and perishable. They need to realize the value of data collection, avoid duplication, coordinate actions and promote decentralized updating. They may benefit substantially by starting after others have gained experience.

Considering the technological trends, the basic needs of countries covered by the second GIS wave may be overcome with geographic information resource policies that are defined with the participation of the main pre-GIT producers (or coordinators) of cartography and statistical data. This participatory policy development approach should help avoid institutional resistance to the new technology. Discussions on data and GI strategies must be promoted in open forums with users and data producers. In Portugal, and perhaps in other countries as well, a national cartography plan, minimizing overall waste and inefficiency, a National Data Sources Catalogue, and national directory of streets and place names are priority investments for GI management.

The growing availability of simpler to use GISs and the awareness of the benefits of the technology, are leading more and more Municipalities to use GIT. Depending on how they start, their locally collected information may or may not have a regional and national value. Thus, there is a need for institutional champions to develop and drive a global information resource management policy.

The production of local statistics with local geocoding may be included in the national statistical systems . This would allow for the production of timely geographic information. However, automated systems must start with traditional geographic information such as toponymic plaques, road and street information, names and numbers on doors. Desktop mapping tools, fed by census and administrative data could serve as starter kits for learning about and using GIT, particularly at the local level. These might then be complemented by micro based GIS demonstration projects.

Data availability, data sharing initiatives, cartography, statistical and GI standards, cost/benefit analysis and demonstrators for the main interrelated application areas are key actions to be considered in planning and driving the diffusion process. The dissemination of digitized administrative limits at a small scale (1: 250 000) at low cost could foster a number of applications from the desktop mapping level of use.

A multinational research initiative aimed at bridging national boundaries, such as the effort the European Science Foundation is launching (Masser 1992), will certainly result in a faster and wiser diffusion of this powerful emerging technology. Decision making and public participation (Dangermond 1991) will change as GIT and related technologies are more incorporated into our daily lives. Those countries able to take advantage of the second GIS wave yet to come may be able to use GIT to help lessen socio-economic gaps for their citizens. Whether they are able to do so will depend substantially on their nation's information policies.

References

Ackoff, R. (1974) *Redesigning the Future*, John Wiley, New York.

Antennucci et. al. (1991) *Geographical information systems: A guide to the technology*, Van Nostrand Reinhold, New York.

Arnaud, A. (1987) Integration des Systèmes d'Information Locaux dans un Système National d'Information Géographique, *12th Urban Data Management Symposium*, Blois, France, May, 1987, 371-82.

Arnaud, A. (1991) Municenso 91, Take Away Census Data, *URSA-NET Fifth Patras Seminar/Forum*, Patras, Greece, June 7-9, 1991, *Computers in Planning* 8: 61-72.

Batty, M. (1992) Position paper, *ESF Workshop on Socioeconomic applications of GIS*, Sintra, Portugal, June 10-14, 1992.

Chorley, Lord (1987) *Report of the Committee of Enquiry into the Handling of Geographic Information*, Department of the Environment, HMSO, London.

Cubitt, R. (1992) The Role of Eurostat in GIS, *Mapping Awareness 92, Proceedings*, London, February, 1992, Blenheim Online, London, 403-411.

Dangermond, J. (1991) Commercial GIS will be driven by diverse applications, *The 1991-92 International GIS Sourcebook*, GIS World Inc., Colorado, 390-91.

Francis, K. and Norris, P. (1985) The Census. In *Information Systems for Policy Planning in Local Government*, Longman, BURISA, Bristol, 20-37.

Goodchild, M. (1991) A taxonomy of GIS systems, *The 1991-92 International GIS Sourcebook*, GIS World Inc.

Goodchild, M. (1992) Position Paper, *ESF Workshop on Geographical Databases*, Aix-en-Provence, France, March 12-14, 1992.

Heath, D. (1991) Eurostat Interest in GIS, *Workshop on European Research in Geographical Information Systems*, Davos, Switzerland, January 24-26, 1991, 105-107.

Masser, I. (1988) The Regional Research Laboratory Initiative, a progress report, *International Journal of Geographical Information Systems* 2(1): 11-22.

Masser, I. and Salgé, F. (1992) European Science Foundation Initiates GIS Research Programme, *GIS Europe* 11(5): 26-27.

Melgoza, C. (1991) Desktop mapping: The next hot techonology? *The 1991-92 International GIS Sourcebook*, GIS World Inc, Colorado, 392-94.

NCGIA (1989) The NCGIA Research Agenda, *International Journal of Geographical Information Systems* 3(2): 117-136.

Onsrud, H. and Pinto, J. (1991) Diffusion of geographic information innovations, *International Journal of Geographical Information Systems* 5(4): 447-467.

Østensen, O. (1991) Infrastructure for Geographic Information Management, *14th Urban Data Management Symposium*, Odense, Denmark, May 29-31, 537-554.

Parker, D. (1991) Introduction: Watching GIS Mature, *The 1991-92 International GIS Sourcebook*, GIS World Inc., Colorado, 1-4.

Rhind, D. (1991) The next Generation of Geographical Information Systems and the Context in which they will Operate. *Workshop on European Research in Geographical Information Systems*, Davos, Switzerland, January 24-26, 1991, 31-37.

Scholten, H. (1991) Geographical information systems for public and environmental health: value-added research and decision-making,. *Workshop on European Research in Geographical Information Systems*, Davos, Switzerland, January 24-26, 1991, 89-104.

Wegener, M. and Junius, H. (1991) 'Universal' GIS versus national land information traditions: internationalisation of software or endogenous developments, *Workshop on European Research in Geographical Information Systems*, Davos, Switzerland, January 24-26, 1991, 167-174.

GEOGRAPHIC INFORMATION TECHNOLOGY AS APPROPRIATE TECHNOLOGY FOR DEVELOPMENT

TIMOTHY J. CARTWRIGHT
Faculty of Environmental Studies
York University
Toronto, Canada

ABSTRACT: The purpose of this paper is to discuss the special case of the use and diffusion of geographic information (Gl) technology in third-world countries. In this paper, Gl technology is presented as advanced technology that is becoming steadily more "appropriate" for use in physical and spatial planning and management. However, the continued evolution of Gl technology and its use and diffusion in developing countries will not be automatic or inevitable: they will depend on a number of key factors, including better training and understanding of the cultural context, lower prices, and more user-friendly software.

1. Introduction

Geographic information (GI) technology is the technology of acquiring and using spatial information. Thus, it includes both remote-sensing and geographic information systems (GIS). As such, GI technology is a good example of an advanced technology that is steadily becoming more "appropriate" in terms of planning and management, in both industrial and developing countries. However, the evolution of GI technology from advanced to appropriate technology is neither automatic nor inevitable: it needs to be planned and implemented. The purpose of this paper is to discuss some of the factors involved in promoting the use and diffusion of GI technology as appropriate technology in third-world countries.

2. The Nature Of Appropriate Technology

Appropriate technology is ultimately technology that is suitable and accessible— suitable for the job to be done and accessible to the person doing it (Schumacher 1973). There are two kinds of myths associated with this concept. One myth is the assumption that appropriate technology is invariably "tried and true" rather than "leading-edge". In other words, appropriate technology is, by definition, low-risk technology. Then there is the opposite kind of myth: i.e. the assumption that today's state-of-the-art technology is by definition more appropriate than yesterday's. In short, high technology means better technology.

Both these views, however, are over-simplified and unhelpful. It is true that there have been cases (in agriculture and construction, for example) where old, even centuries-old, technology

I. Masser and H. J. Onsrud (eds.), Diffusion and Use of Geographic Information Technologies, 261–274.

has been successfully revived. But that does not mean appropriate technology must of necessity be out-of-date—nor, of course, that all out-of-date technology is appropriate. Similarly, advances in technology are just as likely (at least in the short run) to make the new technology less appropriate as more appropriate. If new technology is more expensive and more difficult to use than the old technology, for example, then it is likely to be less appropriate for most applications and most users. "Appropriateness", in other words, is a function not just of the technology, but of what Eric Trist and others (Trist and Bamforth 1951; Emery and Trist 1963) have described as a "socio-technical" relationship between the technology, its users, and its uses.

Take the case of information technology. Clearly, computers have not always been appropriate technology. In fact, for the first forty or so years after they were invented, computers were probably quite inappropriate for many applications related to physical and spatial planning. But the technology—and its relationship with users and uses—changed dramatically with the development of the microcomputer. For it was the microcomputer that created what is now called "end-user computing"—that made it possible for ordinary people to do their own computing (Levy 1984). Thanks to the arrival of low-cost microcomputers, users no longer had to rely on access to hardware over which they had no control. Thanks to off-the-shelf, "generic" software packages (like spreadsheets and database-management software), users could write their own programs, or at least readily understand programs that were written for them. Thanks to the explosion of books and popular magazines on computing, users were no longer dependent on computer "experts" who had little knowledge of or interest in specific applications. Thus, information technology is not inherently appropriate or inappropriate. It became appropriate in the 1980s, when computing became accessible—physically, financially, and intellectually—to the ordinary person.

In his delightful book, *Hackers*, Steven Levy (1984, p. 157) writes that, because of microcomputers, "the practice of computing would become so widespread that the very concept of it would change". Microcomputers, he continues (p. 337) "were amazing not only for what they did, but also for their accessibility.... people totally ignorant about computers [could learn] to work with them and gain in confidence, so that their whole outlook on life had changed. By manipulating a world inside a computer, people realized that they were capable of making things happen [in the world outside the computer] by their own creativity. Once you had that power, you could do anything."

In short, in the 1980s, computing became one of the mass media. Moreover, the revolution has to some extent been self-sustaining. When ordinary professional people began to see how computers could facilitate what they were then doing by hand, they naturally began to think of how computers might also allow them to do completely new things—things they had hitherto not been able to do at all, because they were too expensive or too time-consuming to do by hand. To give just one example, it was obvious that computers could automate registration of land titles; it gradually became apparent that computers could also make it possible to detect possible cases of tax evasion, to model alternative systems of land taxation, to forecast changing patterns of land use, and so on (Yeh 1990). Economists refer to this kind of growth as Say's Law, which states that "supply creates its own demand." Or, to paraphrase Winston Churchill, people shape technology and technology shapes people.

Of course, developments during the 1980s were not entirely unambiguous. For example, the advantages of local-area networks (LANs) have been offset by the inevitable standardization and re-centralization of control that networking brought with it. Similarly, there has been a tendency to confuse the benefits of graphical user interfaces (or GUIs) with ease-of-use or user-friendliness. Third, while hardware prices seem to have fallen, even as the technology has advanced, software prices have not. On balance, however, there can be no disputing the fact that computing in the early 1990s is radically and fundamentally different from what it was in the early 1980s. In the process, information technology has become appropriate technology.

3. GI Technology and the Micro Revolution

Geographic information (GI) technology has been part of this revolution, a point noted by the Chorley Committee in the UK (DOE 1987). GI software became available for microcomputers as early as 1980, almost as soon as they were invented and well before the appearance of the IBM PC (Robinson and Coiner 1986; Cartwright et al. 1988). Now, with microcomputer-based software packages like ARC/INFO from Environmental Systems Research Institute (ESRI) in Redlands (USA) or SPANS from Interra Tydac Technologies Inc. in Ottawa (Canada)—and even low-cost, university-developed programs like IDRISI from Clark University in Boston (USA) or CHIPS from the University of Copenhagen (Denmark)—today's GI technology on microcomputers is pretty close to state-of-the-art.[1]

At the same time, GI technology cannot be said to have taken the microcomputer world by storm, at least not in the same way that word processing, spreadsheets, or database management systems have done (Cartwright 1990; Cartwright 1991). The reasons for this are complex and, to some extent, interrelated. But ultimately they can all be found in the lessons of the 1980s: namely, accessibility to the user. If GI technology is to become appropriate technology, accessibility is the fundamental issue that has to be resolved.[2]

Accessibility has at least three dimensions: physical, financial, and intellectual. As far as physical accessibility is concerned, a user has to feel that he or she can physically use the technology, whenever and wherever it is needed. The more you use a technology, the more you become dependent on it. Users are hardly likely to embrace a new technology, if it is difficult to get to or if there is a risk of being cut off from it. Second, accessible means financially accessible or affordable. Naturally, users have to be able to afford to buy the technology (including hardware and software) in the first place, if they are going to use it. Furthermore, users also need to feel they can afford to keep up with future improvements. Users, especially new users, need assurance that they will get their money's worth out of the product they buy, before it becomes obsolete, especially if improvements in the technology may make it not only more powerful but also easier to use.

Third, accessible means intellectually accessible. GI technology has to be easy to learn and use (or "user-friendly"). But what is also important is that it is easy to re-learn after an absence. Many users will not need to use GI technology on a daily basis; so they will want to be able to "pick it up" again quickly and painlessly when they need to. Similarly, they will want to be able to transfer information quickly and easily between their GI applications and others.[3] At a more profound level, most professionals will be understandably reluctant to rely on a technology they barely understand, particularly for those of their functions that depend on their creative skill and judgement.

The single biggest challenge is probably the last (Nijkamp and DeJong 1987; Goodchild 1991): namely, training. Moreover, there are two distinct aspects to this problem. On the one hand, potential users need to be shown the mechanics of how to use GI technology. But they must also be trained to take advantage of the capabilities that GI technology gives them. Until you have access to a car, for example, you do not normally learn how to drive. But even when you do learn how to drive, there is still another learning process that has to take place—and that is learning how to make the best use of the mobility that driving a car suddenly gives you. Indeed it has been said that the single biggest factor shaping American cities in this century was Henry Ford' s invention of the private automobile.

Much the same is true of GI technology. It is one thing to learn how to use the technology; it is another thing to learn what to do with it.[4] Users have to "internalize" their newly acquired technical knowledge to the point where it begins to shape their conception of their job and how they do it. Indeed, it has been argued (Maguire et al. 1991, vol. 2, p. 4) that the history of the US Geological Survey (USGS) and that of the UK Ordnance Survey are both the stories of such adaptation. Both agencies began as mapping agencies; both moved through a stage where computers were seen as a means of automating an existing manual process; both eventually reached a stage where computers have transformed the map-making process, effectively turning it inside out. Maps are no longer artifacts, merely printouts. "The database is regarded as the central concept and maps are spun off to meet either standard or special user requirements."

In practice, however, there is usually a great deal more attention paid to teaching people the "mechanical" skills needed to use a technology than on what to do with it, once they have acquired the mechanical skills.[6] In fact, one of the great lessons of history is that learning to use power responsibly is often a lot more difficult than gaining the power in the first place. As for the hardware and software, high-end microcomputers are now an adequate platform for at least some kinds of GI applications (Goodchild 1991). Moreover, significant improvements in hardware seem almost certain to continue for some time to come. Best of all, prices seem, if anything, to be falling. Unfortunately, the same cannot be said of software. Most commercial GI software packages in the GI field are still too expensive for them to be accessible on the same scale as spreadsheets and database-management software.[7]

Finally, support for the use of GI technology is not yet widely available. While there is a growing literature on the nature of the technology (e.g. Teicholz and Berry 1983, Aronoff 1989, Peuquet and Marble 1990, Star and Estes 1990, Worrall 1990, and Maguire et al 1991), there are still surprisingly few books on what to do with the technology. As Stan Openshaw recently put it (1991, p. 390), "there is as yet no globally useful text . . . about the complete range of spatial analysis methods that might be considered relevant and appropriate...." Moreover, some of the techniques of using GI technology are fairly esoteric (cf. Samet 1990) and not the sort of thing that one might expect to pick up casually.

4. GI Technology and the Third World

It is sometimes argued that the use and diffusion of GI technology in the third world raises questions not just of practice but also of principle (e.g. Taylor 1991). In other words, quite apart from the practical difficulties of implementing a sophisticated technology in the third world, ought we even to be making the attempt? Is there not something fundamentally inappropriate about using "high tech" in a "low tech" environment?

This, it seems to me, is a very difficult argument to make without being paternalistic. It amounts to saying that there are fundamental or inherent reasons why GI technology is inappropriate for third-world countries, or less appropriate for them than it is for first-world countries. In practice, its use in third-world countries should be subject to the same assessment of costs and benefits as it is (or should be) in first-world countries and the same criteria applied: is the technology appropriate to the circumstances in which it is being applied, given the resources available and the results desired? Any argument, it seems to me, that a country is not in some fundamental sense "ready" for GI technology cannot avoid the implication that there are "stages" of development through which all must pass. To my way of thinking, technology offers one of the few mechanisms by which the less developed countries may be able to "catch up" with some of the more developed ones (Cartwright 1990, p. 269).

Of course, there are risks entailed for any country in adopting a new technology— just as there are risks in other aspects of development, such as improving public health or raising standards of education. There may be unforeseen consequences or undesirable side- effects. For example, it is sometimes argued that technology developed in the first world is ipso facto "foreign" to the third world. So its diffusion to the third world must inevitably represent some form of cultural imperialism.[8] There are many sides to this argument— which is not confined to discussions of GI technology, of course—and we cannot hope to resolve it here. However, there are several points that should be borne in mind. First, we have to be clear what we mean when we say "foreign". In one sense, every innovation is "foreign", even to the culture in which it is developed, because it is "new". That is the nature of innovation. Thus, innovations affect all cultures that adopt them, including the cultures that created them.

Second, when an innovation is "imported" from outside, that may well entail some broader degree of cultural convergence between the importing and exporting countries. In some cases, this may be cultural imperialism by the latter; in other cases, it may be cultural invigoration by the former. But the simple act of borrowing technology from someone else need not imply dependency—as the first world has already learned from some of the so-called "Little Dragons" of Southeast Asia, which have gleefully imported western technology only to re-engineer and re-export it whence it came, to their considerable profit. In cases where the relationship does create a dependency, simply cutting off the flow of technology does not really solve the problem and may in fact contribute to its long-run persistence.

In any case, the contribution of GI technology to current processes of global, cultural integration is scarcely worth worrying about, compared to the impact of television and the other mass media. Even if it were, proponents of GI technology have no more right to bar the access of others than to promote it. Action is not inherently more or less responsible than inaction. In fact, there may be many people in third-world countries who want nothing more than to see the evolution of their own culture in the direction of other, more technologically oriented cultures. Such goals are prima facie no more bizarre than the tendency to idealize the "simple life".

Fourth, it may be somewhat superficial to think of GI technology as "foreign" to third-world countries. Although the engineering that lies behind the technology may not exist in some of those countries, many of the concepts and principles underlying GI technology are far from unknown. Map-making, for example, is deeply rooted in many cultures, from Western Asia to the South Pacific. The Survey of India was founded in 1767, albeit by the British, some eighteen years before the establishment of their own Ordnance Survey! Mathematics, including complex astro-physical calculations, are not unknown outside western cultures. Of course, there are big differences between current levels of achievement in different countries. But there is no reason to view these quantitative differences as creating a qualitative division between those for whom GI technology is appropriate and those for whom it is not.

In fact, it should be noted that the first, operational example of GI technology (the Canada Land Data System) had its origins in a developing country (Coppock and Rhind 1991). Apparently, the idea of digitizing aerial photographs and using a computer to analyze them came to Roger Tomlinson while he was working on a project in Tanzania. In the end, Tomlinson failed to sell his idea to that particular project and succeeded only later with officials of the Department of Agriculture in Canada. Nevertheless, it can be argued that the concept of GI technology actually originated in the third world.

5. "High Tech" in a "Low Tech" Environment

Even if there is no reason in principle to deny the appropriateness of GI technology in third-world countries, there are nevertheless considerable practical difficulties in using high technology in a relatively low-technology environment.

I recall visiting the site only some six or seven years ago of a multi-million dollar mainframe computer, newly installed in the Ministry of Finance (where else?) in one particular developing country. It was, as I recall, the fifth mainframe computer installed in that country and had been in operation for about a year when I arrived. The computer was located in a high-security area of the building and, of course, one had to get passes, sign registers, and remove one's shoes in order to be allowed to approach it. I asked what the computer was being used for and was told that, if I could wait a few minutes, there was a high-speed printer attached which could print out everything they had entered into the computer in about 15 minutes! While this spoke volumes for their attitude towards freedom of information, it did little to inspire confidence about the cost-effectiveness of the computer or its impact on public administration.

Moreover, for the price of the mainframe computer, it would have been possible to have bought about 2,000 microcomputers and distributed them through the civil service. Think what kind of impact that would have had on the standard of public administration, on the development of computing generally in the public sector, and on human-resource development and morale in an underpaid (relative to the private sector) civil service. The point of the story is not that 2,000 microcomputers are better than one mainframe computer, because that naturally depends on what you want to do with them. The point is that the design of a computer system defines its accessibility, and its accessibility determines its impact. If you want people to use computers, you must give them the opportunity to do so. If you choose a single, large, central computer, you are not going to persuade the same number of people to use computers as if you chose many, small, decentralized computers.

On the other hand, I have also seen numerous examples of highly imaginative and devoted applications of microcomputer-based GI technology in developing countries. For example, there is a nation-wide inventory of birds and their habitats, including maps for each species, compiled with a BBC microcomputer. An NGO in India has compiled a complete database of the budgets of all local governments on more than 700 floppy disks. At the Asian Institute of Technology (AIT) in Thailand, a regional-planning system has been developed, based on integration of off-the-shelf software packages for database management and computer-aided design (Yapa 1988), supported by some ingenious geographic modeling programs (Kammeier 1986).

The diffusion of GI technology in the third world will depend ultimately on four factors: motivation, opportunity, training, and support (cf. Brodman 1987, pp. 117-8).

- People need motivating, because learning takes a considerable effort. The rewards may be intangible (excitement, prestige, job prospects, etc.), but they must be real. In the case of something like GI technology, motivation may stem from the fact that technology can provide a means of fighting sexism, racism, nepotism, and other kinds of job discrimination.

- People need to have an opportunity to learn about and to use GI technology, either on the job or through an educational institution of some kind.
- People need training, both in how to use the technology and what to use it for. In some respects, the latter may be easier in third-world countries, since there may be fewer preconceptions that require altering.
- People need help and advice in retaining and applying what they learn. To some extent, this too may be easier in third-world countries.

Calle Hedberg from the University of Oslo reviewed a number of cases in Africa where computerized land information systems have been implemented. On the basis of these studies (1991), Hedberg made twelve major recommendations regarding the use of GI technology in developing countries. These recommendations (paraphrased, re-arranged, and elaborated on here) are as follows:

1. When GI technology is exported from the first to the third world, strategies for implementing and using the technology are invariably exported along with it. Such strategies have not proved fool-proof, even in the country where they were invented: when they are exported, their weaknesses are likely to be magnified.
2. Debate about the technology invariably overshadows any discussion of its organizational and political implications (cf. Barrett and Curtis 1976 and Harris 1987). Yet it is more likely to be the latter than the former which means the difference between success and failure.
3. During selection and installation of the technology, the views of donors, their consultants, equipment suppliers, and other "outsiders" invariably predominate. Yet their interests in the project are all, by definition, short-lived.
4. Ideally, GI systems should be designed to be flexible and adaptable, so that they can respond to feedback from their users (cf. Burrough 1986 and Dale and McLaughlin 1988).
5. The convenience of those who have to install and maintain the technology inevitably seems to take precedence over the convenience of those who will use it. Yet it is presumably the latter who provide the raison d'etre for adopting the technology.
6. Real user-friendliness in GI technology is still a long way off, even in some of the most popular and widely used examples of the technology.[9] Therefore, it is essential to provide adequate training to an initial group of users and suitable materials for training later generations of users.
7. User interaction with GI technology rarely receives sufficiently serious attention from programmers or systems people. Yet if "customers" are unable or even unwilling to use the technology, they will avoid doing so.
8. No matter how highly regarded GI technology may be, "customers" for its capabilities must still be created (cf. Dangermond 1991). The fact is that people get used to not having information. When more information suddenly becomes available, it may be necessary to do more than just tell people about it: you may have to show them what to do with it (and sometimes even persuade them it is worth doing).
9. In deciding what to do with GI technology once it is installed, it is usually wise to set fairly modest objectives, especially in the early days. In most cases, it is

better to achieve a modest objective than to fall short of achieving a more ambitious one (cf. Cartwright 1987, p. 196 et passim).

10. When designing projects in which GI technology is to be applied, it is usually wise to avoid projects with inter-linked objectives and several phases, since complex projects are prone to being "derailed" by events. Instead, it is better to set "reference lines" and "check-points" and to move incrementally from each one to the next, revising your goals and expectations as you go.
11. Information generated by GI technology in the third world is sometimes poorly understood or even ignored. Thus, the problem is not so much lack of use of the technology as an inability or unwillingness to take advantage of it. In most cases, the "dust collectors" and "white elephants" are the reports produced by the GI systems, rather than the systems themselves.
12. Donors must be prepared to take a long-term view of the impact of GI technology in the third world and accordingly to provide long-term funding for it. Outside experts, on the other hand, should be prepared to accept shorter and more fragmented roles on specific projects, so that their influence is not so pervasive.

Of course, it should be added immediately that many of these recommendations may apply to the use of GI technology in industrial as well as developing countries. The impact of the above problems is particularly acute in developing countries, however, because of the relatively high cost of advanced technology and the frequent involvement of a "donor" agency with interests and priorities of its own.

6. The Future of GI Technology in the Third World

Subject to these comments, I think it is reasonable to be as optimistic about the eventual role of GI technology in the third world as in the first world. The promise is certainly there, but it is still far from a "sure thing". In particular, the speed with which the promise of GI technology is realized will depend on how well we deal with a number of factors. For example, there is the risk of what Robert Aangeenbrug (1991) calls "boosterism" in our zeal to sell GI technology. For one thing, there are many important data about the world that are not spatial in nature. As Stan Openshaw (1991, p. 395) has pointed out,

> Spatial analysis in the short term can only be a fairly primitive descriptive science, but maybe for many purposes that is sufficient. The ultimate difficulty is the inherently limited utility of geographical information in studies which attempt to understand process....In geographical analysis also, virtually all the available data are surrogates and proxies for other variables which are either missing from the database or incapable of measurement or not yet identified. It would be sheer folly to go beyond what these sorts of data can sustain. This is not to underestimate the immense potential utility of geographical information, or of the important insights about process that spatial analysis may provide, but merely recognition that there are limits to what can be achieved in a geographical context.... current spatial analysis methods are nowhere near meeting these needs and are a long way from attaining their full potential.

Furthermore, Aangeenbrug (1991) argues, there is altogether too much willingness among proponents of GI technology to gloss over the cumulative (multiplicative) nature of errors that is inherent in any kind of overlay mapping.[10]

Another danger is that of insularity. For example, the editors of a recent survey of the state of GI technology (Maguire et al. 1991) conclude their book with a chapter in which they refer to the technology (quoting Peter Taylor) as "geography's own little bit of the 'high-tech' revolution" (p. 314). They also express the hope that GI technology will one day become a discipline in its own right. But these (it seems to me) are dubious objectives. A more useful goal would be for GI technology to aspire to permeate and hopefully integrate some of the traditional disciplines.[11] The last thing we want in development planning is to have more disciplinary boundaries to cross.

A third threat is that of information overload. We are already collecting vast quantities of information about the planet every day, more possibly than we will ever use (Clark et al. 1991). According to reports, the EOS or Earth Observation System proposed by NASA will alone generate from 1 to 10 additional terabytes (10^12) of data every day. Archiving data on such a scale, let alone analyzing them, may prove to be the biggest challenge of all. In a deeper sense, the availability of data on such a vast scale raises profound questions about the meaning and value of spatial data. If more data do not necessarily lead to better decisions or even to better information, then when do we reach the point of diminishing returns? How many data do we really need?

Thus, the future of GI technology in the third-world seems promising but not assured. Thanks to microcomputers, the technology is becoming steadily more and more appropriate for developing countries. There are still major obstacles to its widespread adoption, including high costs, shortage of training, and lack of user-friendliness. But these difficulties are not insuperable and may, in due course, be overcome.

7. Endnotes

1. For example, SPANS is a quad-tree raster system with integrated vector-to-raster and raster-to-vector conversion (DPA Group 1989), while CHIPS makes ingenious use of 32-bit video RAM to store and manipulate three-band images and overlays (Holm et al. 1988). See also Eastman 1990.

2. It is unlikely, of course, that spatial data-processing will ever rival word-processing in popularity. But the principle of accessibility as a motive for learning to use a technology is just as important.

3. Standardization in the area of spatial data exchange still seems some way off(Guptill 1991), perhaps in part because of the commercial potentialities in the market for GI technology (Dangermond 1991).

4. See, for example, the literature on remote sensing: Barrett and Curtis 1976, Lintz and Simonett 1976, Rosenfeld and Kak 1976, Swain and Davis 1978, Slater 1980, Richards 1986, Curran 1987, Harris 1987, Lillesand and Kiefer 1987, and others.

5. Details of the history of the USGS are provided in Starr and Anderson 1991 and of the Ordnance Survey in Sowton 1991.

6. David Unwin (1991) is quite critical of and pessimistic about even academic teaching in GI technology. He fears that technique is too often taught without reference to theory or applications. Indeed, he sees a parallel between the current commitment to Gl technology and the quantitative revolution in geography in the 1950s. He notes the irony that many of the cartographic principles swept aside by statistical analysis in the last revolution are now reappearing in the current one! Robert Aangeenbrug (1991), on the other hand, is more optimistic. He proposes five different roles that training in GI technology might play in a university: as a specialized graduate course, as an introductory graduate course, as an advanced methods course for undergraduates, as an introduction to undergraduate geography, and as an introduction to the liberal arts.

7. Data, too, still seem unreasonably expensive. In spite of the fact that the entire planet has been photographed by Landsats 4 & 5 every eight days more or less continuously for at least ten years, students and other impoverished researchers are still reduced to begging or borrowing most of their datasets. The United Nations Environment Programme (UNEP) in Nairobi has collected a number of "public-domain" datasets as part of its Global Resource Information Database (GRID) project, which is in turn part of the United Nations Global Environment Monitoring System (GEMS); but its resources are still very limited. See also Guptill 1991 on problems of data exchange.

8. The original formulation of this thesis is usually attributed to Fanon 1963, but it has been repeated by Rodney 1974 and others since then.

9. Even in first-world countries, users complain about the lack of user-friendliness in their systems. In Hawaii, for example (according to Kim 1990, p. 105), "Two problems have arisen: first, GIS technology is currently very expensive, and second, ARC/INFO is not an easy system to learn." Similarly, in the Randstad Holland, Geertman and Toppen (1990, p. 213) write that the system is "by no means ... user-friendly.... Furthermore . . . the analytical functions are limited, particularly in their ability to be integrated with spatial models; the query function demands extensive knowledge of the GIS used; and the on-screen design and evaluation of polygons (building sites for new houses) is cumbersome."

10. But see Goodchild and Gopal 1989 for a comprehensive treatment of accuracy in spatial databases.

11. This presumably is one of the ideas inherent in Aangeenbrug's continuum of roles for GI technology in the university. See also endnote 6 above.

8. References

Aangeenbrug, R.T. (1991) A Critique of GIS. In D.J. Maguire, M.F. Goodchild and D.W. Rhind, editors, *Geographical Information Systems: Principles and Applications*. Longman, London, 8:101-107.

Aronoff, S. (1989) *Geographic Information Systems: a Management Perspective.* Ottawa: WDL Publications.

Barrett, E.C. and L.F. Curtis (1976) *Introduction to Environmental Remote Sensing.* London: Chapman & Hall.

Brodman, J. (1987) Key Management Factors Determining the Use of Microcomputers in Decision- Making in the Governments of Developing Countries, chapter 7. In S.R. Ruth and C.K. Mann, eds., *Microcomputers in Public Policy: Applications for Developing Countries.* Boulder: Westview for the AAAS.

Burrough, P.A. (1986) *Principles of Geographical Information Systems for Land Resources Assessment.* Monographs on Soil and Resources Survey. Oxford: Clarendon, 12.

Cartwright, T.J. (1991) Information Systems for Urban Management in Developing Countries: the Concept and the Reality. *Computers, Environment and Urban Systems* 15(1): 37-53.

Cartwright, T.J. (1990) *The Management of Human Settlements in Developing Countries: Case Studies in the Application of Microcomputers.* London: Routledge.

Cartwright, T.J., M.R. Brown, and H.V. Seaforth (1988) UDMS: the Growth and Development of Microcomputer Software for Planning. *Habitat International* 12(4):171-193.

Cartwright, T.J. (1987) Information Systems for Planning in Developing Countries. *Habitat International* 11(1): 191-205.

Clark, D.M., D.A. Hastings, and J.J. Kineman (1991) Global Databases and Their Implications for GIS. In D.J. Maguire, M.F. Goodchild and D.W. Rhind, editors, *Geographical Information Systems: Principles and Applications.* Longman, London, 50: 217-31.

Coppock, J.T., and D.W. Rhind (1991) The History of GIS. In D.J. Maguire, M.F. Goodchild and D.W. Rhind, editors, *Geographical Information Systems: Principles and Applications.* Longman, London, 2: 21-43.

Curran, P.J. (1985) *Principles of Remote Sensing.* London: Longman.

Dale, P. F. and J. D. McLaughlin (1988) *Land Information Management: An Introduction with Special Reference to Cadastral Problems in Third World Countries.* Oxford: Clarendon. With a Foreword by Shridath Ramphal, Commonwealth Secretary-General.

Davidson, D. A. (1980) *Soils and Land Use Planning.* Topics in Applied Geography. London: Longman.

Dangermond, J. (1991) The Commercial Setting of GIS. In D.J. Maguire, M.F. Goodchild and D.W. Rhind, editors, *Geographical Information Systems: Principles and Applications.* Longman, London, 4: 55-65.

DPA Group Inc. (1989) Sustainable Development & GIS Technology: a Demonstration Project. Vancouver: DPA Group.

Eastman, J. R. (1990) *IDRISI: a Grid-Based Geographic Analysis System*. Users Manual. Version 3.2.2. Worcester, MA: Clark University Graduate School of Geography. November.

Emery, F.E. and E.L. Trist (1963) Socio-Technical Systems. In C.W. Churchman, editor, *Management Science: Models and Techniques* London: Pergamon.

Fanon, F. (1963) *The Wretched of the Earth*. New York: Grove.

Geertman, S.C.M. and F.J. Toppen (1990) The Application of a GIS for the Allocation of Land for Housing in the Randstad Holland, 1990-2015. In Worrall 1990. Chapter 9.

Goodchild, M.F. (1991) The Technological Setting of GIS. In D.J. Maguire, M.F. Goodchild and D.W. Rhind, editors, *Geographical Information Systems: Principles and Applications*. Longman, London, 3: 45-54.

Goodchild, M.F. and S. Gopal. (1989) *Accuracy of Spatial Databases*. London: Taylor and Francis.

Guptill, S.C. (1991) Spatial Data Exchange and Standardization. In D.J. Maguire, M.F. Goodchild and D.W. Rhind, editors, *Geographical Information Systems: Principles and Applications*. Longman, London, 34:515-30.

Harris, R. (1987) *Satellite Remote Sensing: an Introduction*. London: Routledge.

Hedberg, Calle. 1991. *Information Systems for Land Resource Management in Developing Countries*. Development and Environment Occasional Papers No. 7. Norwegian Centre for International Agricultural Development and the University of Oslo. Oslo: Aas Trykk.

Holm, J., H.H. Olesen, H.S. Andersen, I. Sandholt, T. Nielsen, and K. Rasmussen. (1988) *CHIPS Users Guide: Version 2.0*. Institute of Geography. University of Copenhagen. December.

Kammeier, H. D. (1986) Microcomputer-Based Assessment of the Accessibility of Central Place Services. *The Asian Geographer* 5(2): 177-91.

Kim, K. (1990) The Development and Application of Geographic Information Systems in Hawaii. In Worrall 1990 . Chapter 5.

Levy, S. (1984) *Hackers: Heroes of the Computer Revolution*. New York: Dell.

Lillesand, T.M. and R.W. Kiefer (1987) *Remote Sensing and Image Interpretation*. 2nd ed. New York: Wiley.

Lintz, J. and D.S. Simonett, eds. (1976) *Remote Sensing of Environment*. London: Addison-Wesley.

Maguire, D. J., M.F. Goodchild, and D.W. Rhind, eds. (1991) *Geographical Information Systems: Principles and Applications*. Longman, London.

Nijkamp, P. and W. DeJong. (1987) Training Needs in Information Systems for Local and Regional Development Planning. *Regional Development Dialogue* 8(1): 72-119.

Openshaw, S. (1991) Developing Appropriate Spatial Analysis Methods for GIS. In D.J. Maguire, M.F. Goodchild and D.W. Rhind, editors, *Geographical Information Systems: Principles and Applications*. Longman, London, 25: 389-402.

Peuquet, D. J. and D. F. Marble (1990) *Introductory Readings in Geographic Information Systems*. London: Taylor & Francis.

Richards, J. A. (1986) *Remote Sensing Digital Image Analysis: an Introduction*. Berlin: Springer Verlag.

Robinson, V.B. and J.C. Coiner. (1986) Characteristics and Diffusion of a Microcomputer Geoprocessing System: the Urban Data Management Software (UDMS) Package. *Computers, Environment and Urban Systems* 10(3/4): 165-73.

Rodney, W. (1974) *How Europe Underdeveloped Africa*. Washington: Howard University Press.

Rosenfeld, Azriel, and Avinash C. Kak. 1976. *Digital Picture Processing*. New York: Academic Press

Samet, H. (1990) *The Design and Analysis of Spatial Data Structures*. Reading, MA: Addison-Wesley.

Schumacher, E. (1973) *Small is Beautiful: a Study of Economics as if People Mattered*. London: Blond and Briggs.

Slater, P.N. (1980) *Remote Sensing: Optics and Optical Systems*. London: Addison-Wesley.

Sowton, M. (1991) Development of GIS-Related Activities at the Ordnance Survey. In D.J. Maguire, M.F. Goodchild and D.W. Rhind, editors, *Geographical Information Systems: Principles and Applications*. Longman, London, 36: 23-38.

Star, J. and J.E. Estes (1990) *Geographic Information Systems: an Introduction*. Englewood Cliffs, NJ: Prentice-Hall.

Starr, L. and K.E. Anderson (1991) A USGS Perspective on GIS. In D.J. Maguire, M.F. Goodchild and D.W. Rhind, editors, *Geographical Information Systems: Principles and Applications*. Longman, London, 35:11-22.

Swain, P.H. and S.M. Davis, eds. (1978) *Remote Sensing: the Quantitative Approach.* New York: McGraw-Hill International.

Taylor, D.R.F. (1991) GIS and Developing Nations. In D.J. Maguire, M.F. Goodchild and D.W. Rhind, editors, *Geographical Information Systems: Principles and Applications.* Longman, London, 40: 71-84.

Teicholz, E. and B.J.L. Berry, eds. (1983) *Computer Graphics and Environmental Planning.* Englewood Cliffs, NJ: Prentice-Hall.

Trist, E.L. and K.W. Bamforth (1951) Some Social and Psychological Consequences of the Long-Wall Method of Coal-Getting. *Human Relations* Vol. 4.

Tomlin, C.D. (1990) *Geographic Information Systems and Cartographic Modeling.* Englewood Cliffs: Prentice-Hall.

United Kingdom. Department of the Environment (DOE) (1987) *Handling Geographic Information.* Report of the Committee of Enquiry Chaired by Lord Chorley. London: HMSO.

United Nations Centre for Regional Development (UNCRD) (1990) *Geographical Information Systems: Applications for Urban and Regional Planning.* Report of an International Conference. J. Edralin, ed., Nagoya: UNCRD.

Unwin, D.J. (1981) *Introductory Spatial Analysis.* London: Methuen.

Unwin, D.J. (1991) The Academic Setting. In D.J. Maguire, M.F. Goodchild and D.W. Rhind, editors, *Geographical Information Systems: Principles and Applications.* Longman, London, 6:81-90.

Worrall, L., ed. (1990) *Geographic Information Systems: Developments and Applications.* London and New York: Belhaven Press.

Yapa, L.S. (1988) Computer-Aided Regional Planning: a Study in Rural Sri Lanka. *Environment and Planning B* 15: 285-304.

Yeh, A.G. 1990. A Land Information System for the Monitoring of Land Supply in the Urban Development of Hong Kong. In Worrall 1990. Chapter 8.

SECTION 4:
MECHANISMS FOR FACILITATING THE DIFFUSION OF GIS TECHNOLOGY

GERARD RUSHTON
Department of Geography
The University of Iowa, 316 JH
Iowa City, IA 52242-1316, USA

The following four chapters discuss mechanisms for facilitating the diffusion of GIS technology. The first paper by Post and McLaughlin, describes the efforts of the Maritime Provinces of Canada to develop a "spatial information marketplace" building upon the program of geomatics activities which the region has nurtured since 1970. The foundation of their program is a multi-purpose land information system called the Land Registration and Information Service (LRIS) that was developed with government support, during the 1970's and '80s. This provided a spatial data infrastructure based on an extended geodetic control network, property mapping and parcel indexing. They describe how, in the 1980s, the Canadian Maritime Region developed a geomatics business sector of approximately fifty firms. Recognizing the export growth potential of this sector of the economy, a regional consortium of governments provided seed funding for a non-profit organization to prepare a strategy for development. They describe how such an organization was formed and how it currently is assisting in the implementation of the Atlantic Geomatics Alliance--an organization that matches federal funds with private sector funds for geomatics activities with export-growth potential. The paper concludes by describing recent developments in advancing the spatial data infrastructure for this region and provides examples of a wide variety of actual business applications for their information.

In the second paper, Ottens describes the goals and activities of the Netherlands Centre of Expertise for Geographical Information Processing (NexpRI). The Centre has developed a GIS information bank the contents of which are available to academics and the public through a national academic data communication network. The purpose of the information bank is to provide up-to-date and reliable information about the fast developing GIS industry. The focus is on meta-information, thus providing a guide to the sources and availability of GIS information. The bank contains information on software packages, application programs, scheduled education and training courses, and geo-referenced thematic databases. The Centre also provides GIS expert advice to clients on such subjects as system customizing, system maintenance, and data conversion. Recently, activities of the Centre increasingly involve contracts with organizations to provide assistance with their GIS implementation problems. These application-specific contexts are leading to the demand for GIS application-area expertise which NexpRI is working to fill.

I. Masser and H. J. Onsrud (eds.), Diffusion and Use of Geographic Information Technologies, 275–276.

In the third paper, Polydorides discusses the education and training requirements for effective GIS diffusion. He argues that in addition to the technical educational requirements of knowing the concepts and implementation details of specific GIS software packages, there is a need for a broader understanding of the professional-scientific environment in which the new technology operates. Polydorides challenges the conventional wisdom that GIS experts exist who are capable of applying their knowledge to any field. Using the example of urban planning, he argues that education in GIS should be a part of training in urban planning and a part of that education should be how to apply GIS to problems in that field. He further foresees that the systematic application of information technologies to fields such as urban planning will alter the character of the professional field. Therefore, he concludes that the professional groups who should lead GIS education should be the application fields, with support from information technology fields such as computer science. Three types of GIS education and training modalities are described and the types of situations for which each is most appropriate are discussed. The final section of Polydorides' paper describes the activities of URSA-NET "Urban and Regional Spatial Analysis - Network for Education and Training." In addition to organizing short, intensive training courses in Greece and more advanced seminars elsewhere in Europe, URSA-NET has developed Hypertext-based training materials for GIS based, in part, on the NCGIA curriculum text. This new, experimental approach to training is currently being evaluated.

Obviously, another means of facilitating the diffusion of GIS use is through improvement of the technology to the extent that it is useful for solving a much wider range of land-related problems. In the final paper, Douven, Grothe, Nijkamp, and Scholten make the point that the current link between GIS and planning models is very weak. They argue that lessening the gap will greatly expand the uses to which GIS may be applied. Benefits of integrating GIS and spatial modeling are illustrated through two practical case examples; sustainable development analysis for the Sporades Islands of Greece and elderly housing needs analysis in neighborhoods in Amsterdam. The paper concludes with a series of specific research recommendations for achieving greater integration of GIS with spatial modeling and a list of suggested potential critical factors for achieving successful introduction of GIS.

The four papers taken together provide a sampler of university-related initiatives which have attempted to have broad-scale effects in aiding the diffusion of geographic information technologies.

DEVELOPING THE SPATIAL INFORMATION MARKETPLACE: A CANADIAN CASE STUDY

GERRY POST
Maritime Geomatics Board
and
JOHN McLAUGHLIN
University of New Brunswick

1133 rue Regent St., Suite 302
Fredericton, New Brunswick
Canada, E3B 3Z2

ABSTRACT. In this decade we will realize the completion of a primary national spatial data infrastructure in Canada and the evolution of a multi-billion dollar spatial information marketplace. In the process we will see a fundamental restructuring of the public organizations and private companies providing spatial information services.

In this paper we examine how this theme is playing out in the three Maritime Provinces of Canada. Long pioneers in the automation of land records and the development of GIS technology. This region is now shifting its focus to the effective use and application of spatial information. At the same time, emphasis is increasingly being given to the role of private industry.

This paper briefly reviews developments in Maritime Canada over the past three decades, describes recent public initiatives (e.g. creation of the New Brunswick Geographic Information Corporation) and private initiatives (e.g. the creation of a regional strategic alliance), and assess as current policies and concerns aimed at building the spatial information marketplace.

1. Introduction

The topic, 'Modeling the Use and Diffusion of Geographic Information Technologies' is timely. A look at market research shows that GIS has matured and is no longer a technology in search of applications. Market research has indicated that the current (1992) world market for GIS products and services is estimated at more than US$8 billion with annual growth rates well outpacing many technology sectors - between 1992 and 1993 GIS revenue growth is projected to be more than 55% (Market Intelligence Corp. 1992). A large portion of this technology investment will be in digital spatial data. The resulting development of a spatial information market place will have a profound and positive impact on the use and diffusion of GIS into every corner of our community (AMC 1991). This is already evident across Canada and the United States.

This paper presents a case study of how the development of GIS in Maritime Canada is leading to the creation of a spatial information market place. We will outline this

I. Masser and H. J. Onsrud (eds.), Diffusion and Use of Geographic Information Technologies, 277–291.

development and note some lessons to be learned by what has happened in Canada. Although the spatial information market place will be a major catalyst for GIS technology diffusion, we caution that it be developed with care and some restraint - the power of the technology has potential for abuse, particularly in the early stages of market development. It will be very difficult to anticipate the impact of electronic data diffusion on such public policy concerns as 'right-to-privacy'. But the concerns are real, as is evident across Canada and in the United States.

2. Maritime Canada: A Brief Profile

Maritime Canada straddles the eastern seaboard of Canada and consists of three provinces: Nova Scotia, Prince Edward Island and New Brunswick (Figure 1). The region is small by Canadian standards, encompassing an area of approximately 52,000 square miles and a population of approximately 2 million, with a relatively poor resource-based economy in relation to Canadian standards - the average income is about 2/3 the national average.

3. Geomatics Activity in Maritime Canada:

Despite strong conservative traditions and a limited economic base, the Maritimes has been a pioneer in the development and use of geomatics[1], primarily due to the region's economic dependence on resource industries such as forestry and fishing. In the early 1970's it was recognized that an important aspect of improving the productivity of the Maritime resource base was to improve the information about its resource endowment. This resulted in the development of a strategic effort to build a Maritime spatial information infrastructure. As summarized in Figure 2, the effort has gone through a number of stages and diversions along the way:

3.1 STAGE 1 (CIRCA 1960'S AND EARLY 1970'S): CONCEPT, INSTITUTIONAL AND DATABASE DEVELOPMENT

In the late sixties and early seventies the region benefited from a synergy of research within the academic community and leadership within the public sector to begin applying the multi-purpose cadastre concept (parcel land based information systems) which had been explored at the University of New Brunswick. Out of this grew the Land Registration and Information Service (LRIS) - a unique regional partnership between the three Maritime provinces with goals to:

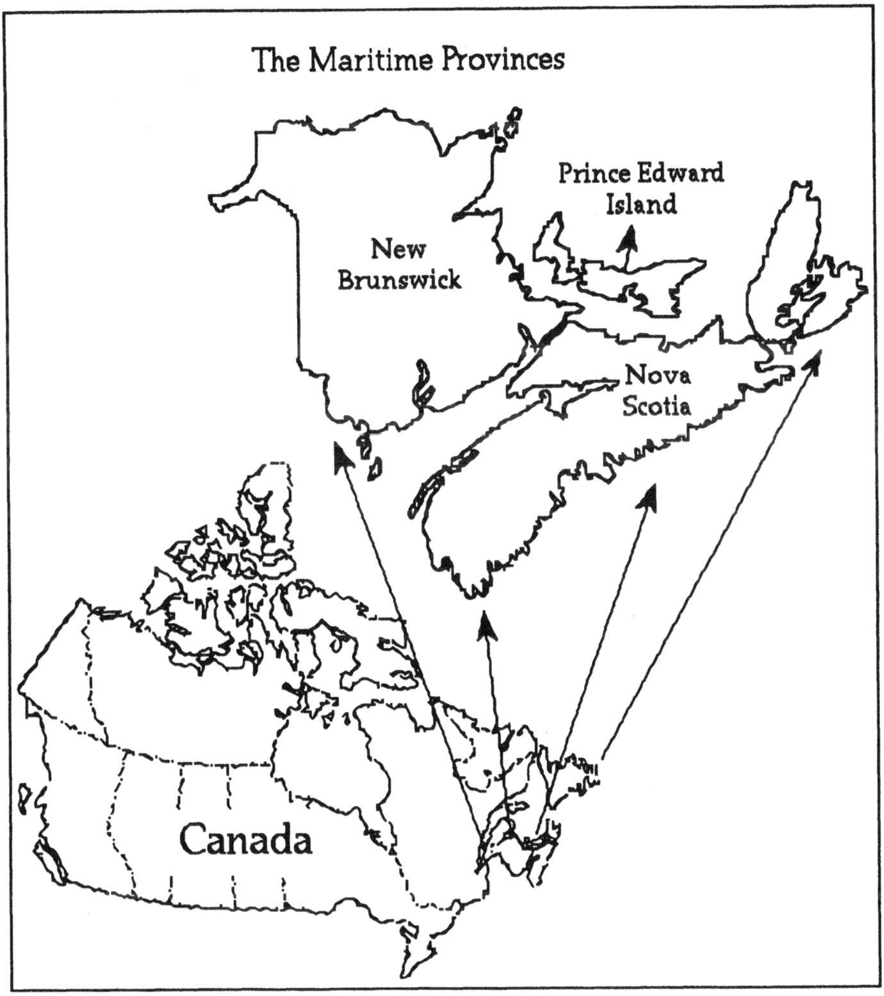

Figure. 1. Maritime Canada

- *extend and densify a second-order geodetic control framework* - this has been completed;

- *develop a large scale topographic and cadastral map series,* a series of conventional topographic maps has been completed and the program converted to digital mapping techniques in 1982 which encompassed the cadastral mapping program which is being completed;
- *convert the existing manual rudimentary deed registry systems in each province to a*

computer-based land titles system, a pilot project in a bilingual (French and English) area was completed in 1987 and is currently operational;

- *build an integrated land records management system throughout the region,* a major initiative is underway to build a Maritime spatial data infrastructure.

Evolution of Geomatics Activity in Maritime Canada 1970-1991

	FOCUS	ACTIVITY
Stage I: 1970's	Establishing a Vision Building an Institution Building Databases	Multi-participant LI Concept Formation of LRIS Agency Control Survey, Topo and Parcel Databases
Stage II: 1980's	Technology Retooling	Introduction of Digital Mapping and GIS Digital data conversion
Stage III: 1987-91's	Institutional Re-organization GIS Applications Development	Establishment of Provincial LI focal points GIS technology becoming mainstream
Stage IV: 1990's	Private sector development in the geomatics sector	Public/Private sector strategic alliances Privatization Government contracting-out
Stage V: 1995 -	Spatial information infrastructure and marketplace development	Establishing networks, standards and public policy to enable the building of the infrastructure and the spatial info market place

Figure 2. Evolution of geomatics activity in Maritime Canada, 1970-1991.

The LRIS organization was unique at the time for a number of reasons:

It had a unique vision: which was to establish a 'foundation' on which to build a multi-purpose land information system. This was well before GIS had entered the mainstream as a land information technology.
It was multi-jurisdictional: involving three provincial governments. LRIS became accepted as the central agency in Maritime Canada responsible for surveys and

mapping not just at the provincial level of government but for municipalities and the private sector as well.

It had centralized geomatics activity under one organization with a sole mandate to deliver surveying and mapping services to the Maritime community. In Canada these functions were traditionally linked to line departments responsible for resource development which often distorted surveying and mapping priorities in favour of the department's principle mandate. Centralizing the agency allowed LRIS to bring to bear its critical mass and expertise to the task of serving a broader community of interests. This aspect of LRIS as a 'community' organization helped spread the basic overhead cost of surveying and mapping to more than one government, and helped reduce duplication of efforts within provincial departments in three jurisdictions and municipalities.

These dramatic changes occurred because the federal government helped fund the establishment of LRIS. Without it, institutional inertia would have killed the concept before it got off the ground.

The primary focus of the LRIS agency in the seventies and early eighties was on building the foundation of the spatial data infrastructure - extending the geodetic control network, mapping the Maritimes and defining the parcel fabric in the region. This included property mapping and parcel indexing, which was computerized from the outset using mainframe facilities.

Lesson to be learned:

In most jurisdictions there must be institutional change to take full advantage of geomatics technologies. Creating change from within is often difficult because of institutional inertia; it took a push from an external force to succeed in Maritime Canada

The clarity of the LRIS vision and its mission helped tremendously in carrying out the agency's mandate (Figure 3).

The Backbone of the LRIS Concept (1972)

A Vision:	***Establishing a multi-purpose land information system***
A Mission:	***Building a strong foundation for the LI system***
A Partnership:	***Use the critical mass of a tri-provincial partnership to accomplish the task***

Figure 3. The backbone of the LRIS concept (1972).

3.2 STAGE 2: INTRODUCTION OF DIGITAL MAPPING AND GIS

From its beginning LRIS had been following developments in digital mapping and GIS. In the early 1980's the agency made a decision to re-tool its mapping operation and go digital. It was an interesting decision because user surveys indicated that there was no demand for digital data. None of the agency's primary users was involved in GIS nor was there an interest in digital mapping. LRIS management persisted, anticipating demand for digital data down the road. They made a significant investment and commitment to the technology. Today LRIS can not keep up with demand and the user community is complaining that they are too slow. Times have clearly changed.

An interesting aspect of the technology investment was that it also accomplished an economic development objective. LRIS's purchase of digital mapping technology supported the development of a local firm, Universal Systems Limited (USL), that was in the process of commercializing digital mapping and charting software developed at the surveying and mapping research laboratories of the University of New Brunswick. At the time of the LRIS purchase, USL employed a few people and currently employs over forty people, exporting software products around the world and continues to enjoy a close relationship with LRIS as a beta test site for software improvements and new products.

Soon after LRIS initiated its digital mapping program some key user agencies began to invest in GIS. Because of the region's resource economy, the forestry sector took the lead - but not until a crisis developed in the region's forestry industry because of a projected shortage of wood fibre compounded by a dramatic infestation of spruce bud worm. This resulted in legislation requiring the forestry industry to prepare and submit to government ten year management plans. This served as a major impetus for both the government and private sector to invest in GIS technology. By the late 1980's New Brunswick became the first province in Canada to have a fully loaded forestry GIS database; the other Maritime provinces have followed. Other sectors have also proceeded with a heavy emphasis on resource management applications.

However, the region also experienced two impediments to widespread diffusion of GIS technology:

•First, those exploring GIS technology quickly found out that the investment in hardware and software was only a small portion of the investment required to employ the technology. Experience in Maritime Canada showed that the cost of building databases to support GIS applications was often a multiple of 3 times the cost of hardware and software (Figure 4). This usually killed projects and it remains a major impediment to technology diffusion.

•Second, until about 1987 the cost of the technology (hardware and software) was clearly out of reach of many users. Only with the introduction of GIS on PC platforms did the community embrace the technology (Figure 5). Hardware and software costs are no longer a major impediment.

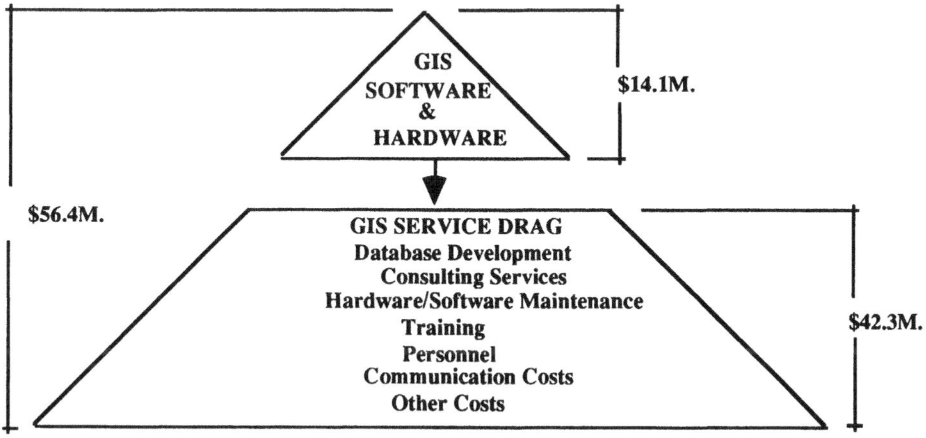

Figure 4. Atlantic Canada GIS Infrastructure Investment 1979 - 1990 (Jan.) (Cabot Group 1990)[2]

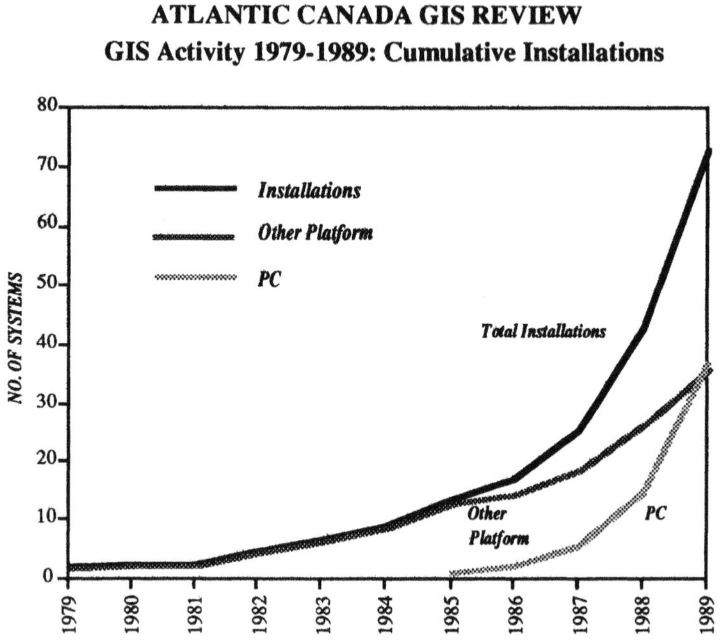

Figure 5 GIS activity 1979-1989: cumulative installations (Cabot Group 1990).

Lessons to be learned:

User surveys do well at defining current requirements and accepted paradigms, but with GIS we are aiming at the future and entering uncharted waters - it often takes leadership and a 'leap of faith' to embrace new technologies

Cost of GIS hardware and software is no longer an impediment to technology diffusion - the major cost impediment remains the cost of database development

The LRIS institutional framework established in the early 1970's served well to spread the cost of building corporate databases for GIS

Legislation can have a major impact on GIS diffusion, as was the case with provincial legislation in New Brunswick which required the forestry industry to adopt an integrated forest management approach which served as a major impetus for GIS technology investment by both government and business

3.3 STAGE 3 (CIRCA 1987 - 1991): INSTITUTIONAL RE-EXAMINATION

In the late 1980's the rapid expansion of the GIS user community caused a re-examination of how best to respond from an institutional perspective. The push for action came from a number of fronts:

- users demanding that government accelerate the building of digital 'foundation' databases;
- corporate planners became concerned about a loss of the Maritime vision to have a multi-purpose land information system operating in each of the provinces; and
- realization that there was significant 'value' in certain digital databases (particularly the parcel database), providing an opportunity to generate revenue to accelerate technology diffusion and perhaps raise net-revenues for public coffers. The notion of a spatial data market place caught the imagination of entrepreneurs in both the public and private sectors.

In response, the LRIS agency proposed a broadened mandate in land information management through the creation of a Maritime Provinces Land Information Corporation (it became known as MAPLINC). The proposal was to restructure LRIS into a crown corporation and provide it with operational responsibility for property assessment and registration, along with its already accepted responsibilities for surveying and mapping. This proposal was not accepted, because of a concern raised by one of the provinces that it would lose control over its assessment and registry functions. The failure of MAPLINC was considered by many a major set back for Maritime cooperation in the geomatics sector.

During this period each of the Maritime provinces also undertook a review of their land information requirements and each came to the same general conclusion - to create a provincial lead agency responsible for land information management. Its interesting to note that each province took a slightly different approach:

Prince Edward Island established a Land Record Centre (LRC) within its Department of Finance. The LRC has operational responsibility for property assessment and the registry system as well as land information standards and technologies within the province (Maritime Resource Management Service, Inc. 1987). The LRC also directs the LRIS activities for the province and was implemented before the MAPLINC proposal. The MAPLINC proposal essentially mirrored the LRC model on a regional basis, but in the form of a crown corporation.

New Brunswick established a Geographic Information Corporation (NBGIC), owned by government, but operating on the principle of cost recovery (the Corporation is able to retain revenues). NBGIC is responsible for property assessment, the registry system and provincial standards and directs the activities for LRIS and was implemented immediately following the failure of the MAPLINC proposal. New Brunswick essentially modelled its institutional structure after the MAPLINC.

Nova Scotia recently established a Land Information Division within its Department of Municipal Affairs. The division has broad responsibilities for *coordinating* land information management within the province but no operational responsibility for land information systems. It is expected that this will change and that Nova Scotia will follow the path established by the two other provinces and merge the property assessment and the registry functions under one institutional umbrella.

These institutional changes have affected LRIS and the agency has entered a period of transition:

- LRIS now has lead agencies to report to in each of the provinces which should improve its accountability to each of the three partners - but will reduce its flexibility and entrepreneurial activity; and

- the database maintenance functions carried out by LRIS are being absorbed by the provincial agencies resulting in a significant downsizing of the organization.

Lesson to be learned:

> *To establish a public spatial information infrastructure requires a strong institutional focus, responsibility can not be fragmented throughout government. Periodic examination of the institutional framework is therefore essential.*

3.4 STAGE 4 (CIRCA 1990 -) : PRIVATE SECTOR DEVELOPMENT:

In the 1980's a geomatics business sector began to emerge in Maritime Canada and by 1990 the region had over fifty firms involved in some way in geomatics technologies (Champlain Institute 1991). Most of the firms were small and became involved in geomatics because of an application that related to their core business, such as environmental consulting. A small number of firms devoted their entire business to geomatics, these could be grouped into two categories: technology firms (such as Universal Systems Limited) - developing geomatics software products; and, geomatics service firms, such as consulting and database suppliers (firms providing data conversion and collection services). Growth of the firms was relatively slow because of the small regional market, although some firms were able to export services to other parts of Canada and abroad. Recognizing the potential export market for these firms, government provided seed funding for a non-profit organization to prepare a strategy for development. This led to the creation of the Champlain Institute which:

- commissioned a market study to identify the strengths and weaknesses of the Maritime geomatics industry and match them with international market development trends; and

- developed an Atlantic Geomatics Alliance, pulling in the Province of Newfoundland, to establish a critical mass within the industry to pursue export market opportunities. The geomatics alliance is based on the following principles:
- to create an investment climate and mechanism to encourage government, the private sector and academia to form alliances and undertake innovative research and development projects which have strong commercial prospects;
- to focus on selected application areas in which the region is strong - three have been targeted: land records management; coastal and ocean information systems; and, integrated resource management systems;
- to establish alliance projects large enough to help build a critical mass within the industry to penetrate export markets.

To finance the Alliance, a $10 million commitment has been made by the Canadian government as an investment catalyst. The conditions of federal funding are that the private sector match the federal investment and that royalties from products and services be used to repay the federal investment. A competition is currently being held to select Alliance projects. The Alliance is a three year development initiative.

This initiative will have significant impact on the use and diffusion of GIS in the Maritime geomatics community in that:

- it has raised the profile of GIS at the political level;
- GIS technology can be used as a vehicle to create wealth and employment;
- it will develop some leading edge geomatics technologies and capabilities in the region to benefit local users;
- it builds a healthy level of cooperation and collaboration between government, business and the academic community which may be expanded to other sectors.

The Alliance initiative has recently been complemented by a Maritime Geomatics Strategy which proposes some significant steps (Figure 6) to build the critical mass within the industry through:

- a greater emphasis by government to 'contract-out' geomatics services;
- establishment of public/private sector alliances
- the privatization of LRIS; and
- the development of reasonable data access and pricing policies.

Lesson to be learned:

> *The synergy of cooperation and collaboration between government, business and academia can have significant benefits for the geomatics community and can be a model to spin-off similar initiatives in other areas*
>
> *An 'honest broker' (the Champlain Institute model), with no vested interest, may be required to build bridges between the government, business and academic sectors*

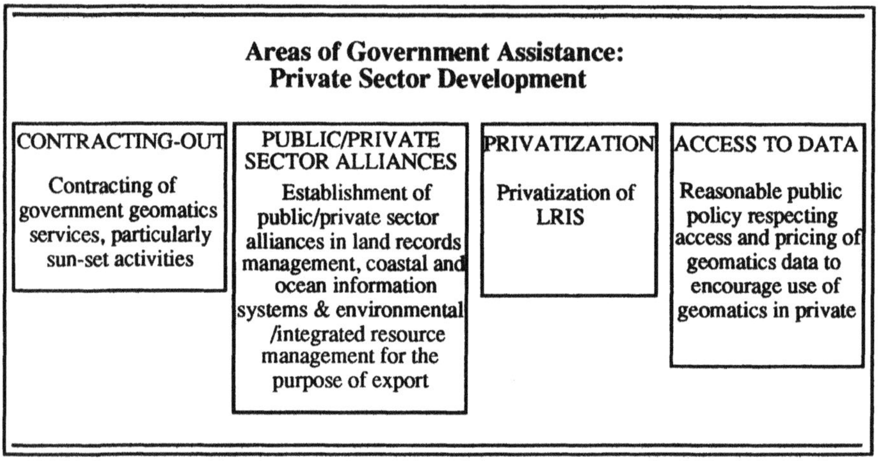

Figure 6. Areas of Government assistance; private sector development (Maritime Geomatics Board 1992)

3.5 STAGE 5 (CIRCA 1995 -): SPATIAL INFORMATION MARKET PLACE

3.5.1 A Spatial Data Infrastructure for Maritime Canada: It is anticipated that by 1995 most of the 'foundation' databases (digital topographic and property records) will be in place in Maritime Canada. The GIS user growth curve should continue to climb which will result in a widespread user community. To accommodate this growth, the region is currently establishing a structure to ensure the orderly development of a Maritime Spatial Data Infrastructure (MSDI - see Figure 7 for a systems perspective) (McLaughlin and Nichols 1991). It is envisioned that the MSDI will be the geomatics electronic highway linking all levels of government and the business sector in Maritime Canada and that it will contribute to a national network. This will enable the full development of the emerging spatial information market place. The MSDI places the emphasis on generating value added information products, enabling users to access different electronic databases and merging them, using technologies such as GIS to create new products .

The main challenges in developing the MSDI will not be technical; rather they will concern institutional issues pertaining to data access and privacy, pricing and liability. Public policy concerning these issues will be one of the main obstacles to fully developing the geomatics market place. This is understandable considering the potential power of GIS technology to merge public data and create new products. Policy makers will be faced with some very tough decisions over the next decade. We anticipate that in some jurisdictions policy makers will overreact because of electronic data abuse. This could have a significant negative effect on the diffusion of the technology. The geomatics community must exercise considerable restraint to avoid situations of abuse (which will be difficult to police), particularly in the early stages of developing the spatial data market place. Signs of problems are already appearing in some parts of Canada, particularly the provinces of Ontario and Manitoba, where major public/private sector partnerships to implement spatial data infrastructure projects have been met with critical media attention and public debate.

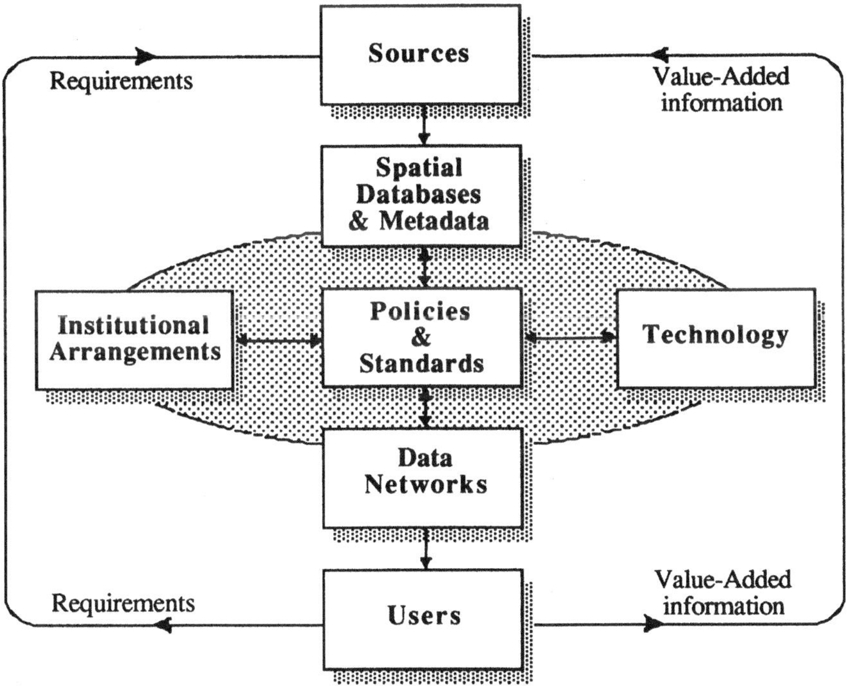

Figure 7. Maritime Spatial Data Infrastructure (Infostructure) - A Systems Perspective (McLaughlin and Nichols 1992)

3.5.2 The Spatial Information Market Place: The geomatics market is eclectic, ranging from traditional uses of the information within government (i.e. property valuation) to mass market applications (AMC 1991). Who would have thought that twenty year old building permits had any value - they are being sold to roofing contractors to target their advertising...or new building permits, which are being used by drapery firms for their marketing. Automobile license plates are being used to define market catchment areas for shopping centres. These catchment areas can be further refined by individual stores using credit card information (ever wonder why some stores ask for your phone number when signing a credit card?). The power of the technology to fuse electronic databases will cause an explosion in the market for geomatics data. The recent issue of GIS World (October 1991) dedicated its entire issue to the business application of GIS, which is a tremendous growth area and will diffuse the technology throughout the community. An overview of the spatial information market place is shown by Figure 8.

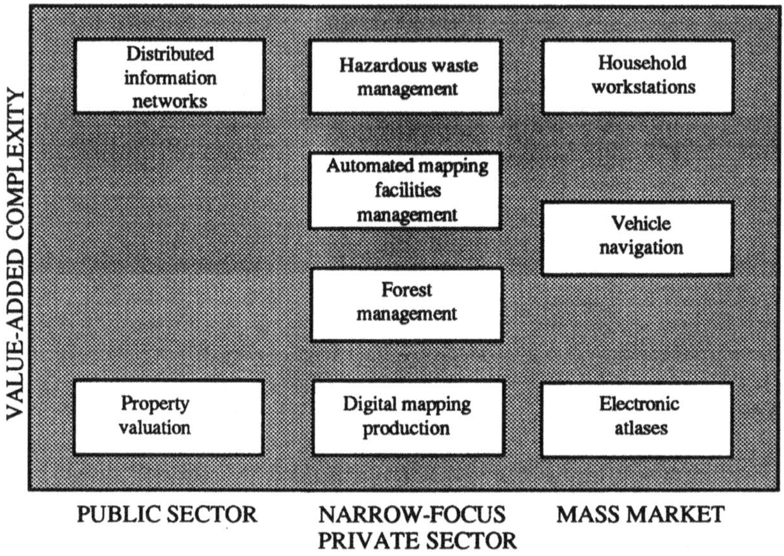

Figure 8. Overview of the Spatial Information Market Place (Coleman 1989)

Challenges we face:

Maintaining a strong focal point to champion the implementation of the spatial data infrastructure

Integration of GIS and communication technologies on a community wide basis

Establishing standards acceptable to the community
Formulating reasonable public policies concerning data access and pricing which facilitates the diffusion of the technology and protects the public interest

Practicing restraint to avoid electronic data abuse, particularly in the early stages of the spatial infrastructure development

Continue building partnerships and bridges between the private and public sectors and the academic communities

4. Endnotes

1. In Canada the term "geomatics" is used to define all activity related to the science, technology, and use of geographically-referenced information, including its acquisition, storage, analysis, display, and dissemination.

2 Please note that these figures are for Atlantic Canada, which includes the Province of Newfoundland. However, the inclusion of Newfoundland has very little impact on the overall figures because of the small number of GIS installations in that province in 1989.

5. References

AMC (1991) *Market Opportunities for the Atlantic Canada Geomatics Industry*. Prepared for the Champlain Institute by Applied Management Consultants and the DPA Group, in association with the Cabot Group.

Cabot Group Inc. (1990) *Atlantic Canada GIS Review 1990*, Publication prepared by the Cabot Group Inc., Halifax, NS.

Champlain Institute (1991) 1991 Survey of the Atlantic Geomatics Industry. *Champlain Institute Review*, August 1991.

Coleman, D. (1989) Geomatics developments, Unpublished presentation slides, Department of Surveying Engineering, University of New Brunswick, Fredericton, NB.

Maritime Geomatics Board (1992) *Maritime Geomatics Strategy*, Council of Maritime Premiers, Halifax, NS.

Maritime Resource Management Service, Inc. (1987) *A Corporate Approach Towards Land Information Management in the Province of Prince Edward Island: Final Report*. Study prepared for the Province of Prince Edward Island, Charlottetown.

Market Intelligence Corp. (1992) Total GIS Software & Services Market: Revenue Forecast (World), 1991-1996 *Computing Canada*, January 1992, p. 23.

McLaughlin, J.D. (1991) Towards a national spatial data infrastructure. In *Proceeding of the Canadian Conference on GIS*, Ottawa 1-5.

McLaughlin, J.D. and S. Nichols (1991) Toward the development of a national spatial data infrastructure. *Report for the Mapping Science Committee*, US National Research Council, Washington, DC.

McLaughlin, J.D. and S. Nichols (1992) Geomatics and the building of a national strategy. *Computing Canada*, January 1992, p. 24.

NEXPRI: A NATIONAL CENTER TO SUPPORT THE DIFFUSION OF GEOGRAPHICAL INFORMATION TECHNOLOGY

HENK F.L. OTTENS
Netherlands Center of Expertise for Geographical Information Processing (NexpRI)
Faculty of Geographical Sciences, University of Utrecht
P.O. Box 80.115
3508 TC Utrecht
The Netherlands

ABSTRACT. NexpRI was founded to promote and support the use of advanced geographical information processing technology in scientific research. Gradually, NexpRI evolved into a communication center for GIS specialists and a meta-information center and consultancy bureau for the application of systems. Meanwhile, the scope of the target and user groups has broadened to include nearly all (potential) GIS users in the Netherlands. In this way the center succeeded in becoming partly self-financing. The goals of gathering, structuring and disseminating GIS expertise are reached in several ways. Conceptual models of GIS activities, of the diffusion of GIS technology and of GIS system planning play an important role in structuring the activities. The NexpRI Information Bank is a key element in nearly all the functions of the center.

Over the last two years, the demand for expertise to support GIS system development projects has grown considerably. Because of the rapid expansion and proliferation of GIS technology the working methods will have to be adapted in the near future. It will become more difficult to serve the increasingly diverse GIS users. Therefore, the general expertise in geographical information technology may become less relevant than knowledge of how to process information specific to a field of application. In order to fit into the new systems of national research institutes and graduate schools, NexpRI's function as a research (support) center needs to be reconsidered.

1. Introduction

Since 1987, some 20 centers of expertise have been established at Dutch universities and research institutes. Most of these centers have been recognized by the Netherlands Organization for Scientific Research (NWO) and are subsidized for a four to five-year period through grants provided by the Ministry of Education and Science. The brochure on the centers of expertise program, gives the following description: "A center of expertise is a national and interactively accessible center, where scientific knowledge (for example: experience, data and software) on a specified field of research is being gathered, regularly updated and made available through SURFnet". SURFnet is a national academic data communication network that connects all institutes of higher education and national research centers in the Netherlands. It also provides access to international data

I. Masser and H. J. Onsrud (eds.), Diffusion and Use of Geographic Information Technologies, 293–305.

communication networks.

The NexpRI Center of Expertise was established to provide computer technology support for scientific research in disciplines like geography, geodesy, landscape architecture, cartography, and environmental and planning studies. Several application fields connected with these research areas were specified: facility management; automated mapping; urban, regional and landscape planning; and natural resource and environmental management (Ottens 1988). Therefore, unlike many other GIS centers, NexpRI is not a research or educational institution but an information and support facility.

Four goals were set for NexpRI:

1. To provide on-line information on (scientific) geographical information processing;
2. To bring together demand for and supply of GIS tools, services and support (non-commercially as well as commercially);
3. To coordinate and promote computer-assisted education in geographical information processing;
4. To function as a platform for all GIS users and developers in the Netherlands and as a focal point for international cooperation.

These goals clearly refer to GIS meta-information (Medyckyj-Scott, et al. 1991). Initially, the GIS expertise handled in the center was primarily meta-information. This information had to be gathered, structured and disseminated. For that purpose, the NexpRI Information Bank was developed on the basis of a model, categorizing and relating GIS activities. The information bank can be accessed by persons interested in GIS meta-information, but it has also been instrumental in producing derived printed publications and services. A model of the diffusion of GIS technology in the Netherlands provided more insight in the types of (potential) GIS users and their meta-information and support requirements. Over the last two years, GIS-experts at NexpRI have become involved in consultancy. The center is mainly approached by institutions and firms contemplating the implementation of a GIS. They seek support in the selection of hardware, software and dataware. These potential users ask for advice on how to organize a GI system and for help with pilot projects. Because of this development, NexpRI has given more attention to aspects like information analysis, system planning and system implementation.

In this paper, the design of the NexpRI Information Bank, the use of GIS meta-information, the diffusion of the GIS technology in the Netherlands, and the need for GIS system planning support will be discussed in turn. The paper ends with some perspectives on the future of NexpRI and a short comparison between NexpRI and national GIS centers in other countries.

2. The NexpRI Information Bank

The NexpRI Information Bank is the prime support tool for all activities of the center. In this section the structure, operation and use of this meta-information system will be described (Ottens & Zandee 1991). The development and implementation of this system has required about half of the staffing budget (about six full-time equivalents) available during the first two years of the center's operation.

2.1. WHY BUILD AN INFORMATION BANK?

The field of geographical information systems is highly dynamic and turbulent. This situation may well last for another ten to 15 years. The technology itself; the market for systems, data and services; and the organizations that adopt the technology for daily use: all are subject to rapid and profound changes. Moreover, the number of players involved increases as the use of GIS proliferates.

At present, it is very difficult to acquire complete, reliable and objective information on the state of the art in geo-informatics. Moreover, it is time-consuming and cumbersome to keep up-to-date on GIS developments, even within a single country.

In the Netherlands, there is no national agency to coordinate and guide the handling of geographic information (RAVI 1990). With respect to large-scale base mapping, the National Cadastral Service (KADOR), supported by the Advisory Council on Land Information (RAVI), plays an important role. But many utility companies, municipalities and other institutions follow their own strategies. The Topographical Survey (TD) is the leading agency in medium and small-scale mapping, while the National Physical Planning Agency (RPD) has been the pioneer and trend-setting user of analytical GIS systems and producer of digital planning base maps. In recent years, the National Institute of Public Health and Environmental Protection (RIVM) and the National Public Works and Water Management Department (Rijkswaterstaat) also have become prominent GIS users. They have served as examples for many provincial and municipal departments of physical planning, environmental management and public works. But also commercial geo-planning and geo-management firms use many different databases and software systems.

Under these circumstances and in view of the aims of NexpRI, it was decided not to start building new geo-databases or to produce GIS programs within this center of expertise. Geometrical and attribute databases are abound, and GIS continues to be a research priority at the institutes participating in NexpRI.

Instead, it was decided to promote and support the use of GIS by providing meta-information and by offering expert services based on this meta-information. A major consideration in choosing this strategy is the fact that demand for GIS technology has also increased outside the geo-disciplines and their application fields. The 'export' of GIS expertise will be beneficial to the geo-science disciplines as well as to the new members of the user community.

2.2. THE ORGANIZATION OF THE NEXPRI INFORMATION BANK

Organizations in which the handling of geographical information is a major activity and where a GI system has been installed form the basic units in the database of the information bank (Figure 1).

A key element in designing the bank, of course is which GI system(s) is (are) being used. These operational GI systems are linked to tables containing information on groups of GIS-related activities. The system is built in a VMS/ORACLE environment on a MicroVAX computer system.

Four types of GIS activities are distinguished: projects, expertise, services, and products. The most straightforward activity is the provision of tangible GIS products by the organization. Information on software packages, application programs (drivers, conversion programs, interfaces, menu applications, etc.), scheduled education and training courses, and

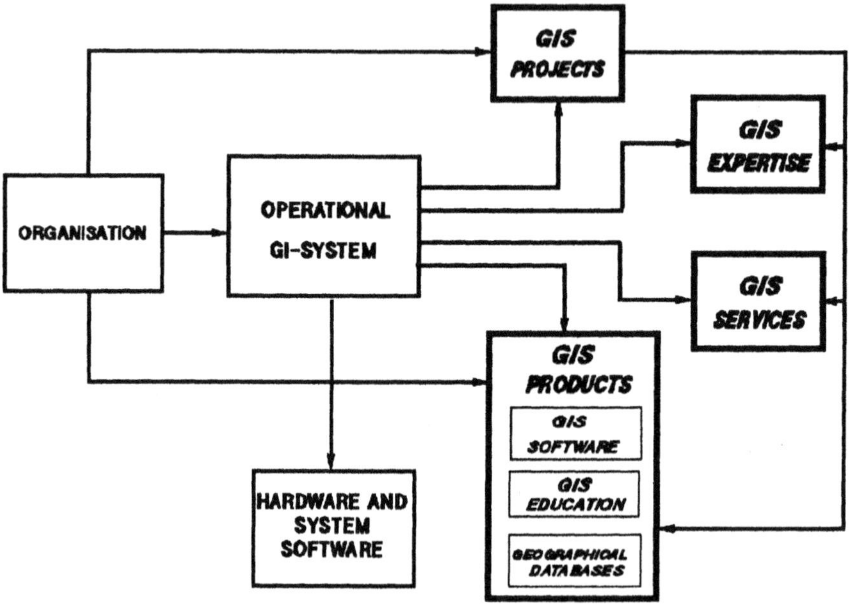

Figure 1. NexpRI Information Bank components

geometric and geo-referenced thematic databases are stored in different tables. Activities like system customizing, data conversion, project support, and system maintenance are grouped together under the heading of GIS services. The table of GIS expertise contains a number of indicators on knowledge and experience accumulated in the organization with respect to GIS. Finally, the database tables on GIS projects store information on research, development, and implementation projects that will result in new or enhanced GIS products, expertise, and services in the future.

2.3. THE OPERATION OF THE INFORMATION BANK

Figure 2 gives an overview of the main information flows of the NexpRI Information Bank. Three different groups of actors are involved in the system. The organizations with GIS activities are the primary sources of information. They are interviewed to get an overall picture of their GIS activities, to record their general characteristics, and to trace contact persons. Subsequently, those contacts are asked to fill out questionnaires with more detailed questions about products, services, expertise, and projects. Once the data have been entered into the information bank, the informants receive a complete listing of the contents of the database for their organization. This allows them to propose corrections before the information becomes available for external use. The same listings are also used for updating in later years.

The NexpRI mediators control the data-gathering process and are responsible for the management of the information bank. They also perform dedicated queries. Since

December 1990, it has been possible to access the NexpRI Information Bank through SURFnet. The system now holds information about more than 150 organizations with a GIS, while detailed characteristics of some 50 packages, 130 courses and over 150 geo-databases have been made available.

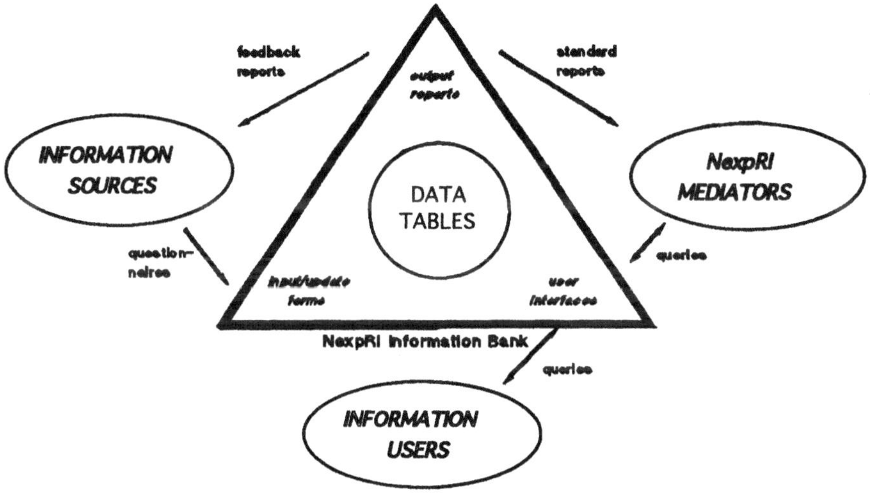

Figure 2. Information flows of the NexpRI Information Bank

Figure 3 gives a more detailed view of the use of the NexpRI Information Bank. Initially, the system was only to be accessible through the SURFnet research network. A user-friendly interface was built allowing users to query the database. Most standard questions can be posed and will be answered with the help of this interface. However, many users still find querying data banks via networks to be difficult and cumbersome. At present, the information bank is only accessed a few times per day. There is no charge for using this on-line system. Moreover, many persons interested in the information have no connection to SURFnet at all. Therefore, more emphasis has recently been given to the production of printed information material and to the use of the information bank through the NexpRI mediators. Because of their knowledge of the structure of the database and of the SQL query language, the mediators are better equipped to perform non-standard queries in an efficient way. Also many persons visit the center to obtain information.

3. The Role of NexpRI and the Use of its Services

The impact of NexpRI on the diffusion of geographical information technology and on the appropriate and efficient use of systems in the Netherlands is not easy to assess. NexpRI surely has brought together the major GIS-related scientific research and education centers in the Netherlands. But so far, structural cooperation remains limited to a few areas, mainly related to activities at the national level: joint workshops, national programs for investment in scientific equipment, proposals for inter-university graduate schools etc. Full-fledged joint GIS research projects and courses have not yet emerged. But the observation that GIS has attained the status of an established field of research and education in Dutch academia may also be attributed to the fact that a center of expertise in this field was recognized and funded. Research and education policies of Dutch university departments continue to be determined primarily at the local level and focused on departmental programs. Infrastructure and specialist support for research continues to be organized and financed locally, making it difficult for interuniversity

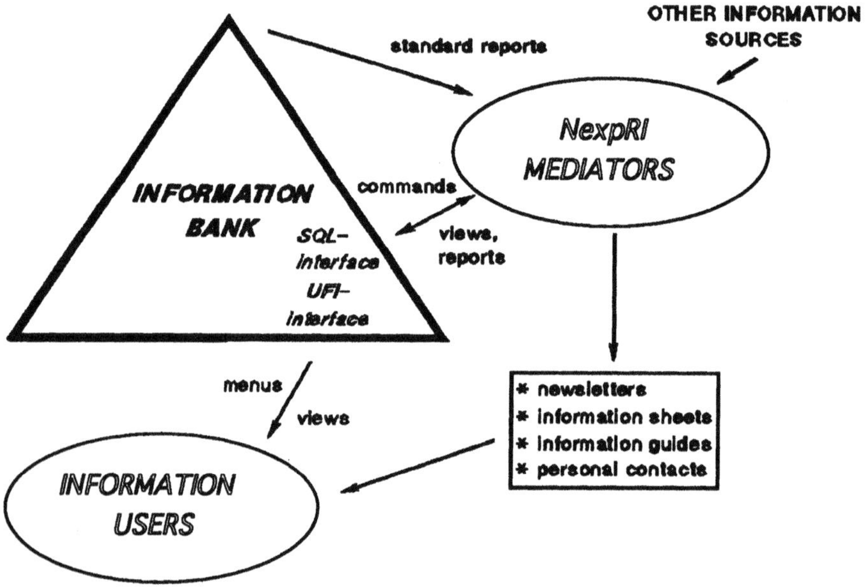

Figure 3. The use of meta-information

centers of expertise to operate in this area. Moreover, research infrastructure costs are covered in general budgets and seldom directly charged to research projects or research groups. The existence of NexpRI has stimulated relations between the geo-sciences and other scholarly fields. Through the center, expertise on the application of geographic information technology could be 'exported' to disciplines like archeology and environmental studies.

NexpRI has become well known to GIS businesses and the center is often used as an entry into the scientific community. The yearly surveys for the Information Bank keep the contacts alive. At the European level, NexpRI has promoted and coordinated Dutch input into the EGIS European GIS conferences and exhibitions, thus facilitating many international relations and forms of cooperation.

The direct use of NexpRI's services is easier to measure. These figures are presented in the annual reports of the center. Currently, the Information Bank receives about 40 on-line queries per month. Half of them are sessions of over two minutes. Last year, there were about 40 formally scheduled visits to NexpRI. Most visitors requested general information on NexpRI or on GIS-related activities in the Netherlands. The circulation of the semiannual newsletter exceeds 600 copies. NexpRI guidebooks (on software, courses and databases) are sold at a rate of a few hundred a year. The yearly GIS Research Workshop organized by NexpRI draws about 150 participants. Although initially identified as the prime target group, only one-third of all contacts are with the scientific community. Government organizations and private businesses each account for about one-third of NexpRI's external relations.

By 1991/92, consultancy had grown to embrace about 50 percent of all activities at the center. The majority of the contracts deal with the selection of appropriate soft-, hard-and dataware for new users, or with system planning studies including pilot projects. Further, a number of geo-database projects have been completed. Both government organizations and private businesses are among the customers.

In conclusion, for the geo-science community, NexpRI functions primarily as a center for national-level communication and a focal point for international cooperation. For other disciplines, NexpRI is a GIS information and advisory center. Outside the academic world, the image of NexpRI as a relatively independent, geo-science-based information and documentation center and system planning consultant seems to be reasonably well established. This profile deviates quite substantially from the original intentions. The necessity to become, at least partly, self-financing over a four-year period caused the center to shift its orientation towards external demand.

4. The Diffusion of GIS Technology and the Need for Meta-Information

The practical application of geographical information systems has increased remarkably. Not only has the number of systems installed risen sharply but the nature of the applications and the type of users have expanded (Meer 1991). This development has had a profound impact on the functioning of NexpRI. This center of expertise monitors the use of GIS in the Netherlands. With the data in the information bank, it is possible to construct a model of the diffusion of the technology among different categories of users. This model is not the outcome of a rigorous scientific survey and analysis. Rather, it is based on an interpretation of the results of the NexpRI inventory among GIS-using organizations and on a listing of contacts with persons and organizations since the center was opened. Figure 4 gives a schematic overview of the diffusion process.

Five stages of GIS development and diffusion can be discerned. As is often the case with generalized models, the stages overlap in time and quite a few exceptions to the postulated stages exist.

During the 'innovation' phase, interest in GIS is restricted to a few university institutes specialized in geographical information processing. Work on GIS mainly deals with fundamental aspects of the technology (concepts, data structures, basic algorithms etc.). The current state of development in hard- and software allows some serious practical pilot applications. Major national agencies that handle large amounts of geographical data become interested and a 'pioneering application' phase starts. At that point, software becomes commercially available, while hardware prices drop to levels affordable to large organizations. Hardware vendors discover this new niche in the market for graphical database management and analysis. They are thus willing to promote the use of GIS, often through a few very specialized software houses. These software houses are the first distributors of the commercial GIS packages. When awareness of the technology has spread sufficiently and competition among vendors has led to lower prices and software that is less user-unfriendly, conditions for a 'commercialization' phase are fulfilled. Especially the consultancy sector becomes active. Various types of consultants enter into the GIS services market: firms specializing in physical planning research contracts; engineering consultants in areas like public works and landscape, water and environmental management; but also general information technology consultants. These firms are instrumental in proliferating the 'pilot use' of the technology. CAD/Mapping users gradually move to GIS. At the same time, a range of organizations new to graphic computer applications, join the growing GIS user community. The final phase is characterized by a widespread and ever more differentiated 'operational use' of GIS by a large variety of organizations.

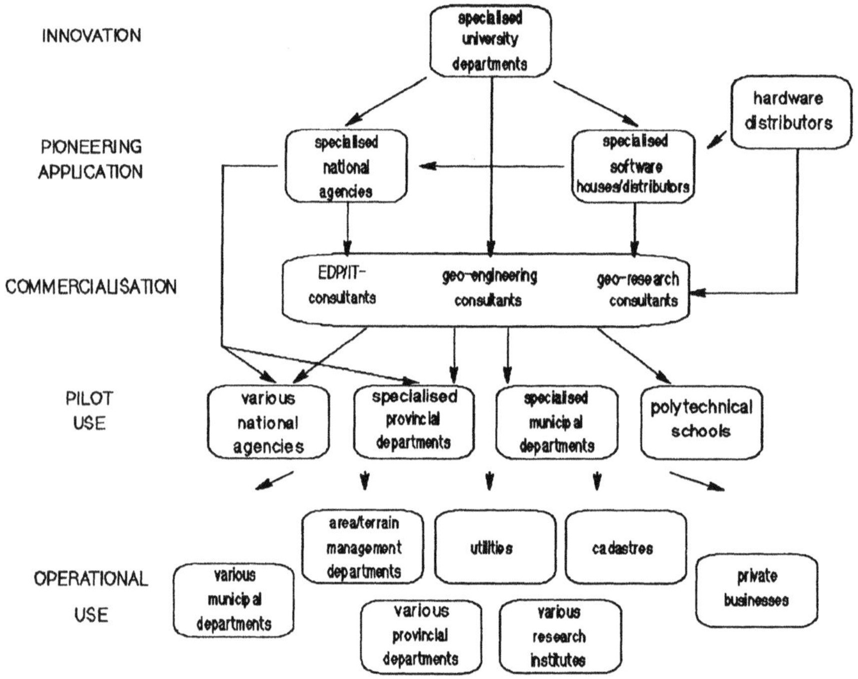

Figure 4. The diffusion of GIS technology in the Netherlands

In the Netherlands, the first phase started during the 1970s, and the second phase followed around 1985. 'Commercialization' at a significant scale started some four to five years ago. Currently, the 'pilot use' phase is gradually being replaced by the 'operational application' phase.

This development has a number of implications for a GIS center of expertise:

- The demand for services and support from NexpRI increases.
- The differentiation among (potential) GIS users also increases, from dedicated full-time GIS specialists with a solid geo- and computer science background to ad-hoc users in policy environments with little or no GIS background knowledge.

- Because GIS users no longer form a homogeneous group, application-specific problems and solutions become more important than general GIS issues.
- Many persons and organizations move from an orientation toward GIS to effective implementation of systems, which dramatically increases the need for expertise on GIS system planning and development.

5. Expertise on GIS System Planning

The current state of the GIS diffusion process more or less compels a national GIS center like NexpRI to exploit GIS meta-information and expertise for system planning. Within organizations, system development nearly always takes the form of projects. So far, about ten system planning projects have been commissioned to NexpRI. The contract partners were mainly medium-sized organizations or agencies. The GIS system development methodologies (e.g. Aronoff 1989, Dangermond 1990 & Korte 1991) and experience gained in the system planning projects, formed the foundation for a stepwise model and check list for this type of project (Figure 5). The model pertains to small and medium-sized GIS implementation projects that run on off-the-shelf software. Ideally, this kind of implementation should only require limited consultancy and customizing efforts. In the event of uncertainty and organizational problems, external advice (second opinions) and support is often sought.

Of course, the GIS system development model has to be adapted to each particular project. The most important elements of the projects were support for choosing the most appropriate hardware and software, surveying the availability of suitable databases, and help with pilot projects that can lead to an appropriate organizational setting for GIS. Accordingly, meta-information still plays an important role. Through the direct provision of information for consultancy work, the meta-information in the NexpRI Information Bank contributes significantly to GIS system planning expertise. But within each project, the available information has to be updated and complemented because the scope, goals, and structure of the projects are never the same.

6. Discussion

Two aspects of NexpRI are addressed in this concluding section. First, the functioning of NexpRI is compared to other national GIS centers. Second, some perspectives on NexpRI's future are given.

National GIS centers have been established in several countries. Their mission is to promote and support the development, application and diffusion of geographical information technology. However different approaches have been taken.

In the United States, the National Center for Geographic Information and Analysis (NCGIA) is a fairly large operation in with three universities participate. The center has received substantial structural funding: $ 5.5 million for a five-year period. This grant is provided by the National Science Foundation. The main objective of the center is to perform fundamental scientific research organized in specially designed initiatives. Its secondary activities relate to GIS education (especially curriculum development) and advisory and consulting services (NCGIA 1988a, 1988b). The NCGIA has developed into a renowned network for advanced GIS research.

FIRST INVESTIGATION/PROJECT SUPPORT
- * global survey and project planning
- * management/user/technical support (internal/external)

ORGANISATION AND INFORMATION ANALYSIS
- * organisational structure
- * information infrastructure and policy
- * persons and tasks (cuurent/future)
- * information problems and requirements
- * data requirements (internal/external)
- * data processing requirements
- * requirements for (presentation of) output

SYSTEM CHOICE (HARDWARE/SOFTWARE/DATAWARE)
- * functional requirements
- * budget limitations/technical constraints
- * look-and-feel evaluation
- * evaluation of offers

IMPLEMENTATION PLAN
- * system organisation and management plan
- * education and training programme
- * customisation, operation and maintenance schemes
- * database, application and presentation design
- * pilot project(s) design

OPERATIONAL USE
- * installation, operation and management
- * monitoring, audit and adaptation

Figure 5. System development strategy for small and medium-sized organizations

The Regional Research Laboratories (RRL) in the United Kingdom encompass a slightly more modest program: nearly $ 4 million of basic funding for eight laboratories over a four and a half year subsidy period, including a trial and a main phase (Masser 1990). The objectives of the RRLs were based on the integrated data resources strategy of the funding agency, the Economic and Social Research Council (ESRC). The RRLs show a wide variety of activities in the fields of research, data handling, database development, education and various contract projects. Their research focus is both fundamental and policy-oriented. The regional dimension of the initiative was primarily shaped by the development of regional geo-databases and GIS application and training projects. Parallel to the RRL initiative, a three-year joint research program on geographical data handling was set up by the ESRC and the Natural Environment Research Council (NERC). The three year program includes fundamental and applied research and has a $ 1.7 million budget (Mather 1992).

In other European countries, different routes were followed. For example, in France, the Physical Planning Ministry (Ministère Chargé du Plan) established the Conseil National de l'Information Géographique (CNIG), an advisory body with several thematic committees. In the French academic world, the CASSINI (Coordination pour l'Analyse Spatiale et les Systèmes d'Information Géographique Intelligents) research network represents a joint initiative of social and computer scientists. It has recently been recognized by the CNRS (Centre National de la Recherche Scientifique). The intention is to let it develop into a GIS research center (Cheylan 1992). The Portuguese Centro Nacional Informacão Geográfica is another example of a national government initiative. In Lisbon, responsibility rests with the Secretary of State for Science and Technology. The Portuguese CNIG is primarily a national geo-data center with local nodes. Additional activities include GIS research, education and training (Gonçalves Henriques 1992).

NexpRI shares many characteristics with the British RRLs. While lacking the research dimension, NexpRI more explicitly acts as a GIS-information and communication center for a regional, in this case the Dutch, GIS community. As explained above, the existing decision-making and funding structures in scientific research in the Netherlands have severely hampered the performance of its role as a service and support center for scientific research.

As an information, documentation and support center, NexpRI is useful as long as geographical information technology is rapidly developing and diffusing. This will probably remain the case for the next five to ten years. Only after that time can GIS become a stable and mature technology.

Should scientific research funds be allocated to finance a general information and support center for a technology that is largely developed and applied outside the academic community? A case can be made for a university-based center. First of all, the services offered by NexpRI are primarily based on scientific research and development initiatives. Moreover, most of the support is provided in response to demand by university departments. Furthermore, the center has proven able to facilitate better and more intensive relationships between the academic and the non-academic GIS R&D and user communities. New initiatives are underway to add research tasks to NexpRI's mission. The establishment of new, integrated institutes for both advanced research and post-graduate education ('onderzoekscholen') will give NexpRI an opportunity to participate in the research programs of these institutes. A number of the proposed institutes to be organized at the inter-university level are expected to have an interest in GIS research or the application of the technology.

In any case, NexpRI will inevitably have to serve a gradually changing target group. Users having limited experience with relevant programs and having little geo-science knowledge will become more important. The demand will increase for implementations on small computer platforms and in small and medium-sized organizations or departments. And a greater need will arise for applications facilitating information management and geographic query and display. These changes in demand have implications for the presentation of GIS information and documentation, as well as for the consultancy services offered. GIS expertise will have to be more oriented towards the casual high-end PC and low-end workstation user and less toward the dedicated and highly specialized GIS user.

Nevertheless, a need will remain for up-to-date meta-information on readily available GIS software, relevant hardware, geo-referenced databases etc. A major part of that meta-information is international in nature. Therefore, it is worth considering provision of this type of information at a supra-national level. Initiatives for an international GIS information and coordination center are underway in Europe. Meanwhile, transatlantic cooperation, which is already effective with respect to commercial trade journals, should be explored as well.

7. Acknowledgement

Thanks are due to Nancy Smyth van Weesep, WCI Interface, for correcting the English text.

8. References

Aronoff, S. (1989) Geographic Information Systems: A Management Perspective. WDL, Ottawa.

Cheylan, J-P. (1992) French GIS Research. *Mapping Awareness* 6(4): 22-24.

Dangermond, J. (1990) How to Cope with Geographical Information Systems in your Organization. In H.J. Scholten and J.C.H. Stillwell, editors, *Geographical Information Systems for Urban and Regional Planning*. Kluwer, Dordrecht, 203-211.

Gonçalves Henriques, R. (1992) Implementing a National GIS: The Portuguese National Geographic Information System. Paper presented at the EGIS '92 conference, Munich.

Korte, G. (1991) GIS 1991: A Practitioner's Handbook. Design Systems Strategies, Chamisal.

Masser, I. (1990) The Regional Research Laboratory Initiative: An Overview. ESRC/RRL, Sheffield (Discussion Paper No. 1).

Meer, A.M. van der (1991) A Profile of Geographical Information Handling in the Netherlands. In J.J. Harts, H.F.L. Ottens and H.J. Scholten, editors, *EGIS '91 Proceedings*. EGIS Foundation, Utrecht, 717-727.

Mather, P.M. (1992) The UK ERSC/NERC Joint Program on Geographical Data Handling: Progress Report. In J.J. Harts, H.F.L. Ottens and H.J. Scholten, editors, *EGIS '92 Proceedings*. EGIS Foundation, Utrecht, 474-482.

Medyckyj-Scott, D., I. Newman, C. Ruggles and D.R. Walker, editors (1991) Meta-Data in the Geosciences. Group D Publications, Loughborough.

Ministère Chargé du Plan (no year) Conseil National de l'Information Géographique. CNIG, Paris.

NCGIA (1988a) The National Center for Geographic Information and Analysis: Research Plan and Call for Participation. NCGIA, University of California, Santa Barbara.

NCGIA (1988b) The National Center for Geographic Information and Analysis: Education Plan and Call for Participation. NCGIA, University of California, Santa Barbara.

Ottens, H.F.L. (1988) A Center of Expertise for Geographical Information Processing in the Netherlands. Paper presented at the Fifth International Workshop on Strategic Planning, Enschede.

Ottens, H.F.L. and A.H. Zandee (1991) The NexpRI-Information Bank: A Meta-Information System on GIS Activities in the Netherlands. In Medyckyj-Scott, D., I. Newman, C. Ruggles and D.R. Walker, editors, *Meta-Data in the Geosciences*. Group D Publications, Loughborough, 139-151.

RAVI (1990) De grootschalige topografie in beeld. Voorlopige Raad voor Vastgoedinformaie, Apeldoorn.

TECHNOLOGY TRANSFER AND TRAINING NEEDS: THE URSA-NET EXPERIENCE

NICOS D. POLYDORIDES
University of Patras and URSA-NET
18, Pythagora str.
15562 Holargos, Athens
Greece

ABSTRACT. The paper discusses the role of training as an extremely important element for a successful process of technology transfer related to GIS. It further examines, as an illustration, issues involved in a specific case, the URSA-NET program.

1. Introduction

GIS is a very successful new information technology. It has been successful both in the speed by which it has been diffused, as well as in the width of such diffusion, in a variety of scientific and professional applications. GIS is, of course, an exciting tool, enabling us to do a variety of jobs in innovative, more efficient, and more rewarding ways. This workshop deals with a wide variety of aspects in the process of diffusion of these information technologies.

In another paper (Polydorides 1989), I have argued that there are four organizational aspects, which are extremely important ***prerequisites***, for the successful application of information systems:

- a rational planning process
- piecemeal development of hardware and software
- a continuously relevant system
- training and education needs

Of these, probably the most important prerequisite is the necessary trained personnel to create and support the system. This paper discusses the role of training, as a critical element for technology transfer related to GIS. The training element is examined from three complimentary aspects:

- what kind of training is necessary?
- training for whom?
- training how?

I. Masser and H. J. Onsrud (eds.), Diffusion and Use of Geographic Information Technologies, 307–316.

To illustrate these points, the paper further examines issues involved in a specific multi-faceted training programme, the URSA-NET experience.

2. What kind of training:

The term training usually signifies teaching of a particular "how-to" methodology or system. In dealing with training issues for GIS, one includes some basic elements:

- learning what a GIS is and how a GIS may be set up
- learning what it can do for the organisation,
- learning how to use a particular software tool and how to develop and administer a GIS.

In addition to these fundamental elements, I will assert that the training process should also include a broader understanding of the professional-scientific environment within which the new technology operates. Such understanding should not be simply a quick introductory exposure to the field, of the kind used by a successful and aggressive salesperson of the technology. From my experience, such salespeople always start with sweeping and impressive reviews of the whole field they is dealing with, including a concise list of its current stimulating debates, before proceeding to prove, in the next hour or so, how almost miraculously, their product solves all these problems.

Introductions like these serve more the need of the salesperson to show how well he or she knows the field, rather than the need of the recipient to understand how the new technology fits in with the old tradition and what the ramifications will be. There is a clear and distinct knowledge base required by planners if they are to use GIS in a Spatial Decision Support System [SDSS] role. Decision-aid and group decision making, commonly used notions in decision-making situations, are not so simple as they appear on the face of it. To be an efficient decision facilitator or a "decision-making process" information provider requires serious training. And, as these roles are important aspects in the potential use of GIS, extensive knowledge related to them should definitely be included in the training package.

In a more general way, a necessary ingredient of the training process for GIS, is a thorough professional understanding of the particular field. In other words, the training for GIS applications in urban planning should include, automatically and simultaneously, training in urban planning proper. This combination is of course achieved in two ways: either the trainee follows a professional programme in urban planning with a strong training element on GIS, or the trainee is a qualified planner who follows the GIS training as an extra, post graduate programme.

This approach to the concept of training for GIS, is based on an important assertion: that there is no such thing as "the GIS applications expert". In other words, I do not accept as valid the model of the professional expert who acquires a deep knowledge of GIS methods and applications in vitro. And, afterwards, applies such knowledge to whatever professional field or application that happens to demand his expertise. Rather, I prefer to have a planner, who is also an expert on GIS and knows very well how to apply this knowledge to problems in his field. There is also the environmental scientist, who might be an expert in using GIS as an extremely efficient tool in modeling environmental pollution. Or the transportation planner, who discovers the flexibility of GIS in analysing his networks.

The discussion makes it apparent, I think, that this issue, of the proper content of GIS training is actually a collorary of another, more important issue: which professional group is going to lead the process of technology transfer and diffusion of this new technology.

The diffusion of new information technologies in the last few years has been taking place with a very quick pace and in a rather haphazard way. This is particularly true in less developed scientific environments. There we frequently witness the effort by the ever growing and much-demanded professional group of computer scientists, who assert that it is they who should be the leading experts in GIS technology [as well as of course any other information technology]. The assertion is that the important knowledge is that dealing with efficient algorithms, memory capacities, and design of proper software. If these are valid, then any application of the technology is a matter of simply imposing such expert technical knowledge on the particulars of the case.

This is unacceptable logic. A knowledge of GIS and corresponding training, are relevant only if they compliment a previous thorough professional knowledge of the particular field. I argue that information technologies are so powerful, so efficient, and so thorough, that their systematic application in a particular field tends to alter the character of that professional field. For example, an urban planning system which introduces GIS and other information technologies in its processes will never be the same again. Soon, planners will discover that they approach urban problems in a different way and they view potential solutions more broadly. Thus, it is extremely important that the person handling this new technology be a planner rather than a computer scientist.

This assertion is considered true in its generality as a starting point. However, it is fairly easy to visualize cases where, particularly in a group situation, somebody will be needed to take care of the more technical aspects of using/running a GIS (especially if the work has shared access of a large database and a multi-user arrangement).
In that case, one of the experts from the particular field might become more expert in the techniques and know-how of the GIS and therefore become a GIS expert. Or the team might decide to include a technical expert who knows a lot about computers, software and hardware issues, and particularly about GIS, but does not necessarily know anything about land-use, regional planning, or transport planning. This technical expert should understand enough about the problems faced by the corresponding subject expert, in order to participate with that expert to help create the appropriate digital solution to problems within a particular GIS environment. In these cases, the role and the position of GIS expert is quite relevant and acceptable.

3. Training for Whom, and How?

A second element of the training issue, concerns who will receive training. We have already asserted above, that the trainee should be professionally trained in the corresponding application field. Therefore, the two possible profiles for a potential GIS trainee are:

- the newly graduated professional
- the mid-career professional

The case of the graduate student, or new graduate, is the simplest of the two. He or she, is still fresh from professional studies, and can simply follow a further post-graduate programme, that should provide a thorough education in computer applications in planning, or more specifically in GIS. Programmes like these have started to appear, usually in Graduate schools of Planning, Geography, or Surveying. I will not attempt a review of these. But I would like to make some general comments on some of the issues involved in the design of a successful graduate training programme.

First, is the issue of length. It is usually agreed that one year (or two semesters) is the minimum for a comprehensive professional programme on GIS. But is one year enough? I am willing to argue that two years is a more proper duration because the programme should be as integrated as much as possible to the mainstream of the corresponding professional field. Also, I think the programme should include a strong element of applied training, possibly with a one-semester placement to an enterprise or public authority using GIS.

Second, is the issue of content. It is very important to decide exactly what mixture of theories, methodologies, and skills, is the correct one. My general opinion is that the curriculum should include "equal" portions of all three. This is in line with the whole model presented above of the integrated expert in a professional field, who also possesses a deep knowledge of GIS tools.

When we turn to the mid-career professional, things become somewhat more complicated. A first issue is the probable inertia or suspicion of the trainee towards this new technology. It demands from the trainee a possibly painful effort to adjust to new ways of doing the old job. Therefore, the training process in this case should be:

- more convincing
- and more tuned to the practicalities of the application.

How is GIS training to be achieved? There are a variety of possibilities: The first model, and also the simplest, is what I would call the "exposure" seminar. This is a seminar intended to give a broad and introductory view of the field. It should include a discussion of the fundamentals of GIS, i.e. purpose, content, structures, processes. It should also include a roughly equal dose of applications. The basic objective is to describe and illustrate possibilities very clearly and to erase misconceptions. The duration of such a seminar should typically be anything between two and seven days.

The second model is the advanced, or specialised, seminar. It deals with a particular topic or application and is addressed at people who are involved already in the field and are receiving training in order to do the job in a new way.

A third model, is the longer training programme, usually of a few months, done in an institution of higher education or similar organisation. The content and clientele of such a seminar is very specific and exclusive. A typical example is the training programme designed for a group of officials and engineers of a particular municipality in order to introduce GIS in their work. Such training programmes might be offered by software vendors in order to be in total harmony with their particular software package.

Finally, the longest possibility, is arranging for fellowships, of one or two years, usually for middle level personnel of the organisation, in order to follow a graduate programme like the ones discussed above.

Designing a strategy for the training of particular personnel is critical. The strategy used may depend on many parameters, such as:

- the type of functions and processes the organisation is involved with, and particularly on their time horizon.
  A body such as a building permit department in a city which carries out primarily routine tasks will have a completely different approach to training from, for instance, the planning department that may be preparing the comprehensive plan for the next century.
- the size of the manpower of the organisation.
  It is obvious that small departments cannot afford to loose a single member for extended periods of time.
- the "age" of the organisation.
  Old departments, doing the same work for years, tend to be a lot different from "younger" departments, created recently, and usually with younger personnel, eager to experiment with new technologies.

All the above simply illustrate the importance of training, in order to achieve a successful GIS application in an organisation.

4. The URSA-NET experience

URSA-NET is an interdisciplinary project organised within the COMETT programme of the European Community. COMETT was launched in 1987, with the explicit objective of bringing together universities (and other institutions of higher learning) and enterprises for the purpose of creating training processes and materials for new technology.

4.1 THE COMETT PROGRAMME

COMETT is a successful programme. In the 5 years of operation (it has already entered its second phase), hundreds of cooperation programmes have been supported by the European Commission, which are essentially of four types or "strands":

4.1.1 Strand A: This strand deals with the creation of a University-Enterprise Training Partnership (UETP). This partnership is an agreement among its member universities and enterprises, to provide the support structure necessary in order to carry out all other programme activities. UETPs are of two types:

- Sectoral: partnerships of a transnational character within a given technological field or an industrial sector.
- Regional: partnerships at a regional level (within one member-state) bringing together within a particular geographical area groups of universities and groups of enterprises engaged in joint training with the support of relevant public and private authorities in the region.

4.1.2 Strand B: This strand deals with transnational exchanges. One element is the system of integrated transnational student placements in enterprises. The placements offer students a period of structured professional training, familiarising them with technology-related issues and career prospects. Priority is given to placements integrated in the student's initial or post-graduate training.

Another element in Strand B, are personnel exchanges between universities and enterprises. University staff participate in training activities in enterprises, to enrich their teaching activities, and to extend their practical experience in a European context. For the staff of enterprises the aim is to share training responsibilities at the university and to contribute to the diversification of teaching activities.

4.1.3 Strand Ca: This part of the programme promotes crash training courses designed and offered by UETP's in co-operation with universities and/or enterprises. Their purpose is to provide rapid dissemination of research and development results in the field relative to new technologies and their applications and to promote the transfer of technological innovations,.

4.1.4 Strand Cb: Finally, Cb projects are multilateral arrangements aimed at new training technologies such as open learning, distance learning, and multi-media training. Their purpose is to improve the quality and delivery of teaching and learning at advanced levels of training, resulting in transferable training products.

4.2 URSA-NET

URSA-NET is a project within COMETT that covers all the strands above. It was inaugurated in 1987, with the first operation of the COMETT programme, and has received financial support from the European Commission ever since. URSA-NET stands for: Urban and Regional Spatial Analysis - Network for Education and Training. It is a sectoral UETP, covering the technological and training area of "Computer Applications in Planning".

The nucleus of the project is the ***URSA-NET*** Network, currently consisting of 52 institutions: universities, enterprises, and local and regional authorities across all member-states of the European Community, as well as affiliated members from Czechoslovakia, Hungary, Japan, and India.

It is a flexible network, based on the common interest amongst its members to exchange and share information and experiences in the field of Computers in Planning. Currently, there are no fixed bylaws or rules for the Network's activities, and there is no formal legal status for URSA-NET. All related work is carried out by the Secretariat, located in Athens.

URSA-NET has been engaged in a series of parallel activities, during this entire period. However, before describing these activities, and in order to provide a better understanding of the impact of this programme on the overall process of diffusion of GIS technology and computer applications in Greece, it might be helpful to make at this point a quick digression in order to describe the current situation concerning GIS development in Greece.

4.2.1 GIS in Greece. Only a few years ago, the people involved with Geographic Information Systems and computer applications in general, in the fields of cartography, facilities management, or urban planning, were very few They consisted essentially of members of academic institutions, or specialised personnel in the one or two State installations (like the Military Cartographic Service). Things have advanced rapidly in the last few years. There are two reasons for this:

- Greece has been well connected with the general international computer market, from an early point in time. The "information revolution" with its new technologies, and particularly the personal computer, arrived in Greece early and in full force. Therefore, it was only a matter of time for the more specialised and technologically less-developed fields, like cartography and spatial analysis, to start catching-up with international developments.

- There was recently a proliferation of young university graduates, who selected the field of GIS for their post-graduate studies abroad. Therefore, when they started returning to the country and were employed in a variety of public authorities and organisations, they were able to argue in favour of the application of the new technologies to traditional tasks.

The above two parameters taken together with the obvious fact that sooner or later GIS had to be introduced in Greece created in the last two or three years an environment quite inducive to the development of GIS applications. On the other hand, currently there are no autonomous local hardware or software developers in Greece for GIS-related products. Only vendors of established foreign products.

Currently, two types of organizations experiment with GIS technology: government-related organizations and private engineering/consulting firms. Among government agencies, most prominent is the Military Cartographic Service, which is responsible for the production of most base-map work in the country. In addition, other government agencies working with maps, like the National Cadastre Service, the Institute for Geological Research, and the Centre for Oceanographic Research, are currently making their first investments in GIS technology.

A second group of organisations seriously considering GIS are local authorities. The City of Athens has not yet decided to go ahead with any significant GIS application. However, more than a few smaller municipalities of the Greater Athens area are already rigorously developing GIS applications for their urban structure and cadastre. Most use PC Arc/Info.

In addition, other municipalities across the country are gradually acquiring and using GIS. Three prominent examples are:

(i) the Municipality of Kalamata, in Southwestern Greece, which established a GIS centre immediately after a devastating earthquake, in 1986, in order to better organize the post-earthquake reconstruction planing.

(ii) the Municipality of Rhodes, in the Aegean, has experimented with the development of a GIS for the historic medieval part of the city, as part of a restoration effort.

(iii) the Municipality of Patras has recently awarded a two-year research contract to the University of Patras, to develop the framework of a comprehensive GIS for the City, as well as to provide the necessary training and technical support.

As far as private consulting firms are concerned, we are witnessing recently a gradual but nevertheless quite systematic introduction of GIS technology in the array of tools used by private firms. A characteristic example is the extensive area of projects financed by the Ministry of Planning and Environment, dealing with rezoning of urban areas in most Greek cities. Many private consulting firms that were awarded such contracts are using GIS and have developed interesting user-oriented interfaces in order to handle the large volume of data manipulation and cartographic work involved.

It is within this environment and in parallel with these activities that the URSA-NET project began offering its seminars and other activities.

4.2.2 Short intensive courses. A training seminar was developed at URSA-NET and carried out repeatedly in Greece. The seminar addressed mid-career planners in local administration and regional authorities. It is a typical "exposure" seminar and consists of two integrated elements: a series of introductory lectures, by the Greek training team of the University of Patras, and a series of presentations of applications, by European experts, who are members of the network.

The ***First Patras Seminar/Forum*** was carried out in May 1988 in Patras. About thirty planners, from local, regional, and central government authorities participated. The first seven working days of the Seminar were covered by lectures (on databases, GIS, modeling, and the like) while the last three days were devoted to an International Forum. In the forum which URSA-NET partner experts presented a series of lectures on applications. The same model was repeated in the following three years 1989-91 at the same place and time with similar audiences and future forums are planned. Overall, the seminars have been instrumental in creating, and sustaining a process of gradual technology transfer and diffusion.

4.2.3 Advanced Seminars in Europe. A series of ***advanced intensive seminars***, on a variety of topics, were held in different European cities and organised by a different partner each time. The first advanced seminar was held in Lyon, France in November 1989, on ***"Multi-media Urban Databases"***. A second advanced seminar was held in Delft, in October 1990 on ***"Data Acquisition for Spatial Information Systems"***. A third seminar was held in Prague, Czechoslovakia, in November 1991, with a general "exposure seminar" character. Five advanced seminars are planned for the coming year.

4.2.4 Training materials. The network has developed a comprehensive set of ***training materials***, including a series of textbooks, manuals, collections, bibliographies, and other relevant material in the field of computers in planning. There are already eight volumes published in the series. Further, a ***database*** was created of bibliographic records (articles, conference papers, etc.) as well as of people, institutions, software programmes, and all other relevant information in the field. The database currently includes more than 2800 bibliographic records, up to the end of the year 1991.

4.2.5 Student Placements. A ***fellowship / training system*** for students and fellows to spend time abroad for in-house training was established. In the period 1990-1991 four student placements and one fellowship were offered. In the period 1991-1992 the system has expanded to a total of 96 student months available for placement in enterprises for training periods of 3-6 months.

Additionally, planning is under way with partners from other universities in the European Community, for the development of a multidisciplinary international post-graduate training programme. The programme should be along the lines described above, of a two-year professional course with strong emphasis on applied training. Further, the network provides systematic support to members of partner institutions, in order to allow representatives to participate at related international activities such as the ***UDMS, Symposia***, the ***EGIS Conference***, or other international seminars and conferences.

4.3. URSA-CAPS: TRAINING WITH AND ABOUT COMPUTERS IN PLANNING

One last activity of the URSA-NET network, of particular importance for training, is the development of a software programme which both supports the planning process and trains the planner on how to plan with the aid of the computer. The idea is to combine in a single comprehensive workstation environment the following three functions:

(i) using the computer as an efficient information provider:
Planning Information Subsystem.
(ii) using the computer as a decision-support tool for planning:
Planning Analysis Subsystem.
(iii) and, at the same time, using the computer as a training medium, in order to teach somebody how to plan with the computer:
Planner Education Subsystem.

The Planning Information Subsystem is an integrated system that includes three types of information: documentary, statistical, and cartographic. It communicates with the user through friendly menus and allows the user to:

- create a particular information file,
- store information in the file,
- retrieve information, based on given search procedures,
- present information in a custom-tailored way.

The modules within the Planning Analysis Subsystem allow the planner to apply various subroutines for his analysis and evaluation tasks, e.g. descriptive statistics and statistical inference, cost-benefit, and other economic analysis techniques, qualitative evaluation schemes, spatial analysis, descriptive and presentation graphics, and the like.

Finally, the third role and function of the programme is training. All topics and modules included in the planning subsystem, have their counterpart in the training subsystem. In other words, corresponding to the statistical analysis subroutines, there are files/chapters of reading materials (with examples, problems, questions, and all other necessary training elements) which present the topic of statistical inference. Similarly, parallel to the automated cartography subroutines, there are lessons on cartography and thematic mapping.

The Planner Education Subsystem is based on the Hypertext format. ***Hypertext*** is an extremely interesting new way of organising large sets of information. The idea has been around for years, after been coined by Ted Nelson in the sixties. Essentially, it is the ability of the computer to handle text-based information in a non-linear, nonsequential form. Within URSA-CAPS, the Hypertext principle is applied to a large set of planning-related information. The idea is simple: A set of reading materials, dealing with all aspects of the planning process, together with exercises, question-and-answer modules, examples, and the like, are organised in one huge textual dataset through Hypertext.

For illustration purposes, we have integrated the whole NCGIA core curriculum text in the GIS training module. In addition, the user always has access to the other collections of documentary, statistical, and cartographic data. The whole set, organised in a user-friendly way (meaning the use of "menus" for selection among various choices, and a dialogue with the user in plain language rather than "computerese"), becomes a self-supporting computerised training system. The planner-trainee is seated in front of a workstation and makes an initial choice of a topic to read about, e.g. Geographic Information Systems. While reading a condensed and concise summary of the topic, the user is able, if desired, to wander through a collection of related articles that are cross-referenced through Hypertext. In addition, the user may explore specific examples of graphics applications and study a variety of thematic maps. As studies advance, the user may ask to see a step-by-step procedure on how to set up a geographic information system, complete with videoclips on digitising, editing the data, and plotting.

5. Conclusion

We have argued above the importance of training in creating a successful environment for GIS diffusion. Training should be of the largest possible breadth and depth. It is not simply a matter of learning the nitty-gritty of a particular software package. It goes much further; than that, to a whole new way of thinking about spatial problems, about possible solutions, and about using this new technology.

6. References

Bogaerts, M.J.M., editor (1990) Data Acquisition for Spatial Information Systems, *URSA-NET Computers In Planning Series*,vol. 7, Athens.

Laurini, R., editor (1989) Multi-media Urban Information Systems, *URSA-NET Computers In Planning Series*, vol. 1, Athens.

Masser, I. (1991) Promoting GIS Awareness: The European Dimension. *EGIS '91*, Brussels, April 2-5, 1991, 2: 700-706.

Polydorides, N.D. (1988) Planners' Education "for" and "with" the Computer. *Ekistics* 328-330: 52-55.

Polydorides, N.D. (1989) Computer-Aided-Planning-Systems: The Human-Machine Interaction in the Planning Process. Coordinated Information Systems for Improving Urban Functioning and Management, OECD Seminar, Copenhagen, 11-13 October 1989.

Polydorides, N.D. and N. Milionis, editors (1990) Computers in Urban and Regional Planning, An annotated Bibliography, *URSA-NET Computers In Planning Series*,vol. 2, Athens.

Polydorides, N.D., editor (1990) Proceedings Forum '90, *URSA-NET Computers In Planning Series*, vol. 6, Athens.

URBAN AND REGIONAL PLANNING MODELS AND GIS

WIM DOUVEN, MICHEL GROTHE, PETER NIJKAMP
and HENK SCHOLTEN
Department of Economics
Free University
Amsterdam
The Netherlands

ABSTRACT. Geographical Information Systems (GIS) have in recent years become an important part of regional and urban research. The combination of large data handling capabilities and computerized visualisation techniques explain the rapid growth path of GIS. A weak point is formed by its insufficient link with urban and regional modeling. The gap between modeling techniques and GIS needs to be bridged soon in order to 'reap the fruits' of this modern technology.

The paper will first give an overview of various caveats in the use of GIS and spatial modeling. Next, it will deal with two case studies which aim to overcome the above bottlenecks. The first one deals with a hybrid GIS and evaluation model for sustainable development of the Greek Sporades Islands. The second case study deals with GIS applications to elderly housing and neighbourhoods in the city of Amsterdam. The paper is concluded with some retrospective and prospective remarks.

1. Introduction

Information has become an important trademark of our modern society. Effective and accessible information systems are nowadays vital to both economic performance and strategic decision-making of nations, regions and cities. The rapid development of digital and electronic technologies, for instance, in the form of digital recording and transmission of sound and pictures, optical fibres for high speed transmission of information, super-fast computers, satellite broadcasting and video transmission, offers a new potential for sophisticated voice, data and image transmission of both private and public actors. From a geographical viewpoint the trend towards advanced information systems has in the past decades led to the design and use of geographic information systems (GIS). A GIS serves to offer a coherent representation of a set of geographical units or objects which - besides their locational position - can be characterized by one or more attributes (feature, label or thematic compound). Such information requires a consistent treatment of basic data, from the collection and storage stages to the manipulation and presentation of such data, in order to enhance the quality of decision-making (see Burrough 1986, Scholten and Stillwell 1990).

All such information systems may be highly important for the planning of our scarce space and of our living environment, not only on a global scale (e.g. monitoring of rain-

I. Masser and H. J. Onsrud (eds.), Diffusion and Use of Geographic Information Technologies, 317–337.

forest development), but also on a local scale (e.g. physical planning). Within this framework, spatial information systems are increasingly combined with spatial interaction models, pattern recognition, systems theory, topology, statistics, finite element analysis and evaluation analysis. The past twenty years have witnessed the development of various computer-based applications of information systems which have changed the activity patterns of people and the decision modes of private and public actors. Especially in the industrial sector, the design, manufacturing and quality control of products have been altered significantly with the introduction of Computer Aided Design and Computer Aided Manufacturing (CAD/CAM). In a similar manner, GIS tools are currently modifying our perception of geography and planning and our ways of spatial and environmental decision-making.

The conventional perception of GIS is that it is able to generate the following functions: handling geographical data entry, storage, analysis and mapping more rapidly, inexpensively and accurately than traditional non-automated methods allow. While effective record keeping, data analysis and information management are important benefits of GIS in daily operations, the greatest benefits of GIS are those resulting from its application to decision-making. A GIS can quickly and accurately provide decision-makers with information in the most concise form: a sophisticated polyvalent picture. A GIS produces maps and supporting tables of information that for many years used to be very difficult or time consuming to produce. A noticeable advantage of GIS is that it generates a multitude of relationships (using information from various sources) which can be made immediately accessible to decision-makers, who can then spend their time more efficiently by considering strategic policy issues rather than extracting data from volumes of written information (see also Fischer and Nijkamp 1993).

Especially the capability of modern GIS tools to answer 'what-if' questions makes them appropriate as tools in decision support systems (DSS). In particular, interactive GIS-DSS methods are likely to offer a great potential for improved decision-making, as such methods enhance significantly the flexibility and choice freedom in relation to the knowledge and expertise of the user. The blend of GIS and DSS can be used in a wide range of modeling activities, such as: statistical and exploratory data analysis, financial accounting models, spatial behavioural models, evaluation models and visualisation techniques. A good illustration of the architecture and structure of a spatial GIS-DSS can be found in Armstrong and Densham (1990) (Figure 1).

A major research challenge rests in linking the traditional GIS functions (measuring, recording, storing, analyzing and visualizing) to spatial analysis (selection, information manipulation, data exploration and confirmation) so as to improve the quality of decision-making (see Anselin and Getis 1992). Clearly, error analysis is an important complementary tool for judging the validity of inferences (see Fisher 1991), while also limitations in data modeling and flaws in software design have to be recognized (van Beurden and Scholten 1990).

In conclusion, GIS has demonstrated its great potential in a wide range of applications. Its ability to store and combine geographical data from different sources in a logical, coherent and structured way, to carry out in a flexible manner various analytical manipulations and to present the results in an appealing visual way (e.g., sophisticated overlay maps) enhances its popularity. Many organizations and research laboratories in various fields gather spatially related data for planning and policy purposes and use increasingly the GIS methodology to

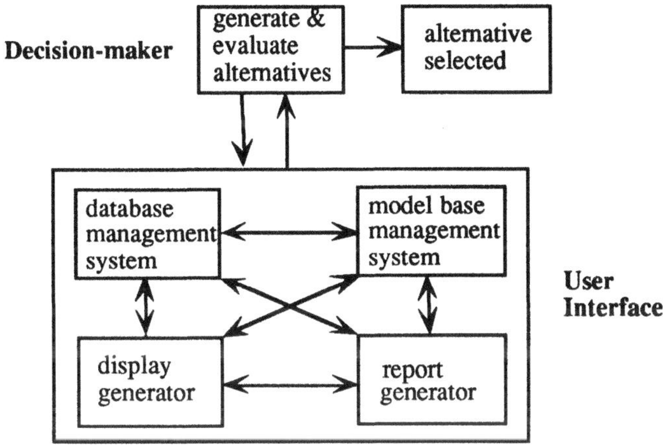

Figure 1. Architecture of a spatial DSS

store and handle such data. GIS is nowadays successfully implemented for planning, monitoring and maintaining spatial objects, while it has the potential to play a role in structured spatial research (see Douven and Scholten 1993). Nevertheless, there is still a major discrepancy between GIS as an instrument for handling spatial data and GIS as a tool for exploratory, explanatory or decision-supporting spatial analysis (see Openshaw 1991). Despite the popularity of GIS, its role as an ancillary tool for balanced decision-making still leaves much to be desired. This paper will first discuss the critical success factors for a further penetration of GIS. The main focus will then be on the linkage between GIS and spatial analysis, while the final part of the paper will provide empirical illustrations of recent GIS applications, viz. sustainable development of the Greek Sporades Islands and elderly housing and neighbourhoods in Amsterdam.

2. Critical Success Factors of GIS Adoption

As with all new products, GIS has to overcome various barriers in order to be adopted and to reach a stage of full market penetration. This product life cycle approach is well-known from the diffusion literature. It should be emphasized here that the new GIS products are polyvalent in nature - such as: efficient data base management, support of model calculations/simulations, flexible geographical presentation, possibilities for locational simulations, flexible geographical scale adjustments etc., - so that these functions for improved decision-making have to be judged from multiple angles. High product quality (i.e., sophisticated GIS technology) and market orientation (e.g., user-friendliness and cost-effectiveness) are a sine qua non for a market success of new GIS products.

In a recent paper (see Maggi et al. 1992) attention has been called to the critical success (and failure) factors of the introduction of new information and telecommunications technologies using the so-called Pentagon prism. In the authors' view, the successful introduction, adoption and penetration of such new technologies depends in general, on five critical success factors:

- hardware (engineering e.g.)
- software (information, e.g.)
- orgware (management, e.g.)
- finware (costs, e.g.)
- ecoware (environmental impacts, e.g.)

These five factors which hold also for a successful introduction of GIS technology, can be incorporated in the so-called Pentagon prism (see Figure 2), which has been used as a systematic way of screening on necessary conditions for market penetration. All five factors are simultaneously and intrinsically playing a critical role.

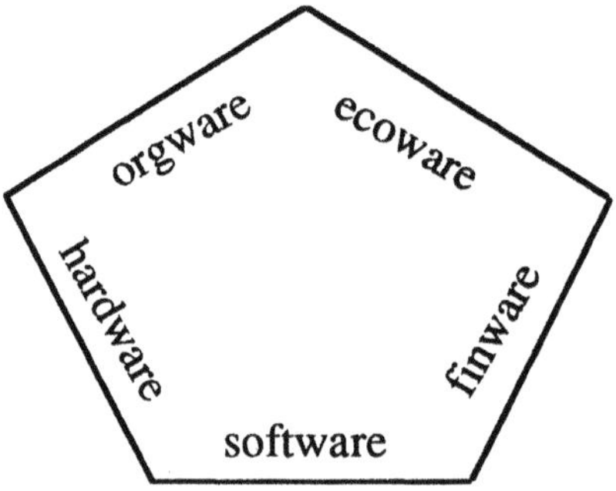

Figure 2. The Pentagon prism

A parallel reasoning can be adopted for GIS. The success of GIS is critically dependent on:

- hardware: sophisticated workstations, computer equipment, multi-colour computer cartography, visualisation technologies, etc.
- software: availability of reliable data, accompanied by advanced data recording and treatment techniques, integrated GIS and spatial analysis methods (e.g., GIS-based spatial interaction models), or educational and training schemes for GIS based spatial analysis.
- orgware: user-friendly DSS-GIS interfaces, interactive visualisation methods, well structured spatial decision-making and AI, administrative responsibility for maintenance, up-dating and back-ups, and interwoveness with decision making agencies etc.
- finware: efficient and inexpensive data mapping techniques, or polyvalent GIS methods.
- ecoware: significant contribution to urban and regional quality of life, balanced conflict resolution/evaluation methodology.

All such factors are of critical importance for a successful implementation and use of the GIS methodology. Lack to overcome the above barriers will inevitably lead to biased or insufficient use of the potential of modern GIS. The intricate synergy between GIS and spatial analysis is one of the crucial developmental factors of spatial informatics and will be discussed in the next section. Next, in the sequel of this paper we will give two interesting examples of successful applications of GIS.

3. Spatial Analysis and GIS

There is an increasing need for integrating GIS with spatial analysis in order to design a new policy-relevant strategic framework for effective and balanced decision-making on spatial and environmental systems. Therefore, the linkages between GIS and spatial analysis/modeling deserve closer attention.

Both GIS and spatial analysis serve to systematize a complex reality by focusing in a structured way on key elements under investigation. Thus abstraction and reality breakdown into objects and relationships (either deductively or empirically) serve to assist human comprehension of reality, although a considerable amount of error will always remain. Mapping (by means of GIS) and modeling (by spatial analysis) are essentially two sides of the same coin, but each offers different perspectives and insights. Following Anselin and Getis (1992) we may distinguish four main groups of important functions to be carried out for increasing our knowledge of spatial systems: selection, manipulation, exploration and confirmation. These functionality groups will now be concisely described.

3.1. SELECTION

Selection is in itself a fairly simple operation. The basic operation is a classification: only units with pre-determined characteristics are selected as members of a group. This is sometimes an important step in visual interpretation and therefore prone to errors (e.g. Middelkoop 1990), which are not always easy to cope with.

Depending on a value x of variable X a given unit (any spatial object) is a member of the class. Such a dichotomic segmentation can be represented by a Boolean membership. However simple the operation, its result is important, because all subsequent processing will be based on these results. The main problem is clearly the demarcation of the boundaries of relevant sets. For instance, determining whether someone is a teenager is simple: if $10 \leq x \leq 19$ is true, the person is a teenager. A problem might be the determination whether someone is young or not. The class boundaries are then not very clear. This case may call for a more continuous kind of class boundary, e.g., a fuzzy approach. In general fuzzy classifications are based on recent theories of fuzzy logic and are able to encapsulate statements such as: "a person is fairly young." For an application in the area of land use evaluation we refer to Munda et al. (1992).

In classification and subsequent evaluation, the kind of method used can be very important. Examples of cases where possible errors have not been taken into account causing a disturbing impact on the results are numerous (see e.g. Draaijers et al. 1991).
The previously defined selection or classification is mainly used for attribute data. Spatial selection may also occur in a broader analysis. The best known examples in a GIS are clipping or its complementary form (often related to overlay). If a fuzzy classification is used in a fuzzy selection environment, the result is a set of units similar to an ideal area, where especially locational uncertainty (border errors) is incorporated.

3.2 MANIPULATION

Manipulation functions are the tools usually termed 'analysis' in GIS software. We will not adhere to this practice here, although most GIS tools are part of these functions. Manipulation of the objects stored (as observational units) is usually also related to the next two groups of functions, i.e., exploration and confirmation; it might sometimes even be difficult to separate them. In manipulating the information one may just analyze the data, but more often one will use some kind of hypothetical model of reality, which forms the basis of the analysis. Main techniques in this functionality group (to be elaborated subsequently) are:

(a) coordinate transformation, like rubber-sheeting, generalization and projecting (although this is very much related to the input stage, it is often necessary to change from, for instance, equal area to equal distance);
(b) partitioning of observational units (or aggregation to larger ones), while also changes in representation may be made here (e.g.. from lines to buffer zones);
(c) overlay, in order to create new observational units, which in themselves are assumed to be relatively homogeneous;
(d) interpolation, usually used to change observational units (from points to areas or lines) or to derive information about a unit's variable, which for the unit is not known, but whose values are known for surrounding objects.

3.2.1 Transformation. Coordinate transformation is rather familiar. When all forms remain intact this is done by just recalculating the corners and using the same function for all other points. This may, for instance, be a transformation from rectangle to rectangle, but also from rectangle to parallelogram (resulting in a 'perspective' view).

A more intricate transformation occurs when the transformation function used differs in the map (spatial variation of the function). This is called 'rubber sheeting', which is a mapping technique, that primarily should be used to match edges of different thematic layers in a map; unfortunately it is often used to get the data in the 'right' position, as part of the input stage. Resulting errors are usually discarded.

The transformation best known to cartographers is projection: the problem of mapping a spherical surface on a flat surface with the least distortion (see e.g., Robinson and Snyder 1991). Projections appear in three forms: true to distance (equidistant), true to area (equal area) and true to form (conform). A projection can have only one of these three properties. Obviously, transformation by projection would not be necessary when a true three-dimensional technique is used; however, most techniques are based on classical techniques dealing with flat maps, therefore processing distortion caused by the projection.

3.2.2 Partitioning and Aggregation. Surveys often yield figures for larger units, for instance, population counts or the total area of crops per administrative unit (a municipality, e.g.). The given amount is most certainly not spread evenly in this unit, but is instead concentrated at certain spots within the unit. If possible at all, some partitioning can be performed, in which the total amount is sub-divided by some rule (either deterministic or empirical), while the resulting figures are attributed to smaller units which together form the larger unit we started with. This means that some knowledge about the large unit must exist: where are concentrations to be expected? This technique puts some constraints on the input data: area recalculations may only be performed in equal area projections; densities or fluxes must first be transformed to absolute amounts. And, of course, the error in the result depends largely on the quality of the pre-conceived knowledge or hypothesis about the component areas of the large unit. Sometimes some kind of area interpolation may precede the step of partitioning.

The process can be performed the other way round as well: aggregation. Summations of attribute values are fairly simple; one only has to take care of true area representation (equal area projection).

3.2.3 Overlay. Overlay is the best known GIS tool in relation to spatial analysis. Its reputation has two major causes. It is one of the oldest and most commonly used techniques by cartographers and geographers. Secondly, an overlay in itself is very simple: only coordinates have to be matched and possibly lines intersected. The result is a new map with new observational units with of course different positions from the units in both input maps.

Projection plays an important role in this technique. Any projected map is drawn on a flat piece of paper, upon which a Cartesian coordinate system can be positioned. This allows easy coordinate matching and line intersection. Things would be more difficult on a spherical object (like the earth). The three dimensional notation (degrees) needs certainly more intricate mathematical transformations than Cartesian coordinates do in an overlay.

An overlay assumes some homogeneity in zones; a condition not often met in reality. Accuracy problems may therefore arise in an overlay.

3.2.4 Interpolation. Spatial interpolation is another 'old' technique to change from one observational unit to another. The problem can be stated as follows. "Given a set of spatial data either in the form of points or for subareas, find the function that will best represent the whole surface and that will predict values at other points or for other subareas" (Lam 1983). Choosing the 'right' interpolation method depends largely on the type of information involved and the degree of desired accuracy. Furthermore, some methods will require much computational effort (of either the mapper or the computer). Basic to the method is the assumption that each interpolation method is a sort of hypothesis of the surface, and this hypothesis may or may not be true. Elaborate descriptions will not be given in this paper; some general remarks can be made here.

Point interpolations are used to derive area information from point data (or derive information at an infinite number of points), which can be collected at a point (isometric). Exact methods preserve values of the points. Most functions pass through all points, sometimes totally unconstrained (and therefore sometimes resulting in absurd extremes), or more constrained, but with more assumptions (like distance-weighting). These techniques can easily be used to smooth the data surface, thus introducing some error. Furthermore, assumptions will lead to ambiguity when the actual surface characteristics are unknown, and uneven distributions can distort the method's functioning.

Splines are functions with an appropriate power degree, so that the derivatives of the functions are continuous everywhere. For computing the function, the surface may be divided by drawing lines between points, to form quadrangles or triangles. Especially in choosing the triangles the process is prone to errors. Nevertheless, this technique works piecewise, is very closely related to the value being interpolated and it may be very accurate (depending on the surface modeled). Finite difference methods assume that the desired surface obeys some differential equations, both ordinary and partial. These equations are then approximated by finite differences and solved iteratively. The method is simple, though the solution to a set of difference equations may be time consuming. Interpolation beyond the neighbourhood of the points is usually poor and flat areas may pose serious problems.

Kriging is perhaps the most distinctive type of interpolation method. Kriging is based on some simplifying assumptions, like stationarity in simple Kriging, which means that the probability density function and the autocovariance matrix can be estimated. A variation is needed for most natural phenomena, where this assumption does not hold. The variogram then forms the basis, representing the relationship between the mean-square difference between sample values and their intervening distance. One might also assume that the increments of the regionalized variable have only some properties of stationarity within a neighbourhood and that the trend or drift for a neighbourhood can be described by a polynomial function: this is universal Kriging. It uses a different set of equations for each point estimate in different neighbourhoods. Choosing the size of a neighbourhood and matching a polynomial function with the experimental variogram may pose problems when using Kriging. The advantage is clearly the ability for statistical inference, which also provides an estimate error. Most of the elements described precede the actual Kriging interpolation.

Approximation methods are concerned with determining a function f(x,y), which assumes values at the data points approximately but not generally equal to the observed values, therefore resulting in an error or residual at every data point. These errors can be minimized (or optimized), for instance, by least squares polynomials, sometimes added with distance weighting or fitted with splines. Choice of weights or uneven distributions may cause problems with these methods.

The areal interpolation problem is more common to geography than to other fields. Its main use is in isopleth mapping. The change from one set of boundaries to another is not part of the interpolation (like overlay), but is an essential step, which is why overlay is also mentioned here. Most generally, nonvolume-preserving methods use point interpolation methods to derive values for small areas (for instance, grid cells), for which the value found is assumed to be a representation. The choice of the control point or the way areas give weights (or even a decisive value) is very important and may have a large effect on the result.

Volume-preserving methods take the topological aspect of areas into account. Overlay is the basic tool for this, since area values of the target zones are determined from sizes of overlapping zones. Derived areas and comparison with input zones may result in weights when attributing a value to a new target zone. This is often accompanied by some regression function of several variables (attributes), of which some may be absent for new target zones. The important assumption of homogeneous zones is in reality often found not to be true.

3.3 EXPLORATIVE DATA ANALYSIS

In exploratory data analysis the data are used in an inductive fashion to gain new insights. The first step in discovering a kind of pattern or cluster in the data available is to derive any information whether patterns exist or not. Pattern indicators have been established with dispersion factors (using nearest-neighbour distances) or, more recently, fractal geometry. The equations will however not yield any spatial result in the form of, for instance, maps (Berry 1991).

A fairly simple approach is to limit the analysis to pure description. Examples of possible solutions are recently appearing (see e.g., Openshaw 1990; Openshaw and Charlton 1990), where computation intensive routines elicit pattern information from the data, but usually no significant indication or information about the fact that the pattern may be spurious is given. Anselin and Getis (1992) conjecture that this is not really exploratory data analysis, as this be of high dimensionality. Nevertheless, the 'new' routines show examples of univariate and bivariate situations.

Clustering is a fairly old method to group individual data according to similarity or dissimilarity. Most clustering methods use attribute values to discern similarities, but in some cases also distances are be used. There are, depending on the clustering method used, different constraints for this distance aspect in computing clusters. The most simple clustering only looks at distance (or, put in another way, point density). Clustering is getting more difficult if attribute values are also involved in the clustering, mainly because of strict constraints and assumptions.

Typical exploratory analysis is of a statistical nature, while most statistical methods are a-spatial. Spatial aspects might be incorporated in some methods, but this usually means a set of straitjacket constraints on the spatial data. If such conditions are not fully met, wrong or false conclusions may easily result; the best known example is spatial dependence (and/or heterogeneity).

Interactivity between maps and other (statistical) exploratory information offers interesting insights into the relationships within the data. Examples of this approach developed in a computer environment are interactive graphics (Haslett et al. 1990a,b and c; Bradley and Haslett 1990; Unwin et al. 1990).

Explorations will most certainly elicit some pattern or structural relationship in the data. Such results should be considered with caution, because causal relations may very well be absent. Therefore, after an exploratory analysis there will be a need for further research into the nature of these relations and processes.

3.4 CONFIRMATORY ANALYSIS

Anselin and Getis (1992) point out that very little has been achieved in the field of automating confirmatory analysis, which can be seen as the main functionality for predictions, forecasts and filtering tests. Most applications are non-spatial, like regression analysis, which fail to make use of the spatial aspect of the information (topology). Some links with pure statistical methods can be made, but it proves difficult to work with the space component. Some model based techniques may use spatial elements, such as Kriging and Kalman filtering when they are explicitly concerned with space (see e.g., Christensen 1991). A well developed field of spatial analysis is that of spatial interaction or, in a broader sense, any kind of spatial dynamic modeling (e.g., diffusion). Spatial interaction and gravity ('potential mapping') are pure models, which are tested after formulation. The test contains both comparisons with observations and statistical features. Unfortunately, the fact that the model is only an approximation of (part of) reality means that statistical tools are indispensable (see for details amongst others Nijkamp and Reggiani 1992; Scholten and Stillwell 1990)

In cases of space-time clustering it is often necessary to apply some preconceived notion (model) to the data, for instance, by stratifying the data (a kind of classification problem). The result is a clear mix of both exploratory and confirmatory techniques (see e.g., Melse

1991).

After this rather lengthy discussion about the connections between GIS and spatial analysis (including modeling), we will in the next sections offer a presentation of two recent applications of GIS modeling so as to illustrate the power of GIS modeling in a regional and urban planning content.

4. A Case Study on Greek Islands

Here a concise description of a GIS-based regional model will be given which serves to analyze sustainable development (SD) for the Greek Sporades Islands, with particular emphasis on the island of Alonnisos (see for details Giaoutzi and Nijkamp 1993). The study deals with conflicts between regional economic development and environmental protection for an island economy dominated by tourism and fishery. Especially the seasonal peaks in tourism exceed the carrying capacity of the islands and may result in abrupt high resource demands (e.g., water, food, etc.). This may in turn cause irreversible processes in the resource stocks. This means that the SD conflicts on the study area are dynamic phenomena. Some of these conflicts can be also expressed in spatial dimensions directly (e.g., the expansion of urban land at the expense of forest areas). But only through the use of a spatio-temporal hybrid GIS-SD dynamic model we can satisfactorily describe and simulate the mechanics of our ecosystem SD conflicts. For such a purpose, the model concerned should focus on dynamic simulations in space and time rather than being an optimization model; should assist planners in examining the dynamic economic and ecological conflicts of the study area in their physical dimensions (three dimensions for the spatial reference together with one dimension for the dynamic reference over time); and should provide alternative strategic solutions by scenarios which should be based on SD constraints for the development of the region at hand.

The data requirements for such a GIS-SD hybrid model are determined by (1) the spatio-temporal resolutions selected (i.e., 100m x 100m ground pixel size, and a simulation period of 1 year), and (2) the data needs for a successful GIS-SD model calibration. The data supply may not necessarily be in one-to-one correspondence with the data demand. This means that (1) some digital data may have been collected and processed which finally are not all used as an input to the GIS-SD model, but rather form an integrated digital data base and (2) necessary digital data for the GIS-SD model may not have been available for the period the model was developed and tested. The data collection process focused mainly on obtaining as much information as possible about the socio-economic and natural environment of the study area. The necessary data input to both our GIS-SD model and the digital data base constructed included amongst others:

Non-spatial data

- socio-economic data of the region (productivity and income per economic sector, tourism, fishery, houses, energy used per household, etc.)
- demographic data (population, age pyramid etc.)
- ecological data (ground water, sea water quality, forest fires, wild life data etc.)

Spatial data - terrain elevation and sea depth data
- land use data
- distance data from important land uses such as urban land, forests etc.
- road transportation network data etc.

After the necessary spatial and non-spatial data requirements were set, the relevant data were collected from the various (mainly Greek) sources: National Statistical Service, Greek Military Geographic Service, Ministry of Environment, Planning and Technical Works etc. Based on these selected data, a complete GIS-SD model was designed and used as follows. Various scenarios on the future development of the island were developed. The main focus was on land use/transportation scenarios, and on two scenario runs generated by successively excluding and including road transportation on Alonnisos island in order to demonstrate the efficient use of the GIS-SD system for monitoring land changes under different urban attractiveness conditions. It was assumed that urban land use development was strongly dependent on the presence of transportation infrastructure. This was reflected in so-called attraction layers. Using these attraction layers the system was run for a simulation period of 15 years, starting with the year 1985 and ending with the year 2000 for a five year period. The results can be represented by means of systematic colour maps (see Despotakis 1991). Here we will only describe some results based on the policy options of "absence" and "presence" of transportation, respectively.

The effects on urban expansion as a result of road transportation are clearly depicted in the above two figures. For the "no transportation" scenario the urban expansion takes place mainly along the sea shore and the already existing urban areas of the island; for the "transportation" scenario the urban expansion presents clusters spread along the roads of the island, thus eliminating the amount of urban area to be spread along the sea shore. This GIS-SD system provides these results in the form of raster images which may also be used for animation applications, so that more intuitive information may be extracted from the simulation results. These two scenario results can be inserted into a DSS. The evaluation by means of a DSS for the Sporades contains four relevant policy criteria, viz. tourism, nature, landscape and transportation. In addition, the impacts of possible development scenario's on each of these criteria have to be assessed. Based on policy trade-off analysis, in particular multi-criteria analysis, the most desirable scenario can then be identified.

For each criterion (e.g., nature, tourism, etc.) an evaluation map can be generated which corresponds to the selected alternative development scenario for a specific year. Multi-criteria analysis is then used to rank the selected alternatives. Based on the above methodology, five spatial scenario runs generated by assuming certain policies with respect to urban expansion over the island of Alonnisos were carried out. The purpose of selecting these scenarios was to use the GIS-SD model for evaluating the effects of various spatial policies on urban development. These scenarios can be developed for both the 'transportation' and 'no transportation' policy. In the sequel of our analysis only scenarios related to the 'transportation' option will be further illustrated.

Scenario 1: the first spatial scenario encourages urban growth within 200m of the sea. The beaches themselves may be changed to urban areas.

Scenario 2: the second spatial scenario encourages urban growth in the middle of the island. The beaches and a zone of 500m from the sea may not be changed to urban areas.

Scenario 3: the third spatial scenario allows the urban area to expand in the southern part of the island only. The northern part of the island remains untouched.

Scenario 4: the fourth spatial scenario allows urban to expand only in the eastern part of the island. Beach areas are allowed to change to urban functions in the eastern part of the island only. The western part of the island remains untouched.

Scenario 5: the fifth spatial scenario allows urban areas to expand only in the existing urban areas of the island. This means that any type of urban growth is strictly prohibited in horizontal direction, and only urban growth in a vertical direction is allowed.

The effect table for these five scenarios is shown in Table 1. The scoring units are non-dimensional relative units. High scores mean 'good' results, whereas low scores indicate 'poor' results. We may observe from the above table that:

(1) Scenario 2 (urban growth in the middle of the island) has a very low relative score for the 'tourism' criterion, whereas scenario 5 (no urban expansion) has the highest score for the same criterion. This results from the fact that the urban distances to the sea - which determine the "happy tourist" in this scenario - are maximal for the case of scenario 2 and minimal for the case of scenario 5.

(2) The 'best' sustainable scenario for nature, determined by land use changes, is scenario 5. This is due to the fact that, for scenario 5, the natural area for the year 2005 remains almost the same as the natural area for the year 1985. The 'worst' scenario from the point of view of natural sustainability in this case is scenario 2; according to this scenario the scarce natural areas of the middle of the island are destroyed and changed to urban areas. We also observe here that the third scenario (urban expansion at the southern half of the island) results in a lower burden to nature than the fourth scenario (urban expansion at the eastern half of the island). This happens because the existing urban area for year 1985 is already located at the southern half of the island; therefore, a scenario according to which additional urban expansion at the eastern half of the island takes place, will destroy the natural areas more than the southern (already disturbed) areas.

Table 1: The effect table in the GIS-SD-DSS system for the Sporades

	ALTERNATIVE SCENARIOS				
CRITERIA	1	2	3	4	5
tourism	6.219	3.761	6.242	6.543	6.817
nature	4.128	3.841	3.906	3.879	5.516
landscape	4.663	5.742	4.575	4.427	4.353
transportation	6.299	5.635	6.945	5.168	8.691

(3) Scenario 2 is the 'best' in terms of preserving the original landscape of the island: the urban expansion takes place in the middle of the island and therefore the urban areas are not visible from either the sea or other parts of the island. This non-visibility results in high scores for the landscape criterion.

(4) Scenario 5 is the 'best' in terms of road transportation; in other words, this scenario results in the minimum transportation burden required for the local people or tourists to move on the island. The 'worst' scenario in terms of road transportation load is scenario 4 (urban expansion only at the eastern half of the island). This is due to the fact that the distances from harbour, located at the southern part of the island, are taken into account for the computation of road transportation load. The effect table was next evaluated by a multi-criteria analysis, based on a regime method (see Nijkamp et al. 1990).

The ranking results, given the four selected criteria, are as follows (see Table 2) .

Table 2: Optimal ranking for the five spatial alternatives in the Sporades

Alternative scenario	1	2	3	4	5
Rank	3	5	2	4	1

We may thus conclude that based on the above assumptions and ranking methods, scenario 5 is to be regarded as a 'sustainable' scenario, to be preferred to the others. The term 'sustainable' here refers to a development (till the year 2005) which 'best' combines the four sustainability criteria selected: (1) the tourist sector activities are favoured and enhanced; (2) the natural areas are as little as possible destroyed and/or changed to urban areas; (3) the landscape values are preserved, and (4) the road transportation load becomes minimal. Clearly, such a conclusion is expected from development scenarios of the type of scenario 5: if the urban areas are to be restricted to their original locations, then this is the 'best' sustainable scenario, at least in an ecocentric sense. Since such 'no expansion' policies often tend to be unrealistic (and much more difficult to apply and control), one might, alternatively, select scenario 3, i.e., the scenario which was ranked as the second 'best'. All results - including all detailed GIS maps - are documented more extensively in Despotakis (1991) and Giaoutzi and Nijkamp (1993). This illustration has shown the potential of combining GIS with spatial modeling for policy evaluation purposes.

5. A Case Study on Elderly Housing and Neighbourhoods in Amsterdam

The population in Europe, and even more so in the Netherlands, is aging rapidly. An increase in life expectancy and a decrease in birth rates are leading to an explosive increase of persons 55 years of age and older, generally regarded as "the elderly". Within this group of elderly, "double aging" also takes place, and the population of the "old" elderly within the total elderly population increases even more rapidly. Concurrently the Dutch government's elderly care service policy has decreased the number and capacity of elderly care services, such as old peoples' homes and nursing homes.

The consequence of this national policy, with regard to housing, is that elderly need to remain independent in their own residential neighbourhood for more years than they did in the past. However in the urban and suburban areas, elderly who live independently face an increasing number and range of problems. A notable shortage of suitable dwellings exists, especially in older urban areas in Amsterdam. There the quality of dwellings is often insufficient: poor technical quality, poor accessibility due to steep and narrow staircases, and lack of amenities like a shower or bath. Furthermore, a decent residential environment often is lacking. The gradual disappearance of small neighbourhood shops and post offices has a negative impact, especially for less-mobile inhabitants who often are nearby-oriented. Increasing fear of crime, threatening behaviour and violence in the streets diminish the interaction possibilities of the elderly and devalue their daily living environment.

Over the past decades, planning concepts of elderly housing in The Netherlands shifted from the planning of institutional housing - old peoples' homes and nursing homes - towards planning concepts of independent housing with a focus on the design of buildings and dwellings. This new policy emphasised the neighbourhood as the geographical unit to study and monitor the living environment of elderly and ensure continuity of the individual life-styles.

Recently, Dutch planners have developed a planning concept, called "residential zoning for the elderly", which includes a framework to monitor and guarantee the possibility for elderly to live independently in a 'friendly' residential neighbourhood. GIS offers promising possibilities to integrate data to measure the physical and social qualities of a neighbourhood for elderly housing. GIS applications in the urban area of Amsterdam already illustrate the potential of GIS technology to monitor housing and neighbourhood quality for older residents and to support planners in local housing policy development.

Originally, residential areas for elderly were defined as "areas with suitable and affordable dwellings, situated within 500 meters walking distance - or 400 meters as the crow flies - of services for elderly: shops, public transport, medical facilities, post offices and green areas. At the same time, these areas have by no means a homogeneous population of elderly, but are built up heterogeneously". Consequently, in considering housing for the elderly, planners need to focus their attention on both physical and social characteristics of elderly residences (Burby and Rohe 1990). Residential zoning for the elderly addresses this need because it is adapted to the local circumstances and qualifies areas according to their suitability for elderly housing based on both physical and social characteristics.

Physical characteristics of the neighbourhood are measured by means of indicators of accessibility to several important shopping facilities, care services centres and recreational facilities. The availability of suitable dwellings - defined and selected on their physical characteristics such as ground floor accessibility, availability of an elevator, amount of floor space, number of rooms, availability of amenities, rent level, house ownership etc - is the other important objective policy variable for measurement of elderly housing. Social variables include neighbourhood livability and satisfaction, fear of crime and potential for social interaction, which are linked with the location, design and management of housing for the elderly. The actual use of neighbourhood facilities and services, the patterns of social interaction and the perception and fear of crime are indicators of housing satisfaction for elderly who are economically and physically often restricted, more vulnerable and less mobile. Interviews among the elderly population are necessary for planners in order to obtain insight into:

- the actual behaviour of the elderly and their perceptions concerning the use of services; and
- their existing and preferred housing situations.

The decisive factor for the perceived housing situation and actual interaction pattern is the household situation. Differences in household situation, in terms of age, mobility, cultural/ethnic background en socio-economic characteristics, cause individual households to perceive the housing environment differently and to weight objective policy criteria differently. This difference in perspectives demonstrates that residential zoning is a relative concept: a range of friendly to unfriendly zones, related to the demographic and socio-economic characteristics of the elderly population, exists. The analytical framework can be found in Figure 3. The objective policy variables are weighted, using interview results, and combined with other variables for the development of local housing policy.

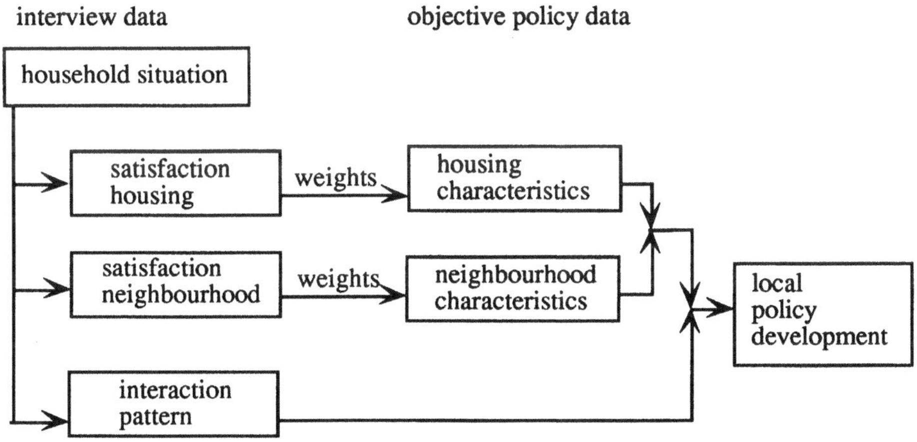

Figure 3. Analysis framework for analyzing elderly housing and neighbourhoods

In several pilot studies conducted in residential districts of the Amsterdam urban area, GIS has been used to define residential areas for the elderly to support policy development for elderly housing. GIS is applied in all aspects of the research framework:

- qualification of neighbourhood areas, based on accessibility of facilities and services; selection and display of suitable dwellings; and
- integration of data on elderly social interaction and attractiveness of collective areas, based on the occurrence of crime.

Data about physical and social variables have been collected, processed and stored using statistical software in combination with GIS. Objective policy indicators for measuring of elderly residential situations in the Netherlands were applied as well. Two types of questions concerning the neighbourhood and housing characteristics had to be answered by the elderly:

1. Which retail facilities and care services are important to the elderly and how is accessibility defined?
2. Which type of dwellings are suitable for the elderly?

From the viewpoint of accessibility, the following retail and recreational facilities and care services appear to be important to the elderly, according to Dutch planning objectives (in descending order of importance):

- daily-visited retail facilities (bakery, green-grocery, butcher's shop, dairy, supermarket and daily market);
- public transport (bus, tram and metro);
- financial services (post office or bank);
- recreational facilities (parks and green areas); and
- care/welfare service centres.

Accessibility for the elderly has been defined as a walking distance of 250 meters for less-mobile elderly and 500 meters for mobile elderly. Suitable elderly housing was defined as:

- being on the ground floor, first floor or in a building with an elevator; and
- dwellings with 3 or 4 rooms, more than 32 square meters of floor space and a low rent level.

Data about the location and type of facilities and care service centres, and the location and characteristics of the housing supply have been collected and stored as point data. A transportation network is used for distance calculations between all residential locations and every single retail facility and care service centre. Using policy variables and weights, this information can be processed and displayed in suitability maps (see Grothe and Blom 1992). GIS analysis and mapping functions produced suitability maps based on the accessibility of several types of facilities and care service centres (in declining planning priority) in the district "de Baarsjes" in Amsterdam.

On the basis of interviews among elderly residents, the weighting was adapted to local circumstances. Elderly expressed their preference and the frequency of use of the different types of facilities and service centres. Using different weighting methods and the weighted summation method of multiple criteria analysis, a classification was derived, based on the product of weights and the distance between residential location and facility location. Shopping and recreation facilities available to elderly in their neighbourhood were measured by counting the number and type of shops within variable distance zones from the residents' dwellings. Using the average distance, a qualification of dwellings according to the neighbourhood service level was generated.

Next, the relationship between the household situation and the spatial component of activities of older individuals was examined based on a behavioural analysis of elderly social interaction. Data about places that elderly regularly visit, so called "habitats", have been collected from individual elderly in order to identify commonly visited areas in the neighbourhood. Data concerning spatial and social interaction by type (shopping, visiting friends etc.), origin-destination and frequency were collected, stored and analyzed in ARC/INFO (see also Shalaby 1991). Using map overlay operations, the interaction maps were combined and collective areas of interaction have been defined and differentiated for the household situation (age, mobility etc.). Measuring the habitat of the elderly is important to understand the actual interaction which was used to correct policy variables of accessibility to the local circumstances. Residential zoning areas were defined based on collective areas of interaction differentiated by age. At the same time, this information is combined with information about "attractive" and "less attractive" areas gathered during

interviews among governmental, care and welfare institutions and elderly inhabitants.
As a result, maps of attractive and less attractive areas, based on the perception and occurrence of crime, can be compiled and stored. These maps reflect the mental and social perceptions of elderly individuals and indicate the parts of the neighbourhood they perceive as decent living environments. In combination with objective policy variables, these maps aid in the development of new local plans for improving the housing environment of the elderly. (see for further details Grothe and Blom 1992)

To apply the concept of residential zoning in a Dutch - or in a broader European - urban planning practice, several integrative elements need to be distinguished:

- a framework for new housing, housing renovation and allocation;
- a structure for the integration and adjustment of housing, care and service facilities to the needs of elderly;
- procedures to handle the flows of elderly residents from large to small(er) dwellings; and
- a system for the management of the residential environment.

Although the concept of friendly residential zoning for elderly is still in a development stage, the Amsterdam applications have clearly illustrated the potential of GIS technology for residential zoning, behavioural analysis and policy development and support. Not only local government authorities but also care service institutions, housing corporations and real estate owners appeared to benefit from this concept and to gain support for their elderly housing policies. GIS use in this regard appeared to be inevitable.

6. Conclusions

The combination of the availability of large numbers of digitised maps with abundant software which allows a wide spectrum of data to be presented by these maps has increased the interest in GIS enormously. One can even speak of an explosion of geographic data that is unprecedented in history (Rhind et al. 1989). The users of such GIS systems are extremely diverse and originate from various scientific disciplines. The increased use of such systems can be explained simply by the need expressed by many planners to attach a spatial dimension to the presentation of their data. Besides, the discovery that the spatial component offers the possibility of integrating data is another reason for the world wide penetration of GIS.

This penetration however, has not yet reached its full benefits to the user, as there are in the operationalization stage still many elements of GIS and spatial analysis to be improved, in particular:

- improvement of models (e.g., the combination of descriptive and evaluative or predictive models)
- orientation towards DSS and AI environments, based on realistic policy information improvement of spatial analytical tools (e.g., spatio-temporal autocorrelation, cellular automata methods, neural networks methods and bio-computing techniques)
- integration with remote sensing data (especially multi-spectral and multitemporal raster data)

- improvement of error analyses dealing with the quality of the data used (including fuzzy set methods for imprecise spatial information or demarcation)
- enhancement of user-friendliness of GIS-based methods (e.g., provisions of source programme documentation, detailed model-base description, presentation of graphical sampling results, menu-driven applications, help-files retrieval etc.).

In a GIS environment the following characteristics can be regarded as important in the determination of relevant spatial analysis techniques, which may favour a further use of GIS models and techniques:

a. The techniques must be explicitly spatial, meaning that they must utilize the specific characteristics of spatial data as distinct from non-spatial data
b. The techniques must be able to deal with large data bases (e.g., more than 10,000 objects)
c. The techniques must be oriented towards the field which are regarded as important and should focus on problem resolution
d. The techniques must be able to have an interface with GIS systems
e. Attention needs to be devoted to the way in which uncertainty is managed in spatial data
f. Attention is also necessary, if possible, for dealing with temporal dynamics in data
g. The techniques must be subject to advanced automation in order to make optimal use of the current technical possibilities and in order to reduce the dependence on the knowledge of the user
h. The techniques need to be oriented in particular towards exploration and simulation instead of hypothesis testing. For this reason it is important that the present use of already available power of computers will be optimised.
i. It is important to focus on issues which allow transferability to results and methods, so that methodological and technical knowledge can be further disseminated .

In conclusion: the penetration, implementation and utilization of GIS in combination with spatial analysis requires both methodological/conceptual progress and applied field work. The critical success factors for a successful introduction of GIS (see Section 2) may offer a meaningful guideline for policy and research strategies in this field.

7. References

Anselin, L. and A. Getis (1992) Spatial Statistical Analysis and Geographic Information Systems. *Annals of Regional Science* 28(1): 19-34.

Armstrong, M.P. and P.J. Densham (1990) Database Organization Alternatives for Spatial Decision Support Systems, *International Journal of Geographical Information Systems* 4(1): 3-20.

Berry, J.K. (1991) Discovering Feature Patterns; Everything Has its Place, Everything in its Place. *GIS World,* October, 42-46.

Beurden, A.U.C.J. van and H.J. Scholten (1990) The Environmental GIS Of The Netherlands and its Organizational Implications. *Proceedings 1St EGIS Conference*, Amsterdam, April 10-13, EGIS Foundation, Utrecht, 59-67.

Bradley, R. and J. Haslett (1990) Interactive Graphics for the Exploratory Analysis of Spatial Data - the Interactive Aariogram Cloud. Department of Statistics, Trinity College, Dublin.

Burby, R.J. and W.M. Rohe (1990) Providing for the Housing Needs of the Elderly. *Journal of the American Planning Association* 65(3): 324-40.

Burrough, P.A. (1986) Principles of Geographic Information Systems for Land Resources Assessment. Clarendon Press, Oxford.

Christensen, R. (1991) Linear Models for Multivariate, Time Series, and Spatial Data. Springer-Verlag, New York.

Despotakis, V. (1991) *Sustainable Development Planning Using Geographic Information Systems*. Ph.D. Dissertation, Dept of Economics, Free University, Amsterdam.

Douven, W. and H.J. Scholten (1993) Spatial Analysis in Health Research. In H.J. Scholten, ed., *The Added Value of Geographical Information Svstems in Public and Environmental Health*, (forthcoming).

Draaijers, G.P.J., W. Bleuten, N.H.S.M. de Wit, W. Hoogendoorn and A.U.C.J. van Beurden (1991) Bodembeschermingsgebieden: Selectie en Bescherming van Abiotisch Waardevolle Gebieden. *Milieu* 6(2): 42-29.

Fischer, M.M. and P. Nijkamp, eds., (1993) *Geographic Information Systems. Spatial Analysis and Policv Evaluation*. Springer Verlag, Berlin

Fisher, P.F. (1991) Modelling Soil Map-unit Inclusions by Monte Carlo Simulation. *International Journal of Geographical Information Systems* 5(2): 193-208.

Giaoutzi, M. and P. Nijkamp (1993) *Decision Support for Regional Sustainable Development*. Avebury, Aldershot, UK.

Grothe, M.J.N. and W. Blom (1992) GIS Measures Elderly Housing and Neighbourhood. *GIS Europe* 1(7): 20-24.

Haslett, J., G. Wills and A.R. Unwin (1990a) SPIDER - An Interactive Statistical Tool for the Analysis of Spatially Distributed Data. Department of Statistics, Trinity College, Dublin.

Haslett, J., R. Bradley, P.S. Craig, G. Wills and A.R. Unwin (1990b) Dynamic Graphics for Exploring Spatial Data, With Application to Locating Global and Local Anomalies, *American Statistician* (to appear).

Haslett, J., R. Bradley, M. Dillon and G. Wills (1990c) Interactive Graphics for the Exploratory Analysis of Spatial Data - Applications to Data of Different Support. Department of Statistics, Trinity College, Dublin.

Lam, N.S. (1983) Spatial Interpolation Methods: a Review. *The American Cartographer* 10(2): 129-149.

Maggi, R., I. Masser and P. Nijkamp (1992) Missing Networks in European Transport and Communications. *Transport Reviews* 12(4): 311-321.

Melse, J.M. (1991) Literatuuronderzoek Betreffende het Gebruik op een Gezondheidsdienst van Statistische Methoden Voor Ziekteclustering in Relatie tot Milieuverontreinging.Gemeenschappelijke Gezondheidsdienst Zuid-Kennemerland, Haarlem.

Middelkoop, H. (1990) Uncertainty in a GIS: a Test for Quantifying Interpretation Output. *ITC Journal*, 1990-3: 225-232.

Munda, G., P. Nijkamp and P. Rietveld (1992) Multicriteria Evaluation and Fuzzy Set Theory. Research Paper, Dept of Economics, Free University, Amsterdam.

Nijkamp, P., P. Rietveld and H. Voogd (1990) Multicriteria Analysis in Physical Planning, Elsevier, Amsterdam.

Nijkamp, P. and A. Reggiani (1992) Spatial Interaction, Chaos and Evolution in Space, Springer Verlag, Berlin.

Openshaw, S. (1990) Spatial Analysis and Geographical Information Systems: a Review of Progress and Possibilities. *GIS for Urban and Regional Planning*, edited by H.J. Scholten and J.C.H. Stillwell, Kluwer Academic, Dordrecht, 153-164.

Openshaw, S. and M. Charlton (1990) Applying GIS to a Small Area Health Statistics System. Paper presented at SAHSU meeting, London, June 22, 1990.

Openshaw, S., C. Brunsdon and M. Charlton (1991) A Spatial Analysis Toolkit for GIS. *Proceedings of the EGIS'91 Conference*, EGIS Foundation, Utrecht, The Netherlands.

Rhind, D., S. Openshaw, and N. Green (1989) The Analysis of Geographical Data: Data Rich Technology Adequate. *Proceedings of the IV International Working Conference on Statistical and Scientific Data Base Management.*

Robinson, A.H. and J.P. Snyder, eds., (1991) Matching the Map Projection to the Need. Committee on Map Projections of the American Cartographic Association, Bethesda.

Scholten, H.J. and J.C.H. Stillwell, eds., (1990) GIS for Urban and Regional Planning. Kluwer, Dordrecht.

Shalaby, M.T. (1991) GIS in Behavioural Analysis and Activity Systems. In *Proceedings Second European Conference on Geographical Information Systems*, J.J. Harts, H.F.L. Ottens, H.J. Scholten, eds. EGIS Foundation, Utrecht.

Unwin, A., G. Wills and J. Haslett (1990) REGARD - Graphical Analysis of Regional Data. Department of Statistics, Trinity College, Dublin.

SECTION 5:
EXTENDING THE RESEARCH AGENDA

HARLAN J. ONSRUD
Department of Surveying Engineering, 5711
University of Maine
Orono, ME 04469-5711, USA
and
IAN MASSER
Department of Town and Regional Planning
University of Sheffield
Sheffield, S10 2TN
United Kingdom

At the outset, the objective of the NATO Advanced Research Workshop was to provide a high-level exchange of methods for analyzing and modeling the diffusion and use of geographic information handling innovations and to establish directions for research. As the workshop progressed, participants met in small working groups to discuss and recommend needed research under four general headings:

- diffusion of geographic information innovations,
- use and value of geographic information,
- institutional issues, and
- cultural issues.

Summaries of their discussions and recommendations follow.

In order to avoid losing the diversity of opinion expressed by the various participants, the editors have made no attempt to summarize common concerns among the various working groups nor have we attempted to highlight or resolve conflicts between reports. Rather, the results offer a sampling of research issues deemed important by each particular group of researchers and any commonalties or discrepancies speak for themselves.

1. Diffusion Of Geographic Information Innovations[1]

The central research question is whether there is any difference between the diffusion process for GIS and for information technology products of similar kinds, especially products that use substantial knowledge bases (e.g. statistical packages, CAD, computer aided education etc.). There are two comparative questions:

1) are the diffusion processes of these products different from the processes described by the general diffusion literature (e.g. Rogers, 1983), and

I. Masser and H. J. Onsrud (eds.), Diffusion and Use of Geographic Information Technologies, 339–343.

2) are there any differences in diffusion between GIS and related products?

We need to make a major theoretical distinction between diffusion to the point of adoption (acquisition) and to implementation. It would be helpful to develop two related models for the two stages. For example, reasons for acquisition or purchase may be different from the actual use that results, the 'champion' needed for acquisition may be different from the 'champion' needed for implementation, and some factors for example, ease of use, may only have a major influence in implementation.

It was felt that we have no need to facilitate adoption per se because it is happening rapidly although it will remain important to understand the factors determining diffusion for different applications, in different nations, in government vs. private organisations, etc. The significant research issue is not what promotes adoption of GIS but what promotes effective use of GIS. As a result of the Workshop there appears to be four different but related requirements for effective use to result and they involve different diffusion rates:

1. Technology
2. Data
3. Knowledge
4. Organisational Strategies

1. Technology. The vendors are and will promote the technology, and the important requirement is for it to become easier to use. Research to assess how 'embedded' or 'transparent' GIS systems can become is important.

2. Data. GISs need large-scale data sets which are mostly controlled by national agencies. The Conference demonstrated how national policies and practices affected diffusion of GISs. Effective diffusion needs reasonable access to these datasets with reference to accuracy, reliability, privacy, ease of data exchange etc. Reasonable access may depend upon the establishment of national alliances. A research project to compare diffusion rates in different countries in relation to reasonable access to data may be instructive.

3. Knowledge. Effective use depends upon a range of knowledge. Vendors may supply the technical knowledge to operate a particular package but knowledge is needed to choose appropriate technology, develop applications and to avoid mis-use. It may be possible to build into the technology some intelligent support to avoid some mis-interpretations of spatial data. However, a valuable research project would be to determine the knowledge required for different applications and to establish the most effective means of delivering that knowledge to those involved in the applications (e.g. intelligent systems vs. education and training).

4. Organisational Strategies. Diffusion within an organisation is perhaps better modeled not as take-up, but as a process of comparative analysis by a variety of actors at different levels. Within the organisation, potential GIS use is one among a number of possible change scenarios and its use will have elements of re-invention. The present scene has been characterised as vendors 'throwing bricks' leaving the client 'to build the house.' An important research theme would be to model what different clients need in order to 'build the house' with the practical target of defining possible sources of such support.

There are in the analysis two broad research strands. The first is the descriptive task of modeling the adoption process especially with reference to comparative analysis of different nations, applications, etc. and referencing it to other information technology applications. The second is the broader research theme of diffusion to implementation for effective use. How one should measure or evaluate "effective use", "optimal use", and "use success" remains as a significant research challenge. Similarly, "success" may be the non-adoption of a GIS. The goal of the second theme is to model the processes which lead to situations where geographical information is being appropriately used. These processes may or may not lead to use of a Geographical Information System.

2. Use And Value Of Geographic Information[2]

The second working group identified four priority research areas that they felt would contribute significantly to increased understanding of the use and value of geographic information.

1. Substantive knowledge should be identified from GIS-related disciplines (i.e. domain knowledge) which if integrated with geographic information technological capabilities would enhance the functionality of the technology and hence its use and value. For instance, pattern analysis combined with crime analysis has the potential for allowing more optimal dispatching resulting in better crime control. Location-allocation analysis methods might be applied in developing decision support systems to improve services in developing countries. Numerous similar knowledge integration scenarios based on real application areas should be postulated and prototype projects developed. Prototypes would allow both users and potential vendors/developers to determine the desirability of each operational system proposed.

2. To date, little knowledge has been gained in the GIS field on purposeful non-adopters of the technology. Comparative studies of adopters and purposeful non-adopters should be pursued in various GIS application areas. Appropriate methodologies for accomplishing this should be designed and tested in a number of different types of GIS application contexts.

3. Methods for assembling information on GIS applications from which costs and benefits may be calculated are still very limited. Methods for developing audit trails in an organization's use of GIS need further investigation. Methods of continuous evaluation of technical, organizational, utility, and cost factors as a GIS matures (i.e. potentially automated tracking) should also be explored.

4. GIS is being used in many public decision-making roles and has high potential for being used in many more. Research directed towards advancing public use of GIS and library access to GIS should be priority research domains. This includes research on group decision support systems which might use GIS in a dynamic mode.

3. Institutional Issues[3]

As stressed at the Workshop, far more work needs to be done on isolating generalizable principles germane to the acquisition, implementation and particularly utilization of GIS. Such principles would be highly useful in aiding the implementation process and in developing prescriptive strategies for gaining greater benefits from already implemented GIS. The focus of such studies should be on factors, structures, and processes which can actually be managed or influenced as opposed to institutional or environmental conditions which managers have little hope of affecting.

How institutions adapt in order to capitalize on technological innovations such as GIS and how the technology is adapted to the organizational context within which it is applied are significant topics of research interest. The community at large could substantially benefit from assessments of the impact of GIS on organizations as well as from the advancement of appropriate methods for making such assessments. Important questions which might be addressed through the results of such work include:

- What organizational and institutional structures allow or enhance effective implementation and use?
- What strategies best facilitate implementation of GIS within and among organizations?
- What factors influence implementation and optimal use?
- Under what arrangements and circumstances can information sharing more readily occur?

In addition to assessment of effects at the organizational and institutional levels, assessments of social policies being pursued in the diffusion of GIS must be carried out. For instance, very different policies are being followed in different countries as well as within single countries with respect to government agency recovery of GIS costs, and with respect to privacy, liability, copyright, ownership and similar issues. Local governments typically have substantial ability to influence or control the policies that will be followed in the use and dissemination of data and information from their GIS. Comparative analyses should be carried out on the actual ramifications and predicted long term effects of the disparate policies being followed in various jurisdictions. The knowledge gained should substantially aid rational decision making in the selection of GIS information policies.

There is widespread recognition that the technical ability to more easily share geographic information and the benefits to be gained by such sharing have been important concepts in diffusing GIS. Yet the promise of sharing has far exceeded actual sharing to date. Organizational and behavioral impediments to sharing appear to far outstrip the technical impediments. Indepth research on options for allowing or enhancing sharing within and among organizations is very much lacking. Why people want to share and are willing to share or resistant to sharing are sub-questions that are ripe for inquiry within GIS domains. Fundamental information relative to both technical and behavioral issues on what works and what doesn't in GIS sharing environments has yet to be confirmed.

The group felt strongly that a cumulative research tradition must be built in order to provide substantive contributions in answering many of the questions raised. A suite of planned research methodologies such as in-depth longitudinal case studies, comparative case studies, and survey-based approaches are necessary to cover major areas of inquiry comprehensively and reliably. Within this suite of methods, selected major GIS implementations should be studied in detail and over time. For instance, the entire range of sharing issues including issues of costs, standards, ownership, access, facilities, maintenance, human behavior, organizational structure, law and the connecting strands between these issues should be observed and analyzed in detail in such a study environment.

Finally, although not a research issue, the need for qualified personnel is significantly impeding the adoption and optimal use of GIS by organizations. Training in the implementation and use of GIS in organizations must be increased for both those within organizations and for those in university programs in order to overcome this very substantial impediment. Teaching materials are still very limited and personnel with the ability to train others are still in extremely short supply.

4. Cultural Issues[4]

Geographic information technologies are being diffused to other than the cultures from which the innovations originated. Of primary concern is that such innovations and the ability to access the power which these information technologies bring should be allowed to disseminate in such manners that are socially responsible.

In addition to knowing how to diffuse and implement GIS throughout a culture, we need to consider the likely social impacts of that diffusion. Key issues for comparative studies on the impact of cultural factors include attitudes in regard to access to information and privacy, styles of government and planning (e.g. top-down versus bottom-up), standards and transfer formats, adoption of specific systems (e.g. UNIX versus DOS) and language (especially for small countries). Similar to environmental impact statements, governments might consider the utility of preparing social impact statements for specific GIS projects which would disclose social benefits of the technology, detrimental effects, hidden costs to society and alternatives. Cross national comparative research studies should be carried out to divulge those policies and strategies which better promote increased equity in the distribution of the beneficial consequences of GIS. Such studies would also provide information useful for accommodating differences among cultures.

Similarly, within a single nation, various levels of government and the private sector represent different "cultures." Coordination among and accommodation of national, regional, local, professional, and political interests and equitable access to GIS present significant challenges. There is a need to evaluate and carry out active participatory strategies for overcoming such cultural barriers in diffusing geographic information innovations throughout the various segments of individual nations.

5. Endnotes

1. Submitted by Thanos Daganis, Ken Eason, Gary Moore, Nicos Polydorides, Gerry Post, Erik Stubkjaer, and Peter Zwart
2. Submitted by Hugh Calkins, Ken Jones, Peter Nijkamp, Gerard Rushton, Henk Scholten, and Lyna Wiggins.
3. Submitted by Heather Campbell, William Huxhold, Jeff Maes, Harlan Onsrud, Jeffrey Pinto and Massimo Rumor
4. Submitted by Peter Burrough, Tim Cartwright, Ian Masser, Alex Hernandis-Mulet, Henk Ottens, and Michael Wegener

Appendix: NATO Advanced Research Workshop Participants
* author or coauthor of a chapter in this book

Prof. Antonio Morais Arnaud*
Faculdade de Ciencas e Tecn.
Universidade Nove de Lisboa
2825 Monte de Caparica,
Portugal

Dr. Peter Burrough*
Faculty of Geographical Science
University of Utrecht, POB 80.115
Utrecht, 3508 TC
Netherlands

Dr. Hugh Calkins
Department of Geography/NCGIA
State University of New York
301 Wilkeson Quad
Buffalo, NY 14261
USA

Dr. Heather Campbell*
Department of Town & Regional Planning
University of Sheffield
Sheffield, S10 2TN
United Kingdom

Dr. Timothy Cartwright*
Environmental Studies
York University,4700 Keele St.
Toronto, Ontario M3J 1P3
Canada

Dr. Thanos Daganis
17A Stournara str.
106 83 Athens
Greece

Dr. Ken Eason*
Department of Human Sciences
Loughborough University
Loughborough, LE11 3TV
United Kingdom

Dr. Paul Goodman*
Graduate School of Industrial Administration
Carnegie Mellon Univ.
Pittsburgh, PA 15213
USA

Alex Hernandis Mulet
Consell Metropolita de l' Horta
46002 Valencia
Spain

Dr. William Huxhold*
University of Wisconsin-Milwaukee
Engelmann Hall, Rm 172A
Milwaukee, WI 53201
USA

Dr. Ken Jones*
Mariesgate 8
0368 Oslo
Norway

Dr. Jeff Maes
Avenue de Triomphe
174-1160 Brussels
Belgium

Dr. Ian Masser*
ESRC RRL Initiative
University of Sheffield
Sheffield, S10 2TN
United Kingdom

Dr. Gary C. Moore*
Faculty of Management
2500 University Drive N.W.
The University of Calgary
Calgary, Alberta T2N 1N4
Canada

Dr. Peter Nijkamp*
Department of Regional Economics
Free University of Amsterdam
De Boelelaan 1105, 1081 HV
Amsterdam, Netherlands

Dr. Harlan Onsrud*
National Center for Geographic
Information and Analysis
University of Maine
Orono, ME 04469
USA

Dr. Henk Ottens*
Nederlands Expert voor Ruim-
telijke Inform., POB 80115
Utrecht, 3508 TC
Netherlands

Dr. Jeffrey Pinto*
College of Business Administration
University of Maine
Orono, ME 04469
USA

Dr. Nicos Polydorides*
URSA-NET, 18b Pythagoras St
15562 Holargos
Athens, GR-155
Greece

Mr. Gerry Post*
Geomatics Board
P.O. Box 2044
Halifax, Nova Scotia B3J 2Z1
Canada

Dr. Everett Rogers*
Annenberg School for Communi
Univ of Southern California
Los Angeles, CA 90089-0281
USA

Ing. Massimo Rumor*
C.E.D., Comune di Padova
Via Municipio 1
Padova, 35122
Italy

Dr. Gerard Rushton*
Department. of Geography
The University of Iowa
Iowa City, IA 52242-1316
USA

Dr. Henk Scholten*
GIS Projectleider, ISC
Rijksinstituut voor Volksgezondheid en
Milieuhygiene
Antonievan Leeuwenhoeklaan 9
Postbus 1
Bilthoven, 3720 BA
Netherlands

Dr. Erik Stubkjaer*
Institute of Development and Planning
University of Aalborg
Fibigerstraede 11
9220 Aalborg 0
Denmark

Prof. George Veis
Department of Survey Engineers
National Technical University
Zografou Campus
157 73 Zografou, Greece

Dr. Michael Wegener*
Institute for Spatial Planning
University of Dortmund,PO 50 05 00
Dortmund 50, D-4600
Germany

Dr. Lyna Wiggins*
Department. of Urban Studies
Massachusetts Institute of Technology
Cambridge, MA 02139
USA

Dr. Peter Zwart, Director*
Centre for Spatial Information Studies
P.O. Box 252 C G.P. O.
Hobart, Tasmania 7001
Australia

SUBJECT INDEX

GPSR Compliance

The European Union's (EU) General Product Safety Regulation (GPSR) is a set of rules that requires consumer products to be safe and our obligations to ensure this.

If you have any concerns about our products, you can contact us on ProductSafety@springernature.com

In case Publisher is established outside the EU, the EU authorized representative is:

Springer Nature Customer Service Center GmbH
Europaplatz 3
69115 Heidelberg, Germany

Batch number: 10816188

Printed by Printforce, the Netherlands